The Man Who Built the Sierra Club

The Man Who
Built the Sierra Club

A Life of David Brower *Robert Wyss*

Columbia University Press New York

Columbia University Press
Publishers Since 1893
New York   Chichester, West Sussex
cup.columbia.edu

Library of Congress Cataloging-in-Publication Data
Names: Wyss, Bob.
Title: The man who built the Sierra Club : a life of David Brower / Robert Wyss.
Description: New York : Columbia University Press, [2016] | Includes bibliographical
    references and index.
Identifiers: LCCN 2015034505 | ISBN 9780231164467 (cloth : alk. paper) |
    ISBN 9780231541312 (e-book)
Subjects: LCSH: Brower, David, 1912–2000. | Environmentalists—United States—Biography. |
    Conservationists—United States—Biography. | Sierra Club—History.
Classification: LCC QH31.B859 W97 2016 | DDC 333.72092—dc23
LC record available at http://lccn.loc.gov/2015034505

♾

Columbia University Press books are printed on permanent
and durable acid-free paper.
This book is printed on paper with recycled content.
Printed in the United States of America
c 10 9 8 7 6 5 4 3 2 1

COVER IMAGE: Courtesy of the Bancroft Library, University of California, Berkeley
COVER DESIGN: Archie Ferguson

Contents

Contents

Illustrations

Chronology

August	Visited Deadman Creek forest, disillusionment began with the U.S. Forest Service
1954, January	Testified against the Dinosaur dams at a congressional hearing
1955, spring	Supervised the publication of first conservation crusade book, *This Is Dinosaur*
November	Dam proponents agreed to drop support for Dinosaur dams
1956, February	Criticized National Park Service $1 billion Mission 66 public-works project
March	Supported law authorizing an enlarged Glen Canyon Dam in exchange for no Dinosaur dams
June,	Arranged introduction of a wilderness bill in the U.S. Senate
October,	Construction began on the Glen Canyon Dam
1960, summer	Published *This Is the American Earth*, the first of the Exhibit Format books
December	Asked author Wallace Stegner to write the "Wilderness Letter," a poetic plea supporting wilderness preservation
1961, April	Journeyed to Rainbow Bridge to draw attention to the threatened landmark
1962, autumn	Published *In Wildness Is the Preservation of the World*, with text by Henry David Thoreau and photos by Eliot Porter
1963, January	Fought unsuccessfully to stop the completed Glen Canyon Dam from backing up the Colorado River and creating Lake Powell
May	Placed in charge of the campaign to stop two Grand Canyon dams
June	Published *The Place No One Knew*, a lament on the loss of Glen Canyon
1964, August	Congress passed the wilderness bill

1966, March	Organized *Reader's Digest* conference on the Grand Canyon dams
May	Failed to object when the Sierra Club board voted to support the Diablo Canyon nuclear plant
June	Published a newspaper advertisement criticizing the Grand Canyon dams. One day later the club's tax-deductible status suspended by the Internal Revenue Service (IRS)
December	IRS confirmed its earlier ruling, which held despite further appeals
1967, February	U.S. government dropped its plans for the Grand Canyon dams
May	The Sierra Club membership in an election voted two to one to back the Diablo decision Sierra Club board members failed to fire Brower or transfer him to New York
1968, summer	Struggled in disputes over book contracts, a new London office, and a new two-part book on the Galápagos Islands
October	Fought charges from three Sierra Club board members seeking to fire him
autumn	Prevailed in campaigns to establish North Cascades and Redwood National Parks
1969, January	Published Earth National Park newspaper advertisement; stripped of access to Sierra Club finances Stepped down as executive director to run with slate to control the Sierra Club board
April	Failed with slate to win any seats on the Sierra Club board
May	Resigned as Sierra Club executive director
summer,	Created new environmental organization, Friends of the Earth

1971, January	Organized what became Friends of the Earth International
1974, January	Focused increasingly on energy in the wake of a global energy crisis
1979, November	Stepped down as fulltime president of Friends of the Earth but remained in control
1983, April	Elected to Sierra Club board of directors
1984, July	Dismissed from Friends of the Earth board
1985, autumn	Joined Earth Island Institute, hired loyal Friends of the Earth staff
1996, November	Campaigned to close the Glen Canyon Dam
2000, November 5	Died in Berkeley, California

The Man Who Built the Sierra Club

David Brower, the Sierra Club, and their quest to save America's great rivers

While David Brower and the Sierra Club had a national agenda to protect wilderness, it was in the American West during the 1950s and 1960s that they achieved some of their greatest triumphs. Brower cherished the wild rivers of the Colorado River basin and campaigned tirelessly to save them for future generations against those who would seek to dam and tame them. The magnificent national park and national monument in Grand Canyon and Dinosaur would not be what they are today without the unfettered rivers that define them. Through his audacious, stubborn leadership of the Sierra Club and his consummate political skill in opposing federal agencies, Brower thwarted efforts to build dams at Echo Park and Split Mountain in Dinosaur National Monument and at Bridge Canyon and Marble Canyon in Grand Canyon. You will find these stories and many others in this book. You will also find the story of the Glen Canyon Dam in Paige, Arizona—the compromise that Brower agreed to that haunted him for the rest of his days.

Introduction

In the spring of 1966, David Brower traveled to a Colorado River canyon he considered beautiful—and increasingly repulsive. The place was Glen Canyon, and it was being drowned. He met *Life* magazine writer Hal Wingo and a photographer, and as they hiked portions of the valley, Brower talked about two of his greatest conservation battles.

One was his ongoing protests over the Glen Canyon Dam, the water behind which was now flooding 186 miles of the Colorado to regulate its use for hydroelectricity production and storage and to supply water to thirsty southwestern states. The hikers reached Cathedral in the Desert, a beautiful, cavernous amphitheater of pink Navajo Sandstone, bathed in sunlight, the trickle of a waterfall cascading into a pool bordered by lush green moss and hanging gardens. Brower had only belatedly discovered the canyon's beauties, and he had been pleading for the dam work to halt and for the flooding to cease. But his desperate appeals to federal officials, even Interior Secretary Stewart Udall, who could be both a friend and an enemy, met with stony rejection. For three years, water had been pooling behind the new dam, and it would soon submerge the cavern. "This makes me pretty damn sick," said Brower. "Glen Canyon is the greatest loss of scenic resources anywhere." Closer to the dam, similar landmarks in this brilliant slip-rock country were already buried under hundreds of feet of water. "I just hate the deception of it," he added. "Down under that water, some of the most beautiful scenery ever created is gone." Brower was also clashing downstream with federal dam builders in the Grand Canyon. Engineers wanted to erect one dam that would back up the raging Colorado for 50 miles and a second that would back it up for 93 miles. "If the Grand Canyon dams really had to be built to ensure the nation's survival, or even a region's, there'd be something to argue about," he told Wingo. "But they are absolutely not necessary."

The Colorado, declared Brower, was already overregulated, overdammed, and overchoked. At least some of it needed to remain free and wild.

The resulting *Life* story hailed Brower as America's "No. 1 Conservationist." The lead photograph depicted the fifty-three-year-old Brower glancing up, his face still ruggedly handsome, his silver hair tousled and whipped by the wind. He was the California city boy who had forged the Sierra Club into a national environmental force. The primary headline touted Brower as the "Knight Errant to Nature's Rescue."[1]

The article failed, as many others had, to capture the complexity and paradox of David Brower. Brower would always believe that failure to stop construction of the Glen Canyon Dam was his greatest mistake as an environmental activist and that the loss of the spectacularly scenic canyon defined the later years of his conservation ethic. What *Life* overlooked, most importantly, however, was that the decision to erect the dam was tied to conservation's greatest victory to date against dam builders. Brower had led an effort that had successfully blocked the construction of two dams on a Colorado tributary, the Green River, in a national park sanctuary, Dinosaur National Monument. This was a triumphant moment for conservationists. It finally established a basic principle for conservationists, one that become so inviolable that politicians would find it virtually impossible to overcome in the future. All federal land, set aside as a national park or monument, was now off limits to dams or other development. That happened in 1956, and Brower emerged as a national conservation leader; the movement flourished and gained strength as a national political force. A decade later he capitalized on that principle and power by fighting and eventually defeating the proposed Grand Canyon dams. It was an even greater victory than Dinosaur. Yet the price that had to be paid was also great. On behalf of the Sierra Club and as the acknowledged leader of the conservation moment, he agreed to trade the two Grand Canyon dams' hydroelectric units for a coal-fired, electric-generating plant six miles east of Glen Canyon Dam. Air pollution soon fouled the skies, extending to the Grand Canyon, evolving into yet another contributor to the twenty-first century's greatest environmental challenge—climate change.

Over the years, David Brower has been called many things—tireless, unyielding, passionate, visionary, bold, influential, uncompromising, handsome, charismatic, opinionated, and articulate. He became a circuit-riding prophet, the environmental movement's conscience who defined conservation and environmentalism from the mid-1950s until his death in 2000. He was an angry trailblazer responsible more than any other for turning environmentalism from hiking and bird-watching into a social and political force.

Those same admirers also called Brower stubborn, contentious, controversial, irascible, impossible, polarizing, impolitic, impolite, and a notorious curmudgeon. He on occasion would willingly stretch facts into falsehoods, was so unwilling to tamp down his views that he destroyed lifelong friendships, and refused to take orders even from those in institutional positions above him. He was frustratingly independent.[2]

And yet he did all of this for one selfless reason—to sustain the earth's natural environment. He wanted to save as much of the planet as possible from humans. He wanted to preserve what remained of the natural world and safely pass it to future generations.

He was forty years old in 1952 when he began this quest as the Sierra Club's first executive director. He was the first full-time employee for this San Francisco–based hiking club that at the time had only six thousand members. When he left the club in 1969, it had seventy thousand members around the nation. He inherited a club known more for its social outings and summer hikes than for its advocacy. The nation's conservation movement was small, splintered, and ineffective. Engineers built dams seemingly wherever they wanted; foresters chopped the nation's trees with little restraint; even in the national parks, the rangers with the funny broad-brim hats were best known not for conservation but for entertaining visitors by feeding garbage to the bears.

3

By 1969, all that had changed.

Dam construction was off-limits in national parks and almost everywhere else; logging was limited in national forests; and new national parks ranged from the California redwoods to the coastal dunes of Cape Cod in Massachusetts. Brower and the burgeoning Sierra Club were instrumental in this political and ecological transformation. Brower played a key role not only in Dinosaur and the Grand Canyon but elsewhere as well. He fought, for example, to protect the wilderness of Washington State's North Cascades and to prevent further logging of coastal redwoods. He espoused the concept of wilderness protection, and he helped win congressional legislation to protect the nation's wild lands, resulting in the landmark Wilderness Act of 1964.

Brower succeeded because he made people care. "Brower's eloquence in conversation, at the podium, or wielding a pen is purely captivating," wrote Stephanie Mills, an acolyte. "Like any great bard's, his voice has a range and wit. He called for the preservation of homely necessities like topsoil, and of conspicuous glories like the Sierra and the slirock country with equal conviction."[3]

He was also a genius in using the media to promote his causes. No commercial book publisher would take the risks that Brower did to produce a series of nature books. Paul Brooks, who published Rachel Carson's *Silent Spring*, told Brower that such expensive books would never sell. Yet some of them sold more than 1 million copies. Brower's friend Ansel Adams produced the first books, filling them with stunning black-and-white nature photographs. Later came books by Eliot Porter, featuring color photography that was so rich and deep that readers found them irresistible even though they would cost the equivalent of $200 to $250 today. Brower produced books, posters, calendars, and films that told stories of nature's wonders while drawing in thousands of new members to the Sierra Club. So did the newspaper advertisements that campaigned for the redwoods and Grand Canyon. Brower did not invent the concept of political advertising by advocacy organizations, but he used it in a way never seen before to win over sympathizers and outrage opponents. When dam supporters claimed it would be easier to appreciate the canyon walls on a power boat atop a

Grand Canyon reservoir, Brower responded with an ad that asked, "Should we also flood the Sistine Chapel so tourists can get nearer the ceiling?"[4]

Brower was shy in personal situations, sometimes even brusque. But his movie star looks, his gift for words, and the charm and charisma that he could flick on were irresistibly seductive. He developed political alliances with such powerful personalities as Supreme Court justice William O. Douglas, writer Wallace Stegner, and Interior Secretary Stewart Udall. They did not always agree with him—especially Udall—but there was mutual respect. He inspired individuals and large audiences. Over time, he developed a special talk for the lecture circuit that was particularly effective at college campuses. In it, he squeezed time into seven days. Life arrived on a Tuesday. Neanderthals came at eleven seconds before midnight on Saturday. The Industrial Revolution happened one-fortieth of a second before midnight. The message—we must all be wary as the earth's stewards: "We ought to learn to ask, before starting a vast new project, what will we gain if we don't build it? What will it cost the earth? One thing it will cost is wilderness."[5]

In 1952, Brower, at the time a registered Republican, a major in the U.S. Army Reserve, and a supporter of nuclear energy, was willing to compromise on a range of conservation issues. He supported a number of dam projects early in his tenure: "Merely to say 'we're against all dams until . . .' does nothing but lump us as aginners [sic] with no specific alternative. It's a sure formula for being ignored by all hands, including other conservationists," he told the Sierra Club board of directors in 1956.[6] Over time, Brower's perspective shifted. There were many reasons for this shift, often tied to his clashes with dam builders, foresters, and park rangers, but it ultimately came down to Brower's realization that protecting what was left of the earth's wilderness was paramount. A loss of wilderness was not just a campaign defeat; it was a loss to future generations forever. But Brower's increasing refusal to negotiate compromises would infuriate opponents. One of them, Colorado congressman Wayne Aspinall, once called him "an utterly unreasonable man who would never compromise on anything."[7]

What especially set Brower apart from others was the fervor and risk that he brought to his task. He was a gambler, a risk taker, a maverick. When he was in his twenties, he often put his life at risk by climbing mountains; he was credited with 130 first ascents of peaks.[8] When he was older, he insisted that others gamble with him. Edgar Wayburn, a friend and colleague who eventually became a critic, said Brower "demanded absolute self-rule. One of his most used expressions was 'follow me,'" a slogan that works only when the leader has absolute control.[9] Brower of course did not. The nature books he produced, for example, were an incredible financial gamble and one that succeeded—that is, until the market changed. Brower stubbornly resisted acknowledging that adjustment. He was never a strong financial manager, and as Sierra Club deficits mounted, many of his friends worried he was bankrupting the organization. But even as the red ink spread, Brower refused to cut back on publishing expensive books if that meant reducing the Sierra Club's mission and doing less to save the earth. Commitment and righteousness can be positive attributes for an advocate. They can also be self-destructive when they become absolutes, especially when they are directed against allies and friends.

For many years, magazine and newspaper writers and filmmakers painted Brower in one-dimensional portrayals. To them, he was a heroic crusader and a fierce advocate for the environment who would not give in to others at the Sierra Club more willing to compromise. Yet to other people he was an irresponsible renegade, willing to sacrifice humanity for the sake of a forest. If only Brower, this driven, enigmatic man, were that simple, his fall from Sierra Club grace so neat, and his efforts on conservation's behalf so monochromatic or short-lived.

Life magazine did get it right in 1966, though, in declaring that Brower was then at the apex of his power. He was the premier conservationist of his generation, perhaps the greatest ever. He was the heir to Henry David Thoreau and John Muir, the contemporary of Rachel Carson and Jacques Cousteau. Having already established the foundation that we call the modern environmental movement, he was on the cusp of greater glory, success, and achievement. Three years later he was gone, removed from the helm of the Sierra Club and the throne of environmentalism that he had crafted. His plummet was fittingly complex, a tale as rich, deep, and idiosyncratic as the man it befell. He lingered another thirty years, more a sage than a messiah.

1. First Fight

David Brower was skeptical and at least a little hesitant. For more than a year, friends at the Sierra Club had been extolling the beauties of the sagebrush canyons within Dinosaur National Monument. He had seen photographs and even a film depicting kayakers running rapids, tranquil beaches, and towering cliffs of the deep river canyon. They were interesting, but none matched the exuberance and the breath-taking descriptions offered by those who had rafted down the rivers. Others contended that the high-desert wilderness on the border of Colorado and Utah was a wasteland, fit only for the two dams the federal government wanted to erect. Such comments threw Dinosaur's defenders into a frenzy. The dams, they said, would flood the roaring whitewater of the Green and Yampa Rivers, the 2,000-foot-deep canyons of twisted and uplifted sandstone and shale, and a magnificent monumental massif called Steamboat Rock. But was Brower or the Sierra Club or anyone in the tight but small conservation community really capable of taking on the nation's dam builders and in particular an obscure federal agency? The U.S. Bureau of Reclamation was flush with cash, tight with the political establishment, and extolled by farmers, ranchers, and the small-business people who had made the West. What was Brower to do? It was early 1953, and he had just been appointed the first executive director of the Sierra Club, one of a handful of fulltime conservation leaders in the nation. As Brower considered his options, he realized his choice to fight or fold was simple. He would raft the Yampa and Green and see for himself whether these river canyons were worth the fight. The trip and his resulting decision would change his life.

Brower's first task was to track down a crusty old river runner, Robert "Bus" Hatch. Hatch was a pioneer and a legend on the rivers of the West. He had been river rafting since 1925, and he became so well known that when Lowell Thomas needed to film a raft running the wild streams of the Himalayas, he hired Hatch. Now Brower asked Hatch to guide Sierra Club members, including him and his two eldest sons, down the Yampa and Green. Brower told Hatch the trips might have to accommodate up to one hundred people.[1]

Hatch hesitated for only a moment before agreeing. He did not tell Brower that he had never run the river on such a scale. No one had. No more than five hundred had rafted the river in the previous ten years.[2] Hatch had only two ten-person rafts. He would have to find more.

By the summer of 1953, when Brower and other Sierra Club members got to Vernal, Utah, and met up with Hatch, the river outfitter had bought several 27-foot-long pontoon rafts. He refashioned them for the river and hired his sons, college kids, and local youth to guide the visitors down the river. The passengers were a "heterogeneous and colorful collection of individuals," according to one account, and they "came in all ages, shapes and sizes with a sprinkling of small fry for seasoning."[3] Brower brought along two of his sons—Ken, who was nine, and Robert, who was seven. They rode old school buses with no suspension upriver to Lily Park in Colorado. After lunch, everyone was fitted with life preservers, gear was stowed on the rafts, and Hatch declared, "Now we're safe, now we're on the river."

They shoved off. The raft ride got bumpy at Tepee Rapids and rougher at Big Joe Rapids, Harding Hole, and the Warm Springs Rapids. The trip took several days, allowing time for the rafts to stop for hikes up Starvation Valley and Meeker's Cave, where pictographs and petroglyphs and other artifacts remained from ancestral tribes. They camped at Rippling Brook,

Anderson Hole, and Harding Hole. Bus Hatch was the evening's star, recounting campfire yarns of Butch Cassidy and his gang, who had hidden here. A grandson, Tom Hatch, remembered that his grandfather "couldn't sing, he couldn't dance, but he could sure cuss and tell stories."[4]

The trip ended at Echo Park, a luxuriant green canyon of cottonwoods and grasses surrounded by steppes of high-desert sagebrush. Here the Green and Yampa Rivers merged, and downstream they boiled down Whirlpool Canyon and beyond for 20 miles. Echo Park featured Dinosaur's most striking geological masterpiece, Steamboat Rock. It was a sandstone monolith that rose abruptly 800 feet over the confluence of the Green and Yampa Rivers. The waters surged around three sides of the rock. For years after the controversy, when editors chose an image to represent the Echo Park conflict, they chose photographs of Steamboat Rock.

The proposed dam would be two miles downstream. The dam builders, the Bureau of Reclamation, described the Echo Park Dam as the "workhorse" that would drive its entire collection of dams and reservoirs in the upper Colorado River basin. It would be the mightiest of the dams north of Hoover and Lake Powell, rising 690 feet, costing $165 million, and creating a 107-mile-long reservoir.[5] It would not bury Steamboat Rock, but it would reduce the monolith to a modest peninsula rising above the blue reservoir. The Echo Park Dam would pay for itself by producing hydroelectricity, which meant that a second dam had to be built farther down the river within Dinosaur at Split Mountain to catch and store the water to regulate flow downstream.

Nearly everyone who floated down the Yampa and Green came out of it with a new appreciation for the canyons. Brower was no exception. Two days after his return, he was still not getting any work done because he was so enthralled, he told one friend. "I have never had a scenic experience equal to that one."[6] Now, finally, he recognized why it needed to be saved.

Others in the conservation community were also beginning to awaken to the dangers at Echo Park and Split Mountain. News coverage of environmental and conservation issues was moribund in those days except for work by iconoclastic writers such as Bernard DeVoto, the first to sound an alarm in the July 22, 1950, issue of the *Saturday Evening Post*. In a story headlined "Shall We Let Them Ruin Our National Parks?" DeVoto wrote that Glacier, Grand Canyon, Mammoth Cave, and Dinosaur were all in danger. Referring to the Yampa and its beautiful canyons, he argued: "The deep artificial lakes would engulf the magnificent scenery, would reduce by from a fifth to a third the height of the precipitous walls, and would fearfully degrade the great vistas. Dinosaur National Monument as a scenic spectacle would cease to exist."[7]

A year later, Sierra Club member and former president Harold Bradley and several of his children rafted the Yampa with Hatch. Near the whitewater of Whirlpool Canyon, Hatch pointed to a series of metal ladders that climbed the sandstone walls and explained that this was the site of a proposed dam. The Bradleys, who had reveled as they careened down the Yampa, were stunned that this canyon could be lost. "For sheer breathtaking beauty, color and variety, there is no other canyon run in the country to equal this one," Bradley declared.[8]

Bradley pestered Brower and the Sierra Club to mount a campaign against the Dinosaur dams. Brower was not the only one who hesitated. John Muir may have been a rabble rouser when he created the Sierra Club in 1892, but his descendants for decades had preferred influence over confrontation, quiet, gentlemanly, closed-door negotiations and compromise as opposed to angry public protests and harsh accusations in public forums. After World War II, a new generation was taking over, but Bradley's proposal was still a formidable change. Routine club decisions were made by the Executive Committee of the board of directors, but the elected board also took advice from several committees composed of members, including the Conservation Committee. Edgar Wayburn remembered one night in 1952 when up to forty members of that committee gathered to discuss the degree to which the club would join the fray. Like many that night, Wayburn was worried. Hardly anyone had been to Dinosaur, hardly anyone knew very much about it. They argued that night and eventually recommended that the club fight. The board agreed. "We realized in this campaign that we had to become national if we were going to succeed in this and other things that we saw ahead of us," said Wayburn.[9]

By the summer of 1953, when Brower was on the river, the new Eisenhower administration and in particular its new Interior Department secretary, Douglas McKay, were deciding whether to back the dam build-ers. Eisenhower had been neutral and had campaigned in 1952 against twenty years of Democratic federal spending, including high-priced dams. McKay, the former governor of Oregon, was also difficult to figure. He had been a surprising choice as interior secretary. The one-time Chevy dealer from Salem did not impress journalist Elmer Davis. After attending McKay's first press conference, Davis declared, "This man does not know the lower outlet of his alimentary system from a hole in the ground."[10]

Brower focused on influencing the president and McKay. As summer turned to autumn, he produced a steady stream of correspondence, com-munication, and even a film. *Wilderness River Trail* was produced by a pro-fessional photographer, Charles Eggert, and it was a slick presentation of the glories of the high, sun-whitened canyon walls and the swift, aquama-rine rivers. To counter claims that the rivers were too wild for any but the most adventuresome, the two Brower kids were seen often in the footage shot. The film, said Brower later, was "the most important thing we did in offsetting the Bureau of Reclamation's propaganda; it was the hardest thing they had to fight."[11]

By now, McKay had ordered one of his new undersecretaries, Ralph Tudor, to investigate the Dinosaur project. An engineer who had designed the Bay Bridge connecting San Francisco to Oakland, Tudor spent three days floating down the Yampa and sleeping under the stars by the river. He was beguiled by the river canyons and, like many, flummoxed by their recent history.[12]

President Woodrow Wilson had created Dinosaur National Monument in 1915, setting aside 80 acres that contained a quarry with an amazing graveyard of dinosaur fossils. Paleontologist Earl Douglass of the Carnegie Museum had discovered them in the summer of 1909, and over the next fifteen years he identified more than four hundred Jurassic period dino-saurs and excavated more than twenty complete skeletons.[13] It was a great paleontological discovery, one that a century later still drew more than 250,000 visitors each year to the quarry and its exhibits in one of the most isolated places in the country.[14] In 1938, President Franklin D. Roosevelt appended another 203,885 acres, including the Green and Yampa River canyons, an area nearly as vast as the five boroughs of New York City. For legal and political reasons, the name "Dinosaur" was retained. The monument also remained nearly unknown because there were no roads into it.

The president and Congress would have to approve any dams in Dinosaur, a condition that many saw as no more than a technicality even if Eisenhower objected.[15] The nation was enamored with dam builders and entranced by the mighty structures with names such as Bonneville, Boulder, and Grand Coulee.[16] Western politicians were besotted by the engineers who could bring back bloom and life to a region of dust and depression. Between 1928 and 1956, Congress authorized the Bureau of Reclamation to build seventy-seven dams and the U.S. Army Corps of Engineers to build hundreds more. The barriers were erected to irrigate the West's sagebrush country thirsty for rainfall; the rivers bottled to prevent destructive flooding on the Columbia, the Tennessee, the Red, and other streams; the water squeezed to supply hydroelectricity. Electricity had become so cheap in the Northwest that no residents there insulated their homes; power consumption in the United States was the highest in the world, and an electric bill in Seattle was eight times lower than in New York City for the same number of kilowatts. For a representative or senator lucky enough to get a dam authorized in his district, it meant a bonanza in campaign contributions from the engineering and construction firms that won the multi-million-dollar contracts. Pork-barrel politics dictated that representatives and senators vote for new dams either in gratitude or in expectation of similar largess. Dams in the first half of the twentieth century had changed the face of America. Historian Marc Reisner calls this change "the most fateful transformation than has ever been visited on any landscape, anywhere."[17]

And who was going to stop this juggernaut now, even if it scarred a national park or monument that had been set aside for its singular natural beauty? The National Park Service was charged with protecting Dinosaur, but it answered to McKay in the same executive-branch department as Reclamation. Whereas the dam builders had a budget of $205 million in

1953, the Park Service made do with $33 million.[18] Parks were beloved, but the Park Service had no clout in Congress. Conservationists were equally unprepared. Since Theodore Roosevelt had left office in 1909, they seemed always to be on the defensive. Muir and the Sierra Club had lost a decisive battle in 1913 when Congress allowed the city of San Francisco to dam the beautiful Hetch Hetchy Valley within Yosemite National Park. Since the end of World War II, conservation organizations had been fighting just to preserve what they thought they had already protected, battling loggers in Olympic National Park and resort developers in Jackson Hole National Monument (later Grand Teton National Park). Ranchers were overgrazing federal land; roads were carving out pristine wilderness; and few heard conservationist's objections. Today there are more than twelve thousand environmental organizations in the United States, many with huge staffs, and together they spend $15.5 billion annually. In contrast, the Sierra Club in 1953 had seven thousand members, one employee (Brower), and a budget of less than $100,000. The rest of what constituted the conservation movement consisted of a variety of local or specialized organizations. Although these organizations often joined under the same banner, many had conflicting missions, serving everything from garden clubs to hunting and fishing lodges and even bird-watchers. Plus, this was the 1950s, populated by what sociologists labeled the "silent" generation, in which conformity trumped protest and acceptance of the status quo overruled unseemly dissidence.[19]

It was no surprise, then, that as much as Tudor had relished his journey down the Yampa, he recommended and McKay quickly accepted sacrificing the Dinosaur river canyons and erecting the two proposed dams. The dams were simply too important as part of what the Bureau of Reclamation was calling the "Colorado River Storage Project." The project was a massive, $3 billion package of reclamation plans, storage dams, and power facilities spread across six western states. Salt Lake City and the Wasatch Range would be funneled water; ranchers in Wyoming, Colorado, and Utah would be able to irrigate exhausted range land; hydroelectricity would allow miners to better exploit natural resources; construction jobs would bolster local incomes, support greater services, and lower taxes from

Wyoming to Arizona. Despite the loss of a scenic but remote river canyon, the project could be a bonanza, the equivalent of what the Tennessee Valley Authority had done for the Southeast.[20] Tudor and McKay emphasized that Echo Park was necessary for the project for one major reason— less water would be lost through evaporation there than at any other alternative sites for dams.[21] They sent the plan to Congress, and hearings were scheduled to begin in 1954.

McKay would later admit that he and other dam supporters were caught off-guard by what happened next. Essentially, the conservation movement awoke in 1954. A range of journalists, environmentalists, and historians have called the year 1954 and the clash that was soon being led by Brower a "milestone" or "watershed" in the American conservation or wilderness movement.[22] In the January issue of the *Sierra Club Bulletin*, Brower warned members that it was the time to stop compromising and to fight over Dinosaur. "There's a touch-and-go election coming up in 1954," he wrote. "We must all make it clear that conservation isn't partisan, it's American."[23] The loose assemblage of garden clubs, fishermen, and bird lovers began to coalesce in what would eventually become an alliance of seventy-eight organizations. The outrage focused on the decision to invade a national park sanctuary; some called the Colorado River Storage Project the gravest assault on the national park system since it had been created in 1916. At a January 4, 1954, press conference in Washington, D.C., representatives of thirty-two of those groups voiced their united support for the campaign. Charles Collison, director of the National Wildlife Federation, declared, "There's going to be one hell of a fight over this."[24]

Dam supporters also readied for the conflict. Backing for the dams was strong in eastern Utah towns such as Vernal, which bordered the national monument. Residents complained that a "fanatical group of nature lovers" opposed a project that would produce growth and wealth for their community. Bus Hatch told Brower that his neighbors were complaining that "those damn Californians are trying to steal our water" and that "we already have too much scenery in Utah anyway." Support for dams throughout the Rocky Mountains had always been steady, and many in the region worried that a loss at Dinosaur could spell troubles elsewhere.

"If we back down on Echo Park," said George Pughe of the Colorado Water Conservation Board, "we'll just get shoved around on other things." Congressman Wayne Aspinall, who represented the Western Slope of the Colorado Rockies and headed a key committee in the House of Representatives that would decide the issue, was even more adamant. Deleting the Echo Park Dam, he said, would be "handing conservationists a tool they'll use the next 100 years."[25]

Hearings were to begin in the House on January 18, 1954, and in the Senate in the spring. Brower had had a year of work already on Dinosaur, and he carefully considered what he should say. Most organizations wanted to stress the beauty of Dinosaur and how the dams would intrude into a national sanctuary. Brower agreed, but he wanted to go further— not just to defend, but to attack. McKay had based his support on the issue of evaporation, so Brower explored that issue for weaknesses. The science of evaporation rates was at best inexact. What was most interesting to Brower was the testimony given three years earlier by retired U.S. general Ulysses S. Grant III. The grandson of the famous Civil War general and president was a former engineer with the U.S. Army Corps of Engineers. Grant now worked for the American Planning and Civic Association, which opposed the Dinosaur dams. Grant had found an error in the Bureau of Reclamation's math on evaporation rates at the various dam sites. The most likely alternative to the two dams at Dinosaur was downstream at little-known Glen Canyon on the Arizona–Utah border, where the bureau also wanted to build a dam. In place of the Dinosaur dams, a higher dam with a larger reservoir could be built at Glen Canyon. Bureau engineers estimated, however, that this second option would cause far more water to be lost through evaporation, measured as an estimated 165,000 acre-feet per year more. But Grant pointed out that the bureau had failed to subtract the amount that would be lost at Echo Park and Split Mountain if those dams were not built. The evaporation rate was low, only 95,000 acre-feet per year. The bottom line was that by eliminating the loss at those two dams, the difference between the two alternatives was only 70,000 acre-feet per year, which made the difference between the two projects negligible.[26]

Brower talked to evaporation experts. They warned him to be cautious. The strongest advice came from Luna Leopold, the son of the late wilderness advocate Aldo Leopold and a nationally recognized hydrologist with the U.S. Geological Survey. Leopold was disgusted with the dam builders, but as a government official he needed to be careful. He was willing to help Brower, but only secretly, and he warned Brower not to argue with Reclamation engineers. They knew far more and would win any technical argument. He especially urged caution on the evaporation issue, calling it "the height of folly to argue with the bureau." He said the odds were that Reclamation engineers were more likely to be right than anyone the Sierra Club could find. "We can afford to be conservative," agreed Harold Bradley, who had initiated the Sierra Club involvement in the Dinosaur dispute. Just showing photographs of what would be lost in Echo Park would be enough to win the argument.[27]

None of this counsel impressed Brower. He felt that he had been told to "stick to your bird-watching."[28] No one quite understood Brower. He had already demonstrated his willingness to take risks, even to gamble when confronted with personal danger. For years, he had climbed mountains. He had been a part of an elite climbing corps in the Sierra Club dedicated to first ascents. He rarely backed down from a challenge.

The House Subcommittee on Irrigation and Reclamation hearing convened in mid-January. It would draw sixty-two witnesses, who testified over nine days in a hearing room crowded and cramped. Tudor, the first witness for the Department of the Interior, argued that the Echo Park and Split Mountain projects made the best economic sense. "The choice is simply one of altering the scenery of the Dinosaur National Monument without destroying it," he said. There were alternative sites, a number of them, and he produced charts listing them as well as their attributes and drawbacks. "In the final analysis the increased losses of water by evaporation from the alternative sites is the fundamental issue upon which the department has felt it necessary to give any consideration to the Echo Park Dam and Reservoir." He continued, citing detailed statistics on storage capacity and evaporation rates to justify his case.[29]

Brower took copious notes, and the next day he reviewed the transcript when it became available. He was amazed by Tudor's testimony. Although some of the numbers on evaporation rates were slightly different, Brower realized that Tudor had come "up with the same kind of error that had been made before and that General Grant had detected."[30] The bureau still had not taken into account that if the dams at Echo Park and Split Mountain were not built, there would be no evaporation loss. It had failed to subtract the numbers from their calculations.

Tudor and other dam supporters spent two and a half days describing the benefits of the dams. Then the opponents had their turn. Many were passionate. Grant testified and again mentioned that the evaporation numbers remained suspect. Subcommittee members ignored the comment. Two of Harold Bradley's adult sons, David and Stephen, showed photos of their river trips. "Echo Park is a temple which has been many millions of years in the building," said David Bradley.[31]

The committee was into its second week of the hearing before it was Brower's turn late on a Tuesday afternoon, January 26. He was nervous. "I did not take my pulse that day, but I knew it was there," he recalled.[32] He began his testimony by reminding the subcommittee of Tudor's comments that the dams would not change Dinosaur. If that is the case, he argued, neither would a dam in Yosemite National Park between El Capitan and Bridalveil Falls. "After all, the ground would still be there, and the sky, and the distant views," he said. "All you would have done is alter it, that is, take away its reason for being. Maybe 'alter' is not the right word. Maybe we should just come out with it and say 'cut the heart out.'"

Then Brower, defying the expert's advice not to argue with Reclamation's figures, questioned the evaporation rates, not only in the comparisons between the projects but also in the estimates of how much would be lost by building a higher dam at Glen Canyon. Congress, he said, would make "a great mistake to rely upon the figures presented by the Bureau of Reclamation when they cannot add, subtract, multiply or divide."

There was very little reaction from members of the subcommittee. Some may have been confused because the numbers Brower was providing in his testimony were different from those in the mimeographed statement Brower had prepared and distributed. Brower explained he had been revising them since he finished the statement the day before. The discrepancies were frustrating to some of the subcommittee members. The afternoon was coming to a close. They agreed to reconvene the next morning. They would see if Brower was really willing to continue with this assault.[33]

The next morning, a blackboard had been placed prominently in the hearing room. In addition, Cecil Jacobson of the Bureau of Reclamation's Denver office was in the room. Jacobson had been summoned to respond to Brower's charges about the evaporation rates.

Brower willingly used the blackboard and scratched out his numbers again. Now his opponents on the subcommittee were more ready, although their amazement remained.

"Are you an engineer?" asked Representative Arthur Miller of Nebraska.

"No, sir, I am an editor," replied Brower, adding that he was using ninth-grade math.

"You are a layman, and you are making that charge against the Bureau of Reclamation?" asked Representative Wayne Aspinall.

Added Representative William Dawson from Utah: "There are some 10,000 employees in the Bureau of Reclamation and 400 engineers in Denver, who have been investigating these sites and working on them. . . . [T]his is like taking the pistons out of the engine if we delete Echo Park, [and then] we must say that those engineers are all wrong."

"My point is to demonstrate to this committee that they would be making a great mistake to rely upon the figures presented by the Bureau of Reclamation when they cannot add, subtract, multiply, or divide," said Brower. "My point is not to sound smart, but it is an important thing."

When it was Jacobson's turn, he admitted to one small error, one so small he would only call it a "misprint." He disputed Brower's math—not the way he had added and subtracted, but the fact that he had even tried to tackle the complex formulas that went into estimates of how water evaporated. These formulas needed higher mathematics than what Brower had used, including a combination of algebra, plane geometry, trigonometry, and calculus. "You cannot use ratios and run the old slide stick and get any answer you want," said Jacobson.

"I do not know of any relationship here except straight subtraction," replied Brower.

Dawson, who often complained that nature lovers were a nuisance, could not restrain himself at this point. "If Mr. Tudor is such a poor engineer as you seem to claim he is, I am surprised he ever got that Golden Gate Bridge down in your town to meet at the center."

Brower replied that Tudor had worked on the Bay Bridge, not the Golden Gate.

Dawson did not care. "It would surprise me if he does not know figures any better than you say he does."[34]

The subcommittee moved on. The many dam supporters were satisfied that Jacobson and the bureau had successfully rebutted Brower's charges. Conservationists felt otherwise. Brower's logic had the ring of truth, and they were sure that it would be persuasive over time. Further, they believed that they had found a new leader for their attacks. Howard Zahniser, who headed the Wilderness Society, was so elated he sent a telegram to the Sierra Club and praised Brower's David-and-Goliath performance. "Salute him well," wrote Zahniser. "He certainly hit the giant between the eyes with his five smooth stones."[35]

Nevertheless, Brower needed further proof of his claims.

He turned to Richard Bradley, another of Harold Bradley's sons. Richard Bradley was a physics professor at Cornell University who knew almost nothing about the esoteric science of evaporation. He spent weeks talking to hydrology experts and reading the technical literature. He concluded that evaporation formulas, especially on a reservoir that had not yet been built, were extremely difficult to calculate.[36] The rate of error, he estimated, could be as high as 25 percent. It was, he wrote to his father, "evaporation-shmevaporation." He was convinced, after his long study, that *no one*—and he emphasized those words—knew what really happened on a real lake. The Bureau of Reclamation, he concluded, had been brilliant in selecting this issue to document much of its case because no one could conclusively dispute the findings.[37] In other words, the Reclamation engineers held all the cards. Brower should not have challenged their numbers. And then Bradley received a letter that changed everything.

The message came from Floyd Dominy, then an acting assistant commissioner at Reclamation and soon to be one of Brower's greatest nemeses. Dominy, who was known to play political games that undercut those ahead of him, wrote a letter that acknowledged that the evaporation numbers were wrong. The bureau had failed, as Grant and Brower had pointed out, to subtract the evaporation levels for Echo Park and Split Mountain. Then Dominy went further. He essentially agreed with Brower's contention that the amount of water that would be lost from the surface of a large lake behind a dam at Glen Canyon had been overstated. The bureau's numbers were now very close to Brower's.[38]

At Interior, Tudor was furious. "The rumor continues that some heads will fall in the bureau," Luna Leopold told Brower. Leopold was so heartened by the victory that he was now openly working with Brower. Tudor had to publicly admit the mistake. Yet, despite the error, Tudor and McKay continued to support the Echo Park and Split Mountain dams, and they seemed to continue to have the backing of a silent Dwight D. Eisenhower. The Dinosaur dams remained preferable to a higher dam at Glen Canyon.[39]

Brower was fairly certain that the engineers had not just been sloppy. Had they lied? "I don't know what's going on in their heads," he replied when asked that question many years later. "I'll use the General Grant line: I am forced to infer that they knew they were lying. I don't think that they could otherwise have been so stupid."[40]

More important was the victory won not just by Brower but by the conservation community. His brilliant, bullheaded, near-arrogant strategy worked. The enemy was wounded, the challengers were heartened. It was still too early to tell who would win, but this issue was now a fight, the greatest for conservationists since Hetch Hetchy thirty years earlier, now being led by John Muir's heir apparent.

Brower and the Sierra Club learned from this exercise. The organization clearly now had the ability to move beyond the high peaks of eastern California to become a national force and organization. Brower had discovered he had talents and leadership skills that until now had been untapped and unrecognized. Even members of his own board, men who had known him for nearly twenty years, were surprised and delighted. They had no idea Brower had such daring, such leadership, such audacity. This was just the beginning of a wild ride, as they were about to find out.

What no one, not even Brower, understood was that in working to save one great natural resource, he would be harming another, that his very success at Dinosaur would lead to a greater conservation defeat elsewhere. But that lesson was still to come, as was a greater understanding of Brower. No one realized just how talented, how complicated, how transparent, and yet how secretive this man actually was—qualities that had taken forty-two years to develop.

2. Mountains

In the summer of 1918, when David Brower was six years old, he took a trip with his family to the Sierra Nevada and Yosemite that was so memorable that he could recall it in great detail for the rest of his life. The journey with his parents and two siblings took days as their two-year-old Maxwell traversed packed-dirt, one-lane mountain roads. In Yosemite, he rebelled on the trail to Vernal and Nevada Falls, refusing to cross the surging Merced River on a fat log that protected crossers with only a rail on one side. He was equally stubborn the next day on the hike to Sentinel Dome, retreating to the Maxwell for a nap. Each night the family slept under the stars, and young Brower would fall to sleep to the wails of train whistles. It was only years later that he realized that one of those engines that he heard on those star-speckled mountain nights was laboring to destroy a paradise. It was toiling to build the O'Shaughnessy Dam, which would flood the Hetch Hetchy Valley. Construction had been authorized in 1913, and laborers then spent nearly a decade devastating a valley that John Muir had failed to save.[1] At the end of the journey, the family returned to the then-modest college town of Berkeley, California, where Brower had been born on July 1, 1912, and where he would make his home for the rest of his life.

His parents had taken different paths before settling in this community on San Francisco Bay, where there were no bridges and the surrounding hills were still wild. His father, Ross, came originally from Bath, Michigan, and had traveled west with his father, Gideon Samuel Brower. The Brower family name in America could be traced back to the seventeenth century with the arrival of the Dutch immigrant Jacob Brower. Gideon was a talented carpenter who built a railroad station in El Paso, Texas, before moving on to Fresno and California's Central Valley. According to his grandson, Gideon was a friend of labor leader Samuel Gompers and ran for governor of California on the Socialist ticket. This, David Brower pointed out, "was not a route to success." He also remembered that his grandfather usually had presents in his pockets for the children when he came to visit.[2] Gideon was married twice, and one of the wives, the formidable Susan Carolyn McKay Brower, would make an indelible mark on the family.

David's mother, Mary Grace Barlow, was born on a farm in Two Rock Valley, just west of Petaluma, California. David never knew his grandfather Barlow, who died seventeen years before David was born. His memory of his grandmother Barlow was also dim; he recalled that she died of a stroke when he was only five years old.[3] The family farm passed to Mary's half-sister Francis, whom the children called Fanny, and David and his siblings were often there. It was primarily a chicken ranch, although it also had some sheep, pigs, and cows as well as an orchard. On the farm over the years, Brower learned most aspects of farming, from pruning and plowing to milking cows and castrating pigs.[4]

According to Joseph, their youngest son, Ross Brower and Mary Grace Barlow met at the First Presbyterian Church in Berkeley and married in August 1906. Ross had enrolled at the University of California in Berkeley and earned his engineering degree in 1900. Since then he had been working as a draftsman and an electrician. Like his father, he was good at fixing things, friendly and attentive. He also had a love for the outdoors, as did Mary. Her degree was in English in 1905, also from the University of California. She got her master's a year later from Stanford University, shortly before the wedding.[5]

After their marriage, the young couple moved to Michigan, where Ross Brower earned a master's degree in engineering. Edith was born in Ann Arbor in 1907. The family returned to Berkeley, where Ross found better employment, first an engineering post at the Union Iron Works and then posts teaching in the Oakland school system and the University of California. A son, Ralph, was born in 1909, followed by David in 1912. For a while, the family rented an apartment in a twelve-unit apartment complex owned by Susan Carolyn McKay Brower, whom young David would call "Grandmother Brower," on Haste Street, four blocks from the Cal campus. Ross, Mary, and their children would rent one of the units off and on over the next few years before acquiring a half-interest in the complex and then becoming its legal owners by 1918.[6]

Susan Brower may have sold the apartment house and moved to Los Angeles, but when she returned for visits, her personality overwhelmed all of the Browers. She was a dowager. She brooked no nonsense, and she never hesitated in expressing her opinion. She also maintained a fierce work ethic. She would arise at 5:00 A.M. and work tirelessly, although never on Sunday. Her religious fervor was fundamentalist Baptist, in contrast to her daughter-in-law Mary's Presbyterianism. David Brower discovered the sheer force of his grandmother's will when she prevailed in having him "dunked in the Baptist style." His mother was "disturbed" by the act. But that was nothing compared to what happened when eight-year-old David went into the hospital to have his tonsils removed: Susan Brower convinced the doctors to wield their scalpels elsewhere, and David awoke from anesthesia to discover he had been circumcised.[7] She was a great apostle against sin and ensured that her son Ross was a disciple. Gayle Brower liked to tell a story about the reception held after she married Joseph, David's younger brother. It was at their new home, and all they could afford to serve was spaghetti, bread, and red wine. When Gayle walked into the kitchen before most of the guests had arrived, she discovered her new father-in-law, Ross Brower, pouring the wine down the sink. Susan Brower had preached against consuming alcohol, and Ross abided by that dictum. The woman, said Gayle, had declared "that there would be no dancing, no whiskey, no drinking, no women, and nothing that was considered fun."[8]

David Brower also reported that neither his grandmother nor his father "had anything to do with alcohol, tobacco, coffee, cards or dancing."[9] But their sermons did not stick. David smoked, giving it up finally in 1959, one year before his father died. He never quit alcohol. His introduction to it came shortly after he was twenty-three years old and still living part of the time at home. Ross, in a letter to his son, said that he had found a jug in a recess of the phonograph cabinet that smelled like wood alcohol. David replied in a long rambling letter, confirming that it was wine. He attempted to justify his decision to drink in moderation. He had never actually gotten drunk from drinking, he said, although he might try it some day. Ross seemed slightly nonplussed by the confession but admitted he could not dictate whether his son drank.[10] During his life, David sometimes did drink heavily, from the notorious two-martini lunches to the 4:00 P.M. cocktail hour at home and the late-night forays until the bars shut down. Some of his associates worried that alcohol too often affected his performance.[11] But as with many aspects of Brower's life, views were mixed on how much his drinking affected him.

Until 1920, the Browers lived a prosperous, comfortable, middle-class life with every expectation that this life would continue. Then it all changed. After the birth of Joe, Mary Brower had trouble recovering. In the hospital, her condition deteriorated. She lost some of her hearing and then her sight. The doctors were helpless. It took a very long time before they finally diagnosed that she had an inoperable brain tumor. It also took a long time before she was able to go home. When she did, she was an invalid.

David Brower's writings and his statements in interviews sought to cast his mother in those subsequent years as "valiant." He poetically described hikes with her in the Berkeley hills. She had been a good walker before she became ill, and she could still get around fine as long as someone led her. David became her eyes on these hikes. He described not only potential hazards on the trail but also the beauty of the natural surroundings. He said he gained his power of description from many of those walks. He remembered other such scenes, too, such as when she sat in the front parlor playing Rachmaninoff and Liszt on the family's Ludwig piano or when she stopped "whatever she was doing, listening attentively, with a pleased smile" while he played (figure 1).[12]

Those days were rare. Her incapacities meant that she could no longer reliably take care of the children, including a new infant, nor could she care for the house or the tenants. Sometimes she would be stricken with spells, terrible seizures, which confined her to her bed. She turned to God and became devoutly religious. She would crank up the family radio so loud that the entire neighborhood could hear "The Country Church" or "The Hour of Prayer."[13]

By then, Ross Brower had lost his teaching position at the University of California. No longer able to leave his wife or small children alone, he became a full-time landlord of the twelve units on Haste Street. He expanded the two-story Queen Anne Victorian on the front of the lot by building dormers in the attic where the children could sleep. Tenants were students, working people, and seniors. Rents were low, but filling the house was always a challenge, even more so once the Depression struck. Ross also got part-time employment at the University of California Extension Division, hawked fire extinguishers for a time, and lost what little inheritance they had on bad real estate investments. He was not a good businessman.[14] A young woman who was a tenant in the house described how Ross once agreed to rent her a room for the ridiculously low price of $10 a month because it was all she could afford. What impressed her the most, though, was not his generosity but his devotion to his wife.[15]

Mary Brower's spells would be hard on the family. When their mother was ill, one of the children often stayed home from school as an extra caretaker. When she was feeling better, she could sometimes do light housework, especially sweeping the stairs. At one point when David was away from home, Edith told him in a letter that their mother had not been well for two weeks and that she had taken a fall that knocked her unconscious. Mary died suddenly in December 1939 at the age of fifty-seven. Joe Brower remembered that his mother was stricken with a seizure in the bathtub and drowned.[16] By the beginning of America's entry into World War II, Ross Brower had married Mae Gabrielson, who had children of her own, and David Brower no longer felt comfortable in the house of his

FIGURE 1 Young Brower playing the piano

Brower's mother would often stop whatever she was doing so that she could listen to her son play the piano. He continued to play throughout his life, sometimes sharing the keyboard with Ansel Adams, who had once studied to be a concert pianist. (Courtesy of the Bancroft Library, University of California, Berkeley) youth with a new stepmother, stepbrothers, and stepsisters. Ross Brower would also die suddenly, but not until 1960, when he was eighty-one. The Haste Street apartments passed out of the Brower family, although locally the buildings are still called the "Brower house."[17]

As for the Brower children, Edith grew into a tall, statuesque beauty who sang, played the piano, and won the California state typing championship. She enrolled at the University of California but did not graduate, and when she was twenty-one, she defied her father's wishes and married her childhood sweetheart. Her oldest brother, Ralph, excelled academically and graduated from Cal with an engineering degree. He was a stickler for details, and in later years the engineer whose task was to build did not always agree with the brother who espoused nature over development. While the four siblings were growing up, the responsibility for watching Joe, who would later become a commercial airline pilot, fell primarily on David, who was eight years older. Joe said David never complained when he tagged along with his big brother on hikes or for neighborhood football games. David was always fair, insisting that everyone be picked for a team, no matter how scrawny or young. "Oh, he was a very good big brother, as good as one could have," recalled Joe. Yet as hard as Ross Brower and his children tried to make Haste Street a normal household after Mary became ill, it still lacked a maternal, feminine touch, according to Gayle Brower, Joe's wife. The Brower children did not learn how to dance; they did not learn the proper etiquette of writing thank you letters; and if they had friends over, those children were rarely invited inside. "Social graces," she said, "were often lacking in the Brower children."[18]

David also excelled academically, eventually skipping three grades in grammar school, though that likely exacerbated his social standing with his decidedly younger classmates. By junior high school, he experienced a series of illnesses that often kept him out of the classroom. He missed nearly a term when he had the mumps, then impetigo, and then burns when gas from the hot-water heater at home blew up when he was attempting to relight the pilot.[19]

In 1925, at the age of thirteen, David discovered butterflies. Two young tenants, Al and Fred Furer, living in the Haste Street apartments with their parents, introduced him to this past-time. Brower began collecting, keeping a journal, where he laborious typed each entry into a small notebook. By the end of the summer, he had captured and mounted thirty species and became enthralled by the science of entomology. The next year he roamed the Berkeley hills and even took a trip to Yosemite to add to his collection. His greatest success came April 17, 1928, marked precisely in his journal, when he found a butterfly so unusual that he could not locate mention of it in any of his guidebooks. Its key distinction was that it had an orange tip. By now, Brower was communicating with other butterfly collectors, and one of them, Jean Gunder of Pasadena, California, paid Brower $10 for the specimen and then had it officially named "Broweri." It was only later that Brower realized that the butterfly probably had been a transition form, a mutation, just an oddball. "It might just be one," he said. "So that was really an endangered species, and I killed it. I wiped it out."[20]

Roaming the hills with a butterfly net could be perceived as effeminate and markedly different from the behavior of his peers. In the hills, when he spotted someone approaching, Brower would combat this self-conscious "oddball" image he had of himself by breaking into song or whistling, preferably a popular show tune of the day. Collecting, said Brower, was lonely, ostracizing, and even embarrassing.[21]

A new neighbor, Donald Rubel, arrived to help end the loneliness. Rubel was everything Brower wanted to be. He was sturdy, good-humored, and even witty at times, and most of all he was a good athlete. Where the Brower family was scrapping for cash, the Rubel family was comfortable. Don's father was employed at the university's Agricultural Exchange; they had a new home, a Steinway piano, and two Dodges. More importantly, though, Rubel introduced Brower to football, installing him as a guard on the Regent Street Varsity, a motley football sandlot squad whose home field varied around the neighborhood. Brower knew so little about football that at first he tried to block and tackle on both offense and defense. But by the time he was in high school, he could toss a football 60 yards down the field and recite the score of seemingly every gridiron encounter between Cal and its archrival Stanford. He would eventually grow to be six feet, two inches tall, but when he entered Berkeley High School, he was two to three years younger than his peers. It was obvious that Brower did not have the physique—in his words he was "too skinny"—to make the famed Berkeley Bees.[22]

What he made up for in improved social standing, however, he lost in the classroom. He had to repeat both Algebra 3 and 4, and he admitted that he did his share of procrastinating on academics in high school. But because of his grade-school advancements, he graduated midyear when he was sixteen. He enrolled immediately at Cal. His academics did not improve. Plus, he continued to struggle socially. He pledged to a fraternity and remembered the day several of the brothers made a quick inspection of his living quarters on Haste Street. They rejected his application. He was younger, poorer, and probably less prepared for college than others, and by his third semester he dropped out. By then, he was working most of the year for a candy manufacturer in San Francisco and at a camp during the summer. An entry in his diary at the time said that it was painful to leave college, but at least he was earning enough money to help the household.[23]

Brower blamed his father for pushing him to start immediately at Cal after high school graduation. Don Rubel graduated from high school at the same time, but Brower believed that Rubel fared better in college because his parents withheld any pressure, and Don entered in August instead of January. The lack of a degree foreclosed some opportunities for Brower but may have opened others. Following World War II, when Brower was approached about taking a position in the history division of the War Department in Washington, D.C., in the letter accompanying his application he wrote that he wished he could have gone farther in college, but only because the degree looked better on job applications. The reply kindly but bluntly told Brower that without a degree he could not be hired. In the 1960s, when he was campaigning to save the Grand Canyon, he would joke that he had graduated from the University of the Colorado River. Brower's future wife, Anne, believed Brower was lucky not to have finished at Berkeley because it might have pigeon-holed him in a lesser career. "David dropped out of school," she would say, "before they could teach him what he couldn't do."[24]

Stymied, Brower discovered mountain climbing. In the summer of 1933, when he was twenty-one, he spent several weeks with his friend George Rockwood hiking in the Sierra Nevada. One night Brower arrived in the small dining room at Glacier Lodge to discover a mountain-climbing legend, Norman Clyde. Since Clyde was having dinner by himself, Brower and Rockwood joined him, and soon they were swapping tales of hiking and climbing. Clyde would eventually summit a thousand peaks in his lifetime and make 130 first ascents in the Sierra Nevada, but by 1931 he had already opened up a new era of alpine climbing as a member of the first party to scale the steep east slope of Mount Whitney. He often climbed solo, and he carried an enormous pack. He had a long waist and short legs, and he dressed in a way that exaggerated that look. He did not present as a world-class climber. Clyde was also quick-tempered and irascible. He had lost his job as a high school principal five years earlier by firing shots over the heads of Halloween vandals. He was on his way to becoming a recluse, often living out of his car or in deserted cabins.[25]

On this night at the Glacier Lodge, a mountain resort at 8,000 feet in the eastern Sierra Nevada, Brower had his own climbing story to tell. He had been climbing nearby in the Palisades, which at 14,000 feet have one of the grandest views in the Sierra Nevada. Brower had left Rockwood behind and was scrambling up a vertical rock column called the Thumb. He saw some holds and grabbed one, but the rock was rotten, and it broke, careening down 75 feet. Desperate and about to fall, Brower reached out with his left hand and with three fingers caught another hold before he could slip. Shaken, he cautiously moved to where he could rest and be more secure. Clyde told Brower he needed to learn more about climbing and especially about the three-point suspension. In difficult situations, Clyde said, Brower should make sure that three limbs were in safe, secure holds before using the fourth to reach for the next hold. It was good advice; Brower used it a few days later in climbing the North Palisades and would use it the rest of his climbing career.[26]

That seven-week trip in 1933 with Rockwood was noteworthy for two other chance encounters. One was with a college student, Hervey Voge, who was hiking the opposite way on the John Muir trail. Voge was to become a good friend and climbing companion, but his advice at this first meeting was especially helpful. He told Brower that if he was that interested in climbing, some of the best mountaineers were in the Sierra Club. Voge was a climber and a member. Brower should join. The second encounter was in Hutchison Meadow, high in the Sierra. Brower in the distance saw a bearded man carrying a tripod and camera. When the man got closer, Brower correctly guessed his identity and addressed him: "You must be Ansel Adams." Brower had already been reading the *Sierra Club Bulletin*, which often featured Adams photos. Adams, Brower, and Rockwood lingered in the meadow for a moment as Adams complained about the cumulus clouds that were too fuzzy to photograph, and then they parted.[27]

Brower's interest in mountain climbing could not have come at a better time or place. The dawn of modern mountain climbing in the American West had begun, and the leading players would include this San Francisco–based hiking club and the glacial granite cliffs of Yosemite. A famous British climber, Robert Underhill, had started working with a cluster of Sierra Club climbers, including Clyde. Over the next decade, this group of climbers, along with late converts such as Brower, would make a name for themselves. Brower would be credited with thirty-three first ascents of peaks in the Sierra Nevada, and he eventually reached every 14,000-foot or higher summit within that range. He and his comrades arrived on the mountain-climbing scene at a time when the technology of mountaineering was undergoing a transformation. Tools such as pitons and carabiners, which were pounded into the rock to anchor ropes and mountaineers, were allowing climbers to reach what in the past had been insurmountable summits.[28]

The Sierra Club climbers, both men and women, trained every Sunday at Cragmont Rock in the San Francisco Bay area. The climbers gave code names to the various cliffs and rock outcrops; the training was rigorous; and it was led by Richard "Dick" Leonard.[29] Leonard was methodical in jotting down notes on attendance, training, and climbs, and he worked out a detailed system to rate various climbers. In one of the scorecards Leonard recorded in 1933, Brower was ranked tenth best out of the seventeen. He had one of the lowest scores for experience, which was understandable considering how fresh he was to the sport. He scored one of the highest rankings for climbing technique and in the midrange for judgment. Brower learned quickly, and soon he asked Leonard and Voge to sponsor his membership into the club. They did.[30]

The next summer, in June 1934, Brower and Voge headed to the mountains to test their climbing expertise. They wanted to "peak bag"— to climb to summits, no matter how low or easy, where there were no records of anyone venturing there in the past. Some of these climbs did require ropes or even a piton or two. At one point, Norman Clyde, who agreed with their goals, joined them. Clyde, said Brower, was a fellow peak bagger.[31] In an account of the trip, Voge described one day when at 5:00 P.M., after a full day of climbing, he elected to return to camp and make dinner, but Brower and Clyde decided to investigate one more peak. They did not return until midnight, having made first ascents on three other peaks. That, wrote Voge, indicated how strong the lure of climbing could become.[32]

Over the next few years, Brower would make Yosemite and its many cliffs his climbing goal. More important than the conquests was the realization that he enjoyed the challenge, the adrenalin rush of the climb, and the gusto and recognition atop the summit. He sought first ascents; he tallied his records and hustled to get them recognized and publicized. He would lie on the bed in his third-floor room on Haste Street gazing up at topographical maps secured to the ceiling, reviewing triumphs and planning future ventures. He would write long letters to friends and relatives detailing his accomplishments. He would sell complete climbing tales to local newspapers and then eventually to major publications such as the *Saturday Evening Post* and *National Geographic*.

His first significant public exposure came in the summer of 1935 as part of a team that challenged Mount Waddington in British Columbia. The 13,260-foot peak was 200 miles north of Vancouver. Brower called

Waddington a "killer mountain" that was "impregnable," especially the last 60 feet to its rock-ribbed true summit. It was pounded by 200 inches of rain a year, and in the summer blizzards would be followed by sun, increasing the avalanche danger. Twelve previous attempts to climb it had failed, and one climber had died. The climbing crew was organized by Leonard and headed by another Sierra Club veteran, Bestor Robinson. They spent months preparing. Brower financed his trip by negotiating a $300 contract to write dispatches for the North American Newspaper Alliance.[33]

Seaplanes in late June 1935 dropped the team off at sea level, more than 13,000 feet below Waddington. They spent two weeks on reconnaissance missions, carrying their gear up through dripping emerald forest to icy-white glaciers. Finally, with only a few days left, they camped on Angel Glacier at 12,550 feet. Brower and four other climbers began a final reconnaissance trip to the snow summit, which they knew would be short of the true summit. Two hours later, the first climber, Jules Eichorn, reached the snow summit and exclaimed, "My God, look at that!" The others scrambled up and stood for a moment transfixed. None of the photos they had studied, none of their telescopic reconnaissance from lower base camps, had prepared them for what was ahead. The true summit was a rocky pinnacle, with knife edges steep and polished. Those rocky faces fell on most sides for thousands of feet to icy glaciers below. The climbers debated routes and finally agreed to retreat and try again the next morning. But that night a violent blizzard arose. They had camped in the lee of the lip of a large crevasse, but "the wind shook the tents with a dreadful fury," Leonard wrote. "The fabric snapped like pistol shots." It was still snowing the next morning, and the team knew that they had run out of time; it would take too long for this latest storm to clear for them to reach the summit. They retreated. The next year Leonard, Robinson, and a new team that included Voge mounted a second failed attempt of Waddington. Brower could not make that trip. A rival climbing club finally succeeded later in the summer of 1936.[34]

The Sierra Club climbers were at the forefront in the climbing community because of their embrace of new equipment, a tactic led by Leonard, who did not tolerate failure. Four years older than Brower, Leonard was born into a military family that traveled often before they settled in Berkeley when Leonard was fourteen and his father had died unexpectedly. He was an excellent athlete and a disciplined student at the University of California. He studied geology, chemistry, and electronics before going to law school and starting his own very successful firm that specialized in business law. When Leonard took up climbing, he seized on the idea of using pitons, which were wrought-iron eye spikes; carabiners; large oval snap rings; and nylon rope that was more flexible than other alternatives. Leonard dramatically demonstrated why this new equipment was so important by assaulting the Cathedral Spires, two massive pillars lining one end of Yosemite Valley. Both were considered unassailable. After two unsuccessful attempts to scale them, Leonard returned for a third attempt in April 1934 and finally prevailed, stunning many in the mountaineering community but scandalizing some because the climbers used pitons unsparingly.[35]

When Leonard attacked the Cathedral Spires, he was also assaulting the sport of mountain climbing. Until this point, the sport was rooted in the philosophy of Victorian Britain that ranked the types of tools that Leonard, Brower, and others were employing as secondary to the climber's wits and ardor. Leonard's gear was not considered "sporting"; some critics even suggested it was cheating. Interestingly, the discussion had more to do with honor and far less with the environmental aspect of placing iron in rock and defacing nature. Leonard and others argued strongly that the reason they used tools such as pitons was primarily to ensure the climber's safety. Obviously, however, it also gave them a decidedly new advantage over climbers of the past. Members of the Sierra Club often argued among themselves about these ethical concerns, and Brower was an integral part of that debate. In a letter he wrote in 1936, he conceded that there was still considerable leeway and judgment on deciding when a mountain route needed pitons and rope and when it was safe to go ropeless. In other words, the line between being safe and having an advantage could be blurry. But neither Brower nor others spent much time discussing how they might be scarring the mountains that they held so dear.[36]

Brower had missed the second Waddington climb because by then he was working and living in Yosemite Valley, doing publicity work for the park. Brower relished living in Yosemite. By his second winter, he had become an adept skier and a winter mountain climber. He notched a number of winter first ascents. But success could give way to overconfidence and poor judgment. One summer day Brower convinced an office worker with no climbing experience, the man's wife, and their four-year-old son to accompany him up the side of a Yosemite cliff. What may be one of the most surprising aspects of this story is not just that Brower made a mistake, but that he did not even seem to be aware of it when he described the journey in letters to Leonard and Voge. Here was another Brower characteristic that would periodically crop up: at times he could make positively brilliant judgments, but at others he could make seriously flawed decisions. And he often could not recognize or acknowledge his error.

The four of them began their afternoon venture by climbing up the rocky Bridal Veil Creek by way of a chute at the Lower Chimney Rock, which is on the south side of Yosemite Valley directly opposite the granite cliff of El Capitan. Farther west the creek plunges and becomes, especially in the spring, the spectacular Bridal Veil waterfall. The family used ropes without a problem, Brower said, although when he got to one large rock, he literally had to push the woman and the child over. He reported that the family seemed delighted by Bridal Veil Creek, and at 4:00 P.M. they needed to head back. That's when problems surfaced. The woman, who had been far more frightened than she had let on to Brower on the way up, announced there was no way she was going to descend by the same route. Brower knew an easier way down beyond Upper Cathedral Rock, but there were no trails, and they would have to go cross-country. He carried the young boy, Stuart, across what seemed like acres of manzanita and scrub oak, up and down ledges, even killing a rattlesnake at one point, until it became quite dark, and they still had a 2,000-foot descent ahead of them. They decided to spend the night on the mountain. Fortunately, Brower had brought some emergency provisions; there was water from the creek, but no flashlight. The next morning they made it down without incident. Brower said that the couple told him that the climb had been unique and enjoyable. He did not indicate that they had any complaints, although he did mention that he had dispensed aspirins to them the night they spent on the mountain.[37]

Brower's most important climb came in 1939, when he and three other Sierra Club climbers were the first to get to the top of Shiprock (figure 2). At the time, mountaineers had labeled this 1,500-foot-tall rocky pinnacle "impregnable." Because of a story in the July 22, 1939, *Saturday Evening Post*, it was also the one summit many wanted to climb. The *Post* had reported on the most recent failure to get up the middle and highest of the three towers.[38] Together, rising abruptly out of the New Mexico desert, the towers resemble a nineteenth-century schooner. The Sierra Club crew that arrived there in October 1939 consisted of Brower, Bestor Robinson as leader, Raffi Bedayan, and John Dyer. They gave themselves five days to mount Shiprock. All four of them would be needed for a successful climb.

The most recent climbers had tried to reach the middle tower by beginning with the north tower, and so did the Sierra Club climbers. But when they reached the place where the others had failed, Brower and his colleagues had a new weapon. Along with the usual supply of ropes, pitons, and carabiners, they brought expansion bolts and stellite-tipped rock drills. It is not clear if they used the drills, but Brower described how the climbers spent up to half an hour pounding in the supports for the expansion bolts. Robinson was clearly sensitive and expected criticism from the greater climbing community for the expansion bolts, which few climbers had used. Their value was that they could be used in wider cracks, too large to hold pitons. Robinson wrote, "We agreed with mountaineering moralists that climbing by the use of expansion bolts was taboo. We did believe, however, that safety knew no restrictive rules" and that if the bolts prevented a fall, their use was justified. When the climbing team reached particularly tricky overhangs, they would pound in three or even four pitons just in case one or two were to let go. They spent the final night up in the rock, the four men crowded into a two-person tent, and scrambled to the top the next morning. At the summit, Brower did not feel the usual surge of elation. For him, the climax had been solving the puzzle during the ascent, not what was on top.

FIGURE 2 Brower mountain climbing

Brower began climbing in the Sierra Nevada when he was twenty-one years old, and he joined the Sierra Club to hone his climbing skills. He became a "peak bagger" intent on capturing as many first ascents as possible, including Shiprock in New Mexico. (Courtesy of the Bancroft Library, University of California, Berkeley)

The Shiprock climb brought national attention after the *Saturday Evening Post* published Brower's account of the climb and color photographs in its February 4, 1940, issue. The $500 check he was paid for the account was appreciated, but so was the note Robinson received from Robert Underhill, who eight years earlier had helped spur what was now being called a golden age of climbing for the Sierra Club. Underhill called the Shiprock climb the "finest thing done in rock-climbing on our continent."[39]

Brower savored the recognition. He had come a long way from the eight-year-old who in Yosemite was too afraid to cross a log bridge over the Merced River and too timid to hike to Sentinel Dome.

3. The Club

The mountains had offered David Brower a lifeline, a way to boost his confidence and to escape from a difficult childhood. It was mountain climbing that led him to the Sierra Club. His admission into what was still a very elite, privileged club would allow Brower, the college dropout, into the salons of the wealthy and influential. The lawyers, doctors, and businesspeople he met at the Sierra Club would offer him new opportunities. He certainly earned their respect, both for his skill on the mountaintops and for his willingness to toil tirelessly for a pittance on behalf of the club and its causes. But his membership in the club was a sinecure, a lucky one, that came just at the right moment and would allow him to become an academic editor, an army officer, and eventually an environmental leader on a national and then international stage. And more than once the closest friendships that he had made through the Sierra Club would aid him at a critical moment.

Brower joined the club in 1933. The organization in those days was very different from what John Muir must have envisioned when he helped found it in 1892 (figure 3). Muir was a radical conservationist nearly a century ahead of his contemporaries. But the Sierra Club by the 1930s had become comfortably establishment and conservative. Its hallmark was leading summer outings, not crusading for nature. It was cozy and insular, with the emphasis on the word "club" in its name. The organization

FIGURE 3 Theodore Roosevelt and John Muir

John Muir founded the Sierra Club in 1892 to promote the preservation of wilderness. While president, Theodore Roosevelt met Muir and was instrumental in setting aside millions of acres of wilderness under the care of the National Park Service. (Courtesy of the Bancroft Library, University of California, Berkeley) was broken up into chapters, and entry could be challenging. Applicants needed recommendations from at least two members to join. They sometimes had to appear in person and face sharp questioning. They might be asked to describe their experience in the mountains or elaborate on whether they would attend and run social events for the club. For years, some chapters also insisted that members be white. The Los Angeles chapter was especially notorious for its racial politics, and in 1945 its policy so angered Ansel Adams that he proposed at a board of directors meeting that the chapter be abolished. He failed to get a second to his motion. In the 1950s, many chapters instituted loyalty oaths.[1]

Members elected their leaders, including the president, who headed the board of directors, but there was no campaigning. Muir was president for twenty-two years until he died in 1914. The charter membership was composed of professors and administrators from the University of California and Stanford University as well as some professionals and businesspeople. Muir's reign was peaceful until he took on the proposed Hetch Hetchy dam project. Muir called the plans for the dam within Yosemite National Park a sacrilege, but many club members were businesspeople with ties to the dam's developer, the city of San Francisco.[2] The schism within the Sierra Club over this issue was great, and after Muir lost the Hetch Hetchy battle, his successors were more cautious. "The attitude in the '20s and '30s was one of almost total ignorance," said Brower. "Whatever lessons John Muir had provided to the board of the directors of the Sierra Club had been totally forgotten by the late '20s."[3]

The club was run by a handful of men: Joseph LeConte, whose influential family would erect a lodge in his honor that is still in use at Yosemite; Francis Farquhar, the scholarly patrician who edited the *Sierra Club Bulletin* for years; and especially William Colby. As a young lawyer, Colby had been the assistant to a man considered the nation's greatest authority on mining laws, Judge Curtis Lindley. Colby succeeded Lindley, acquired his influence, and fattened on retainers from virtually every major mining company in the West. The same pattern followed between the young Colby and the older Muir. "He fell in love with Muir almost as a God; he worshipped him," said Richard Leonard. "He devoted huge amounts of time to Muir and the Sierra Club." Once Muir was gone, Colby, Farquhar, and LeConte would trade the presidency or farm it out to others, with the tacit understanding that Colby was in charge. He would sit on the board of directors for forty-nine years and serve as its secretary for forty-four. Until age caught up with Colby at the end of World War II, the Sierra Club was his.[4] The club's reputation as a conservative institution was so great that it persisted for years after Colby's power waned. Edgar Wayburn, who would become one of the Sierra Club's most prominent leaders in the second half of the twentieth century, once asked a San Francisco–based lawyer with a strong interest in the environment why he never joined the Sierra Club. "Oh, I've been anti-establishment," he replied, "and the Sierra Club was too much establishment for me."[5]

Brower joined the club to find climbing companions, but he also discovered camaraderie and conviviality. It was the Depression, so work came fitfully, and he spent much of his time volunteering for the club. He began by cataloging maps and recording first ascents and other accomplishments, and then he moved on to writing and editing the club's monthly publication, the *Sierra Club Bulletin*. Farquhar heavily edited Brower's first submissions, but he encouraged Brower. In 1935, Brower joined the *Bulletin's* editorial board. With Farquhar's oversight, Brower began editing copy and working with authors and printers. Farquhar was demanding, but Brower responded well and learned quickly.[6]

In 1935, Brower found a job at Yosemite National Park, and he came under the influence of Ansel Adams. The Curry Company, which had been operating as Yosemite's concessionaire since the nineteenth century, hired him to do accounting, but in a few months, with Adams's assistance, the company offered him a post doing publicity.

Yosemite in the 1930s was significantly different than it is today. The National Park Service now tries to limit human impact in its parks. In the 1930s, Yosemite Valley was a small town with hotels, restaurants, swimming pools, stores, golf and tennis courts, and other special amusements to lure visitors. One of the big attractions at night was discarding garbage at the local dump, which essentially meant feeding the bears. It drew a big crowd. Another was firefall, a nightly event featuring a large bundle of live embers and burning logs thrown over the side of the cliffs of Granite Point. Firefall had begun decades earlier, and the 9:00 P.M. event could be seen across the valley because the descent was about 900 feet. In those days, the National Park Service played a secondary role to the concessionaire, Curry.[7] Brower's job was to make the park desirable by taking photos, writing press releases, and designing brochures touting the park's lures. He was good at it, although in later years he would express embarrassment about his huckster role.[8]

One reason Brower flourished was Adams, who was already becoming one of the great photographers of the twentieth century. The two men had much in common. Although Adams was ten years older than Brower, both had been raised in the Bay Area; both had spent summers in their twenties in Yosemite and the Sierra; and both played the piano—Brower was adept, Adams a master. Adams spent years training to be a professional musician and did not give it up until 1930, when he was twenty-eight years old. He was lured away by photography and the Yosemite Valley. He first visited Yosemite when he was twelve, got his first full-time job at seventeen when he managed the Sierra Club's LeConte Memorial Lodge for the summer, and met and married Virginia Best, who lived in the valley. Together they made their first home in Yosemite, running the Best Studio, which her parents had established. His photography career flourished after he met Albert Bender, a San Francisco art patron in 1926, who promoted Adams's work. By 1935, when Brower went to work in Yosemite, Adams was exhibiting and collaborating with such elite artists, photographers, and promoters as Alfred Steiglitz, Georgia O'Keeffe, Dorothea Lange, and Edward Weston. Yet the artistic commissions that Adams garnered needed to be offset by commercial assignments from such clients as Bank of America, Kennecott Copper Corporation, U.S. Potash, and the Curry Company. For a man who loved the wilderness and had a strong environmental conscience, it was not always easy to work with such clients. Adams shared Brower's devotion to the Sierra Club, generously giving the club and the *Bulletin* free use of his photographs. In 1934, Adams was elected to the board of directors. It would finally take a heart attack to get him off the board in 1971.[9]

Brower spent hours with Adams in the photographer's darkroom learning lessons on photography. When they tired of that, Adams would talk about Yosemite and the true meaning of a national park. Adams disliked Curry's commercialism and its carnival approach to Yosemite. He wanted nature preserved for nature's sake, and he tried to convey that through his words and his artist's eye. "He led me to seeing what was behind and within a photograph—his, not mine, and what could happen when words and photographs work their magic together," Brower said.[10] When Brower was not in the darkroom, he might be in Adams's gallery or house. The gallery was split level, with items to buy on the lower level and a row of Adams prints on the upper. There was also a Steinway piano. "When Ansel played the Steinway you almost regretted he had long ago decided to give up the concert stage for the lens," said Brower. Brower saw him play the piano with an orange and do a bump, bump, bump of the "Blue Danube," rising from the bench, turning, and playing the chords with his rear end. Michael Adams, Ansel's son, remembered that Brower was often at the family home and that his father sometimes used Brower as a model. "He was part of the family," said Michael.[11]

Despite Brower's disclaimers about his role as a Yosemite publicist, it may have been the best job he had during the 1930s, and it lasted only eighteen months. The job was eliminated; his photography assignments went to Adams, and another man was brought in to write copy. The decade would be filled with these short-term positions interspersed by summer journeys into the mountains to hike or climb. No one questioned Brower's conscientiousness or his ability; he just had other interests. His bosses at Curry offered him a job the next summer as a ticket seller and guide in their touring cars that shuttled guests around the park. A Hollywood film company hired him to do publicity film work for Curry in Yosemite.[12] It was great training for his future tasks as the Sierra Club's executive director. Then Adams proposed that the Sierra Club hire Brower as an executive secretary to handle various administrative tasks. At this time, the club had only a part-time secretary. The board considered but did not accept the proposal. Instead, in 1938 Farquhar made Brower associate editor of the *Sierra Club Bulletin*, thus significantly increasing his workload.[13] The club also agreed to hire Brower part-time, and from 1939 to 1941 he ran outings in the mountains and produced a film for the campaign to make the Kings Canyon area a national park. By 1941, he was so well known throughout the Sierra Club that he was elected to the board of directors.[14] And yet he was now approaching thirty, and although he loved climbing mountains, it would not pay the bills. He had had a string of jobs that led nowhere, he was single, and he was living at home. His situation did not look good.

And then, through the Sierra Club, a more tangible opportunity was offered. In spring 1941, Farquhar told Brower that his brother Samuel, who ran the University of California Press, had an opening for an editorial position. He urged Brower to apply. With Francis Farquhar's backing, Brower met Samuel Farquhar and Harold Small, the editor. They hired him.

He was to share an office with Small's editorial assistant, Anne Hus. The relationship did not begin well. She showed him around the office on his first day, introducing him. She seemed cold, perhaps resentful. He suspected that she was miffed. He had less experience than she did, yet because he was a man, he was going to be paid more. Plus, she had to give up half of her office to him.[15]

That is how it began, with exchanges in their office that were brief and businesslike. Over time, Brower began to learn more about Anne, and the relationship blossomed. He discovered that Anne had been born in Oakland and raised in a family that encountered some of the same troubles, at least financially, as Brower's. Her father had emigrated from Germany, worked as a dentist, could speak four languages, and was an amateur musician. Then, despite having a wife and two children (Anne's brother, Francis, was seven years older), he suddenly abandoned dentistry for more speculative get-rich schemes. Those schemes failed, the Depression arrived, the family struggled, and eventually they moved in with Anne's maternal grandparents. By the time she graduated from high school, she needed to take a secretarial job with Farquhar at the press to be able to take classes at the university part-time. It took her thirteen years to earn her English degree, and by the time she met Brower she had finally advanced to editorial assistant.[16]

To break the ice, Brower teased her. The story most often told in the family was about the time Anne was working on a particularly difficult manuscript overloaded with footnotes. While she was at lunch, Brower typed out a page that mimicked the style of the manuscript, complete with ridiculous footnotes and an increasingly ludicrous text. She was halfway through the page before she understood the stunt and laughed in delight. However, she forgot to remove the page when she sent the manuscript back to the author. He was not pleased and complained to Farquhar and Small. They liked the joke and sided with Anne.[17] At the time, she had a beau, Paul Gordon, who had given her a small wooden duck that sat on her desk. When Anne was present but others were not, Brower would shoot the duck with a rubber band. The duck would fly into the next office. He was encouraged each time she laughed. But he could not seem to compete with Paul, who, like Anne, had an English degree. He envied how Anne and Paul could get into deep conversations about English literature.[18]

Brower had never had a serious relationship with a woman. He had had a crush on Betty Hillier when he was eighteen years old and working at a summer camp in the mountains. He sometimes dated Charlotte Mauk, and it was clear to some that Charlotte was in love with Brower. But he never returned that affection. Was it because she was less than a beauty or something else? Shyness? Discomfort? "I don't think Dave was ever comfortable with women," said Patricia Sarr, who worked with Brower years later and considered him only as her boss. He seemed to "radiate discomfort" around women, including herself.[19]

The courtship of David Brower and Anne Hus would be as unusual as their fifty-seven-year marriage. They never really dated before the wedding. She did take him to lunch for his birthday, and a couple of times they had sherry at her parents' house. By now, the war was under way, and on October 12, 1942, Brower left Berkeley and his post at the press to join the army. He called Anne to say good-bye, and they exchanged a few letters. Then in December, he received a card from her. She had drawn a traffic signal with a green light. The message: Paul was gone, she was available. He sent daily letters to her and eventually one that included a marriage proposal. "Since I had never kissed him, it seemed a little abrupt," she joked years later. But she accepted. "We went to our first movie when we had been married two weeks," she added. "That was our first date. So I married a total stranger, actually."[20]

They were married on May 1, 1943 (figure 4). The Sierra Club board of directors was to meet that afternoon, and the timing of the wedding became precariously difficult because so many of the guests—including the groom—had to attend the meeting. It was unclear when the meeting would end so that the ceremony could commence. For a wedding gift, Brower gave his bride a membership in the Sierra Club.[21]

After Pearl Harbor, Brower had one goal—to be an army officer and use his mountaineering and skiing experience. Army recruiters told him that being an officer was impossible for him because he did not have a college degree. But what he did have was Dick Leonard and the Sierra Club.

Two of his Sierra Club hiking companions, Dick Leonard and Bestor Robinson, had gone into the army in 1941 to help prepare mountain troops. When Brower appealed for help, Leonard wrote letters of support and even intervened to counter orders that would have meant Brower's expertise was not used. Leonard helped him become and remain an officer in the Tenth Mountain Division, which has been called the most unusual unit in U.S. military history (figure 5). The National Ski Patrol proposed the Tenth shortly before the war and recruited many of its members. Recruits were required to supply three letters of recommendation on their moral standing and skiing ability.[22] Brower applied and then waited as the spring of 1942 turned to summer, growing increasingly frustrated. Leonard's letters were crucial, and the delays actually served as a blessing. By the time Brower went to basic training, the Camp Hale base in Colorado was teeming with old hiking buddies from the Sierra Club and elsewhere. He impressed the brass. Soon he was doing basic training in the morning and avoiding other tasks ranging from 15-mile marches to KP duty. Instead, he was assigned to write a new mountaineering manual for the army. He proved so invaluable for his writing and editorial skills that at one point he worried whether he would be allowed to go to Georgia for officer training. But he did. By the time he left Georgia as a second lieutenant, he had just enough time to return to California in late

FIGURE 4 Newlyweds David and Anne Brower
Brower met Anne Hus when they were book editors at the University of California Press. He did not begin to court her until after he went to basic training in the army in October 1942. They married on May 1, 1943, a union that lasted fifty-seven years and produced four children. (Courtesy of the Bancroft Library, University of California, Berkeley, and the Brower family)

FIGURE 5 Brower in the army
During World War II, Brower served in the Tenth Mountain Division of the U.S. Army, working as a mountain-climbing trainer first in Colorado and later in West Virginia. He was overseas in Italy and returning home when the United States detonated two atomic bombs, and Japan surrendered. (Courtesy of the Bancroft Library, University of California, Berkeley)

April, get married on May 1, and drive Anne on their honeymoon trip to Denver before he had to return to duty. They found a small apartment for her in Denver, and he returned to nearby Camp Hale to begin his duties now leading troops. Brower later transferred to the Seneca Rock Assault Climbing School in West Virginia. Anne then found a job in the historical G-2 division at the Pentagon in Washington, D.C., where she handled soldiers' journals direct from the fields of combat—bloody reminders of where her husband was headed. Her task was to write histories of the war as it continued.[23]

In both Colorado and West Virginia, Brower prepared troops for winter in high elevations, including teaching classes and training up in the mountains. Many troops and officers were unprepared for winter duty in the mountains. Many were from the Plains or the South, and most officers had little or no training in skiing or climbing or experience in harsh winter climates. Those officers did not understand that at higher elevations, packs and marches needed to be reduced to avoid exhaustion. One early two-week training session was especially notorious after temperatures fell to −25°F, and the result was a troublingly high number of cases of frostbite and pneumonia. By the time Brower and other veteran climbers took over, they included both enlisted men and officers in the training, some of whom outranked the trainers. One week would be spent in classrooms and another in the mountains, and the latter sessions could be especially brutal.

It was apparently some time during this period that Brower began having sexual relations with other men. There are no details about the liaisons, only Brower's rare accounts to others that this is when he first began to have bisexual relations.[24] Brower's discovery of his sexual interests in other men would not have been unusual at that time; some historians have suggested that World War II disrupted traditional gender patterns, moving young men out of traditional families into sex-segregated barracks and thus opening opportunities that many closeted gay and bisexual men had never considered possible for them before this point.[25]

In 1944, the army closed the West Virginia program and reassigned Brower to Texas, where his talents were of dubious value. Leonard again intervened and had Brower assigned as an intelligence officer with the Tenth Mountain.[26] Rumors were that the unit was going to Burma, and they were issued summer clothing. They instead landed in Naples, arriving just before Christmas for the final assault to push the Germans out of Italy.[27]

In Europe, Brower was often near or in combat zones. Outside Florence, he watched as waves of German aircraft crossed the skies that were streaked with tracers and high-altitude explosions from antiaircraft fire. In February 1945, he was in the Apennines atop a hill when a shell passed overhead, narrowly missing a tree and detonating. It was his closet brush to death in the war. In late April, he was on Lake Garda, at the base of the Alps, and witnessed fierce fighting between the Tenth and the entrenched Germans. On May 2 at 5:00 P.M., Brower was in San Alessandro when the phone rang, and Captain Everett Bailey answered. Bailey asked to have the message repeated. Then he smiled, grabbed Brower's arm, and exclaimed, "Dave, the war in Italy is over."[28]

In August, Brower and the Tenth were on transport ships on the Atlantic, back to the United States. They understood that any leave there would be short. The Tenth Mountain Division would soon head to the Pacific, for what was anticipated to be a deadly, brutal assault on Japan. One day before they were to arrive in Virginia, they learned about the atomic bombing of Hiroshima. On August 14, 1945, when Japan surrendered, Dave and Anne Brower went dancing at the storied white-turreted Claremont Hotel in the Oakland foothills.[29]

Discharged, Brower returned to the University of California Press, where he was promoted from editorial assistant at $2,100 a year to editor at $2,760.[30] By now, Brower was a good editor, willing to suggest numerous changes to a manuscript when he felt it was necessary. He was weaker on the more technical aspects of publishing, especially in the pressroom. Samuel Farquhar died in 1949, and August Fruge took over. At one point, Fruge made Brower responsible for overseeing the printing of the press's books. However, his inexperience became a problem, and he had to be taken off of that assignment.[31]

The biggest changes for Brower in this period were his marriage and, soon, his new status as a father. Kenneth had arrived in 1944; he was followed by Robert in 1946, Barbara in 1950, and John in 1952. This growing family needed a place to live, and Brower found an empty lot on Stevenson Avenue in the Berkeley hills, with a view of the bay. The two-bedroom house was finished in 1947; it would soon need two more bedrooms. Anne chose not to stay at the press, although she would remain an editor of both her husband's and others' work.[32]

For Brower, even more important than the job or the family was his return to the Sierra Club. It meant less mountain climbing and more toil to save the mountains. Seeing Europe had changed him. Brower's environmental ethos had been slow to develop. He had originally joined the club to hike and climb mountains, not to change the world. But he had listened to Adams's conservation messages; he had read Muir, Thoreau, and others; and he had seen firsthand in Europe how humans could spoil wilderness. It had been one thing to appreciate the wilds of California's Sierra Nevada, another to contrast that mountain range with the Alps of Italy, Switzerland, and Austria. In an article written in 1945 for the *Sierra Club Bulletin*, he described seeing wild places, precipices, mountain torrents, glaciers, and forests ravaged by a conqueror. The culprits were too many people and too much development. "The wilderness," he concluded, "died of over dosage" in Europe.[33]

The beauty of the wild needed to be protected, he was realizing, a viewpoint that others, including Leonard and Adams, shared. This group of what was being called the "young Turks" began to challenge the old guard made up of Colby and his brethren. One of the early battles dealt with roads. For decades, road builders had had carte blanche to go wherever they wanted, including the Sierra range on every available pass, with the full support of the Sierra Club. In 1949, Brower and another prominent Sierra Club member, Harold Bradley, wrote a seminal article on the hazards of road building in the mountains. They picked out Yosemite as their prime example. Access to the valley had inalterably changed it from a national treasure to a recreational resort, they complained. Instead of concentrating on the sheer granite cliffs and lacy cascades of the waterfalls, too many people went there for the swimming, tennis, dining, and dancing.[34] The article was noteworthy because it was here that Brower began laying the foundation for an environmental creed that would grow over the next twenty years.

How tensile that conviction could be was tested around the same time. The National Park Service wanted to build a road into the new Kings Canyon National Park. Ten years earlier Colby had asked powerful political and business forces in the nearby San Joaquin Valley to support creating the new park. In return, he promised that easy access would be provided to get there, potentially opening up new trade for local business. After the park was created, however, the road was delayed by the war. Once the war was over, the new road design plans unveiled in the late 1940s startled Leonard, Brower, and others. The road was wider, more developed, and penetrated deeper into the forest than they had anticipated. But Colby was adamant that it had to be built, that he would not back down on his promise. And Colby prevailed.[35]

It was one of his last victories. Colby still cared, but he was weary, and he had trouble leading. He would come to a board meeting, make a fiery speech, and then, as the meeting settled, fall asleep.[36] In 1949, Colby and his allies relinquished control of the board presidency, allowing it to go to Lewis Clark, a close friend and confidant of Leonard's. Four years later Leonard was president. This new leadership was going to be different—still willing to work with government and industry but also far more willing to question, to criticize, and perhaps even to embarrass the other side when necessary. It was that willingness that led the club to take on the Dinosaur fight.

The army had made Brower an officer and thus began the process of making him a leader; now the Sierra Club and the mountains would complete that task. In 1947, the Sierra Club asked Brower to become responsible for its esteemed high trips. It was quite an honor. Muir had first asked Colby to organize a series of trips into the mountains. He wanted to document that the mountains were being used recreationally and to build support for conservation by making members experience it. Colby had developed these high trips, which could last two, four, or even six weeks. Hikers were limited to forty pounds of gear, which was carried by mules along with all of the food and stock required by the commissary staff. Camp would be established for three or four days, and then the hikers, staff, and mules would move on to another site.[37]

Campers would be awakened by 4:30 A.M. by the cooks calling, "Everybody get up, get up, get up, get up." Some might take a dip in an icy stream, and everyone lined up for breakfast before they were off on the trails, supplied with box lunches.[38] One of Colby's duties, or pleasures, was setting the pace on hikes. The tramps were swift and measured by what was called a "Colby Mile." Hikers on high-trip treks with Brower soon named his measurements "Brower Miles" because of Brower's pace. Early on Brower brought his two oldest sons, Ken and Robert, on these trips. Ken would start the hike with his father, but he could rarely keep up. He soon found that older hikers had the same handicap. "He could really do it down the trail," said Ken. "Everyone of his generation said nobody could keep up with him." The name "Brower Miles" also stuck because it often seemed that the distance traveled was far longer than what Brower had advertised. "The Brower Mile may have been a white lie to keep up the spirits of his flock, or it may simply have been cardiovascular," said Ken Brower.[39]

Back at camp, after dinner came the campfires, the daily highlight of the high trip. The campfire was a time to make announcements, sing, tell stories, skits, and give sermons. Colby was always asked one night on each trip to talk about Muir, and that usually led to discussions about the need to preserve the wilderness surrounding them. It was here, around the campfire, that Brower's reputation for oratory was created, honed, and perfected. Brower had been helping with the high trips since 1939, and that was when he was first asked to speak at the campfires. He had no experience in public speaking. By nature, he was shy. Yet around these campfires Brower discovered his speaker's voice, the excitement and charisma that would charm audiences for decades to come. "He had a talent for influencing people," said Lewis Clark. "He had a great repertoire of songs. He used to take his accordion along, give performances at the campfire and entertain people and inspire them" (figure 6).[40]

Young Ken Brower would sit on the edges of the campfire and watch the audience watching his father. "For the first time, I sensed his way with words and the power of his presentation," he said. "And everyone in that circle of the campfire was moved by what he said." By the time his father was deep into the battle to save Dinosaur, he had become a master of public speaking not just at a campfire but in the halls of Congress. Ken Brower remembered that one time when he was about eight years old and was going to breakfast the morning after his father had given one of these campfire talks, a woman stopped him and exclaimed, "You do realize, don't you, that your father is a great man?"[41]

"He was heroic," said Phillip Berry, who went on his first high trip in 1950 when he was thirteen. To the young teenager, it seemed that Brower had accomplished more than anyone else he had met. "When he said he liked climbing, it was real. When he ate the fish that you brought him, he

FIGURE 6 Brower playing the accordion

Brower became an accomplished performer as he spoke and played music on many camping trips organized by the Sierra Club. For years, he led Sierra Club expeditions in California's Sierra Nevada and beyond, including Dinosaur National Monument on the Colorado–Utah border. (Courtesy of the Bancroft Library, University of California, Berkeley) genuinely enjoyed it and thanked you. He could climb higher, hike faster, sing as well as anyone and play the accordion, the only musical instrument that made sense in the mountains." He even looked the part in his mountaineer pants with the many pockets.[42]

Then in the summer of 1952 everything that Brower had worked for was threatened. He reportedly made sexual advances to a man at the Hearst Gymnasium on the University of California campus. The man not only resisted but also complained to the administration. University officials apparently told the head of the press, August Fruge, to fire Brower.

The most detailed but also probably the most biased account of this situation came from Richard Sill, then a physics professor at the University of Nevada at Reno, who was a Sierra Club board member in the 1960s and 1970s. Sixteen years after the complaint was made, in 1968, Sill made a series of accusations against Brower on why he should be fired as the Sierra Club's executive director, and among them was this statement: "Dave was discharged in an oblique fashion from the University of California for homosexual attacks on a young negro in the Hurst [*sic*] Gymnasium. Dave fought the charge; he was able to keep his job until the end of the contract year, and they did not discharge him in a formal sense; they merely saw to it that no money was available for him to be rehired."[43]

Fruge confirmed in his oral history that Brower was in trouble in 1952 and needed a new job. "There was no future for him" at the University of California Press, said Fruge. He would not explain further. Jeffrey Ingram, who worked for Brower, reported many years later that his boss had told him he had gotten into trouble at the campus for making a pass at another man.[44]

Fortunately for Brower, it was around the same time, according to Fruge, that Richard Leonard told him that the Sierra Club was interested in hiring him and had wondered about his availability. Virginia Ferguson, the part-time secretary for the Sierra Club, overwhelmed by the demands of the club membership, which now numbered seven thousand, had told Leonard she needed help. Fruge wondered later if Brower talked to Leonard about his situation. Board member Lewis Clark remembered Leonard telling him, "I think we can get Dave as our executive secretary."

Brower, however, wanted the title *executive director*, a position that went far beyond that of executive secretary, and the idea gathered momentum. The board approved the appointment on December 15, 1952. Club president Harold Crowe announced it in the January 1953 issue of the *Sierra Club Bulletin*. Crowe said that the idea for the position recently became "both necessary and practical." He did not explain what he meant.[45]

If Leonard, Crowe, and other board members, including Ansel Adams, knew about the incident that had precipitated the University of California Press to force Brower out, they did not talk about it. Nothing positive could be said at the time. Homosexuality in the 1950s (and for years after that) was labeled a perversion in the United States. The Eisenhower administration prohibited the hiring of gay men or women by the federal government or contractors. The Federal Bureau of Investigation spied on the meetings of gay organizations. The U.S. Postal Service traced the mail of those suspected of homosexuality and sometimes passed on evidence to the individual's employer. Vice squads invaded homes and entrapped homosexuals in public places. It was a witch hunt, and the Sierra Club would gain nothing if it became publicly known that its new executive director had sex with men.[46]

Ingram, whom Brower had hired to work for the Sierra Club in the late 1960s, had a number of conversations with Brower on this issue, and he is convinced that at least some members of the board had to know. He believed it was a situation in which the friendships Brower had developed over the years overrode any fears of the consequences of America's homophobic culture. "To me it was an example of the old-boy network where everyone looks out for everyone else," said Ingram.[47]

If so, the decision was an act of both courage and friendship by men such as Leonard and Adams. They believed that Brower had the ability to be a good steward, to manage the outings program, to take care of the financial books, and to assist in conservation campaigns. They would come to realize that they had seriously underestimated Brower and his vision. For the next few years especially, though he would vex and frustrate them at times, they would not be sorry they had selected him to lead the Sierra Club.

4. The Lesson

Conservationists have to win again and again and again. The enemy only has to win once. We are not out for ourselves. We can't win. We can only get a stay of execution. That is the best we can hope for.

DAVID BROWER, IN JOHN McPHEE, *ENCOUNTERS WITH THE ARCHDRUID*

In March 1956, David Brower was in Washington, D.C., about to accomplish the greatest victory so far in his career. He would also consider it his greatest defeat.

The triumph was Dinosaur. After years of fighting, dam builders had agreed to capitulate and not erect dams at Echo Park and Split Mountain within the Dinosaur National Monument. They instead accepted Brower's compromise to build a higher dam downstream on the Colorado in Utah's Glen Canyon.

For as long as he lived, Brower regretted not fighting harder to prevent the Glen Canyon Dam, which began holding back the Colorado River in 1963. In the foreword to the book *The Place No One Knew: Glen Canyon on the Colorado*, by Eliot Porter, he wrote: "Glen Canyon died in 1963 and I was partly responsible for its needless death. So were you. Neither you nor I, nor anyone else, knew it well enough to insist that at all costs it should endure. When we began to find out, it was too late."[1]

He traced the failure back to his decision in March 1956 not to oppose legislation authorizing the dam. "My horrible mistake at that time was to have stayed in Washington," said Brower, "instead of to have grabbed the next plane back and called for an emergency meeting of the Executive Committee or the Sierra Club board to argue why we should have stayed in the battle and stopped the whole thing."[2] Compromise with his enemies, consultation with his friends, these were the hallmarks of Brower's Sierra Club administration in the 1950s. Working with allies, he was building the Sierra Club into a national force. His name was taking on meaning, and as the Sierra Club grew, so did Brower's organizing leadership of the loose confederation of environmental groups. But winning at Echo Park in 1956 by compromising over and losing Glen Canyon in the same year would change him. He would conclude that the policies of compromise and consultation were a mistake. No more would he compromise if it meant losing a unique natural resource. No more would he be bound by anyone who would settle, even if it meant breaking longstanding friendships. Mike McCloskey, who came to the Sierra Club in 1961, saw Brower change. "One reason I realized that he became so disillusioned with compromise was during the '50s he made a lot of them," said McCloskey.[3]

The fight to oppose the two dams in Dinosaur had already been under way for several years when Brower demonstrated in January 1954 the flaws in the Bureau of Reclamation's math on evaporation. It would take two more years to kill the project, and Brower worked tirelessly to accomplish that aim. The Sierra Club planned another series of rafting trips down the rivers in Dinosaur in the summer of 1954, and the publicity would draw more than seventy thousand visitors for a visit to the national monument.[4]

Brower also commissioned a second film and a book on Dinosaur. The film, produced and narrated by Brower, was shot in a single day. Called *Two Yosemites*, it countered the argument that a reservoir at Dinosaur would make the Green and the Yampa Rivers more accessible to more people. It described the beauty of Yosemite Valley and then showed how the rim around the reservoir at Hetch Hetchy had been denuded into a wasteland. Silt turned to dust that turned to scum in the water. "Where is the pulsating heartland of this place?" asked Brower, the narrator. "It is gone." The film was only eleven minutes long, but it was effective and was distributed to groups throughout the country. In the summer of 1955, Howard Zahniser of the Wilderness Society set up a movie projector in the halls of Congress. "You've got to see what this does; it's only eleven minutes," he would announce to anyone who showed any interest. Passing lawmakers paused, intrigued by Brower's scenes. Representative Gracie

Pfost, who was from Idaho and sat on the House Committee on Interior and Insular Affairs and the Subcommittee on Irrigation and Reclamation, cried. Others did too.[5]

Brower convinced the famed Western writer Wallace Stegner to edit a book that would be titled *This Is Dinosaur: Echo Park Country and Its Magic Rivers* and New York publisher Alfred Knopf to produce it. It was ninety-seven pages long, thirty-six of them consisting of spectacular photographs, many in color, with chapters from several authors, including Stegner. In the preface, Stegner wrote that the book's contributors had not chosen "to make this book into a fighting document."[6] Perhaps not, but that was certainly Brower's intention. To make sure that everyone got the message, Brower tucked into the back of the book a photo of the ugly mud flats on the shore of Lake Mead. Its caption, a quotation from the Department of the Interior secretary Douglas McKay, read: "What We Have Done at Lake Mead Is What We Have in Mind for Dinosaur."

Although Brower's name is never mentioned in *This Is Dinosaur*, it was clearly his creation. Said Brower, "I got all the contributors together, I got the photographs together, got the editor—worked with him in laying it out—and got the publisher." It was a small print run, only five thousand copies, and the book was rushed to ensure that a copy was delivered to every member of Congress. No one had used a book as a political tool to save a pristine natural area. Brower pioneered the technique and would perfect it over the next fifteen years.[7]

By the spring of 1955, when the book was published, seemingly every major print media outlet was raising concerns about Dinosaur, including *Life, Time, Readers Digest, Collier's, Saturday Evening Post, Harper's, National Geographic, Scientific Monthly, New Republic,* and *Sunset.* Brower was sending information out to reporters and, as he described it, running up "enormous telephone bills." He accompanied John Oakes, the senior editor at the *New York Times* who wrote a conservation column, on a trip to Dinosaur. They spent days in a car traveling together, and Brower bragged that he knew more specific details about the Echo Park and Split Mountain dams than Reclamation's engineers, including his nemesis on the evaporation issue, Cecil Jacobson. He later claimed often that during the trip, when Oakes asked a question, "Mr. Jacobson would turn to me for the answer. I had boned up on that; I knew more about the Colorado River than I ever want to know again, about anything. It was coming out of my ears." The trip was successful; the *Times* published an editorial opposing the dams.[8]

Although opposition to the dams was growing nationally, the pro-dam opinions in the Rocky Mountains remained firm, and a local public-relations campaign for that side was initiated. The supporters formally created the Upper Colorado River Grass Roots, Inc., but in their literature they called themselves the "Aqualantes." They raised money, sent volleys of letters to local newspapers, ran newspaper ads, mailed flyers, distributed handbills, and produced their own film, *Birth of a Basin*. The film was so bad that some dam opponents suggested making copies and distributing it to help their side.[9]

The Sierra Club launched its own letter-writing campaign, which was so effective that mail sent to the House of Representatives on the issue was running eighty to one against the Echo Park Dam. As early as winter 1954, President Eisenhower had sought in his State of the Union Address to reassure those with concerns about Echo Park.[10] The letters eventually had a boomerang effect, though—dam opponents began receiving propaganda from pro-dam forces. C. Edward Graves of the National Park Association said that a mailing list had apparently been leaked by someone in the Department of the Interior, a step he called "a decidedly unethical procedure." The *Christian Science Monitor* discovered that Senator Arthur Watkins of Utah had asked for the names from Interior officials, and he allegedly had passed the names on to others.[11]

Sierra Club board members were pleased with Brower. When they had hired him in December 1952, he was to work out of the Sierra Club offices in Mills Tower, the handsome, Romanesque-style edifice in San Francisco's financial district. He did not spend much time there. His original responsibilities were to work on membership, run the summer outings, and carry out the club's conservation priorities. He reversed the prioritization of these tasks, spending much of his time on Dinosaur and other conservation issues. When he was not on conservation trips, he was

out each summer on outings. The club was looking beyond California for
its trips, which now included rafting down the Green and Yampa Rivers
in Colorado and Utah, hiking in the Pacific Northwest Cascades and
the Tetons in Wyoming, and boating the Colorado River downstream in
Utah and Arizona. There were now so many trips that Brower could not
make them all, and sometimes he would meet a trip that was concluding
for a day or two and then join a new one at the trailhead, again for only
a day or two. Finally, in 1955 the board, concerned about Brower's long
absences from Mills Tower, reluctantly limited his time on outings to two
weeks a year.[12]

Anne Brower and the children would travel with Brower to the trail-
heads. The two oldest, Ken and Robert, often hiked with their father, but
the other two, Barbara and John, were simply too young. Anne Brower
was a loyal wife throughout their marriage, but living with her husband
was never easy. She once told an acquaintance that when Brower became
executive director, she thought that finally he would use his leisure time
for something else, such as spending more time with the family. Instead,
he was traveling half the year, and being at home simply meant he was in
the area. On the rare occasions that he arrived for a family dinner, phone
calls often interrupted them. He worked virtually every weekend. The
bottom line, she said, was that Brower's work for the Sierra Club caused
damage to the children and the marriage.[13]

Phil Berry, who was thirteen years old when he met Brower and knew
him his entire life, said that much in Brower's lifestyle was difficult for
those who were closest to him, especially Anne. He began to elaborate
and then stopped himself. "It was just sad," he said, "it was just sad."[14]

Despite the hardships, she stayed with Brower. She became an intellectual confidante and an adviser, and together they developed a synergy
that renewed their strengths and their aspirations. Gayle Brower, a sister-
in-law, said the relationship was difficult at times, but it was supportive to
both husband and wife. "Anne helped form Dave, but at the same time
Dave helped form Anne," she said.[15]

Brower had been in the position of executive director for only two
months when he made his first pilgrimage to the East Coast in February

1953. He attended the annual meeting of the Natural Resources Council (NRC) in Washington, D.C. The NRC was an unusual organization. It represented forty environmental organizations, from outdoor clubs to hunting and fishing organizations and professional societies such as the American Planning and Civic Organization and the American Society of Mammalogists. Its diversity was as much a hindrance as a benefit. It struggled to agree on anything. Yet it was a great forum for meeting people and sharing views.[16] At that first meeting, Brower was a bit player, almost overwhelmed by the size of the gathering, which drew 1,229 registrants and 870 participants to the banquet in 1953. At the behest of other conservation leaders, he sat in on meetings with Richard McArdle, who headed the U.S. Forest Service, and even had a private meeting with Conrad Wirth, director of the National Park Service. Brower had no audience, however, with Douglas McKay, the Oregon car dealer turned secretary of the interior.[17]

In a measure of how rapidly his stature was rising, two years later, in 1955, he was elected chairman of the NRC. This time he set up and ran a meeting with McKay. Sierra Club board member Bestor Robinson suggested that Brower not "embarrass" the interior secretary by asking him to reverse his stand on Dinosaur. The meeting was amicable, although McKay rambled at times and complained about distortions perpetuated by the Sierra Club. Brower did not argue with him, and they were able to work out a joint statement indicating that they were striving to work together.[18]

In New York and Boston, Brower attempted to curry favor with editors and writers. Carl Gustafson of the Council of Conservationists told Brower once that he did not know anyone in the conservation movement who spent so much time with the press. "This shocked me, for I have done so little," observed Brower. In Boston, he talked books with Paul Brooks, an editor at Houghton Mifflin, and Brooks mentioned that he was working with Rachel Carson, who was writing a book that Brower would find interesting. In New York, Brower offered story ideas about Dinosaur to Raymond Moley, an influential columnist with *Newsweek*; Jack Fischer, editor of *Harper's*; and Oakes of the *New York Times*.

In Washington, Brower persuaded Gilbert Grosvenor, president of the National Geographic Society, to let him write a long piece on the Sierra Club high trips for the magazine.[19]

In Washington, where the Sierra Club had no lobbyist or even a physical presence, Brower's tasks were more political. He lobbied on Capitol Hill, going up and down corridors, reaching the people in power who affected conservation in the Eisenhower administration. After making two trips to the East, during which he had seventy different meetings, Brower reported that "the influence of the Sierra Club" was well represented.[20] When testimony was needed before Congress, Brower was usually the first on everyone's list to represent the conservation side. He noticed that as time went on, his prepared statements kept getting longer and his time testifying also lengthened. During one hearing, he realized he was on the stand for two hours and fifty-seven minutes without a break. "Senators would go off to the library, back to their office," he said. "They would switch around. They were questioning me quite hostilely for that whole period." He said he did not mind. "It was fun to go after their economics," he recalled. "They were really after an extremely crazy project," the Echo Park Dam.[21]

On that first trip to Washington, he dined at the Cosmos Club, the Victorian mansion and club off DuPont Circle and tony Embassy Row, with the few other colleagues who worked full-time in the conservation movement. The Cosmos was to become Brower's Washington home away from home, just like the Biltmore Hotel in New York City and the Parker House in Boston. Brower's expense account was generous, and he went first class with airfare, hotels, and restaurants. He especially liked lunch. He never wanted to eat alone, and he preferred a dining party of no more than seven because too many diners produced too many discussions. He was the raconteur, the equinox, and, most important, the host at the vortex of each discussion (figure 7). He selected the restaurant, preferably plush and opulent, with overstuffed leather upholstery and a full bar—a man's place. On one occasion in the mid-1960s, several junior members of the staff and friends went with Brower to his favorite Manhattan restaurant for lunch. It was his birthday, and they celebrated. They wanted to pay the tab afterward, but Brower would not hear of it. The struggle over

FIGURE 7 Brower with other conservation leaders

Five leaders of the conservation movement in the 1950s (*counterclockwise from lower left*): Howard Zahniser, Wilderness Society; Carl Gustafson, Conservation Council; Joseph Penfold, Izaak Walton League; David Brower, Sierra Club; and Ira Gabrielson, Wildlife Management Institute. (Courtesy of the Bancroft Library, University of California, Berkeley)

the bill, which had begun jokingly, quickly escalated as Brower became noticeably angry. Everyone let him pay the bill.[22]

Despite his success with the Bureau of Reclamation, the press, and the campaign, Brower throughout 1954 never felt that confident. In June 1954, he was back before the Senate Committee on Interior and Insular Affairs, which was even more hostile than the House committee. In July, it approved a bill authorizing construction of the five dams by a vote of eleven to one. But it was clear that the issue was bogging down in both branches of Congress. Brower, speaking in August to the Associated Sportsmen of California, said, "We've forged together a team for ourselves that has worked together better than any other . . . in conservation history."[23]

Brower decided that the Sierra Club needed to become stronger not only to fight the dam in Dinosaur but to tackle the inevitable clashes of the future. To make it stronger, he needed the support of the Sierra Club's board of directors. In the fall of 1953, he began discussions with them, asking if they wanted the Sierra Club to be a national, regional, or local organization. A good mountaineer, he told them in a prepared statement, always knew when it was time to pause and check the course.[24] In response, board members agreed almost unanimously that the club needed to expand. Even Robinson, more prone to compromise than others, backed him. Only Francis Farquhar, a key member of the old guard, objected, expressing a view that he would hold for the rest of his life. The board agreed to expand, and it set in motion changes that would eliminate the strict membership requirements.[25]

At the time, the Sierra Club had ten chapters. Nine chapters were in California, and one had been created in New York in 1950, although the Atlantic chapter was designed to represent the entire East Coast.[26] Sierra Club members in the Pacific Northwest wanted to create a chapter, and Brower envisioned others throughout the United States. He played a key role in forcing the club to look outward. Until that night, it had been gliding, so Brower gave it direction. In assessing Brower and his achievements, Carl Pope, executive director of the Sierra Club from 1992 until 2010, believed that the changes Brower made to restructure the club marked his greatest accomplishment. He took what was a club, said Pope, and created an organization. "In the process, what he was doing in creating a new environmental organization was that he was helping create what became the environmental movement."[27]

Brower was very much into consulting with the board and communicating exactly what he was doing, especially because he was out of the office so often. From 1953 to 1959, he typed long, detailed reports about his trips, some of them up to forty pages long.[28] What is most remarkable about these reports is Brower's degree of detail. In a few years, however, Sierra Club Board presidents would claim that Brower rarely told them where he was traveling for the club.

Although some in the Sierra Club worried that Brower was too prone to criticize those who did not agree with him, he had very strong support overall in the early years of his tenure. In 1959, one of his longtime champions, Ansel Adams, wrote Harold Bradley, then the board's president, and called Brower "the conservationist of the era" and "a creative genius." Bradley agreed, but he worried that lack of sleep, the travel, and the inevitable frustrations of the work might wear down Brower. At times, he could be difficult to manage. He did not take criticism well, and he occasionally disobeyed orders, such as the time in 1957 he was told not to go to a meeting in Florida but did anyway.[29]

Brower still supported raising the height of the Glen Canyon Dam, although in October 1954 he wrote that the Bureau of Reclamation might be stupid enough to continue to oppose the higher dam and risk losing support for any dam at Glen Canyon.[30] By 1955, however, he had more doubts about Glen Canyon Dam. "I began to look at it with just an editor's mind, not an engineer's," Brower said. "The main thing I found was that they were overdeveloping the entire river."[31] He began suggesting that there were other alternatives to building more dams at Dinosaur, Glen Canyon, or anywhere. He argued that coal-powered electric generation was cheaper and could be built closer to power sources than hydropower. And, like many people of this era, he was optimistic that nuclear power within a few short years would be even more competitive. In a talk in 1955 to the Associated Sportsmen of California, he said, "We're not asking anyone to stop using water, but merely not to store it where it floods dedicated lands, nor to store it for damaging hydro electrical installations that the atomic scientist, with his black magic, will in all probability render obsolete before they can pay for themselves."[32]

The politics of water, especially Colorado River water, in the West could be colossally byzantine, often pitting California against everyone else in the region. Some began to suspect that the California-based Sierra Club was only a stalking horse for California's water barons. In one hearing, Senator Watkins of Utah kept Brower on the stand for ninety minutes grilling him on the Sierra Club's objectives. "What he was trying to do in any way he could was to prove that I was just the patsy of Southern

California," said Brower. It is true that Brower had joined forces with California's water interests, especially their lawyer, Northcutt "Mike" Ely. "I'm not sure whether they approached us or I approached them," recalled Brower. "We certainly saw that we were together. [Ely] was the man that Arthur Watkins was trying to say that I was in collusion with. Indeed we had meetings. We would join anyone who could help us save Dinosaur."[33]

By late fall of 1955, the issue of the two Dinosaur dams had stalled in Congress. Flustered by the stand-off, governors, members of Congress, and other dam supporters agreed to meet in Denver on November 1, 1955, to consider their options. It was a critical meeting. But the conservationists trumped it by running a full-page advertisement in the *Denver Post* on October 31. Five representatives of conservation organizations, including Brower, signed it, and Edward Mallinckrodt Jr. of Mallinckrodt Chemicals, a Sierra Club member, paid for it.[34] The wording in the ad was less important than its message, which the politicians understood immediately. The Dinosaur dam opponents were not going away, and they would oppose all of the dams in the Colorado River Storage Project. Further, they were in for the long term, no matter how long. Their stance was unacceptable to those who wanted the greater water project. The ad and the warning seemed to have an impact. The day after the ad ran, at the Denver meeting on November 1, a longtime supporter of the Dinosaur dams, Senator Clinton Anderson of New Mexico, announced that the Dinosaur dams must be removed from the bill. By the next day, a resolution was offered agreeing that two Dinosaur projects would not be reinserted into the legislation. As the parties milled around before casting votes in Denver, a longtime dam supporter began congratulating dam opponents. He explained that if they did not like the resolution that proposed removing the Dinosaur dams from the bill, they should offer their own version, and it would be approved.[35]

The battle was over. On December 2, Secretary McKay announced that the administration would abide by the wishes expressed in Denver. On March 28, 1956, Congress formally authorized the Colorado River Storage Project without the Dinosaur dams. Everyone had to wait almost two weeks for President Eisenhower to return from a trip to the Master's golf tournament, but he finally signed the bill on April 11, 1956.[36]

By then, however, Brower was already having misgivings. Years later, he revealed that many congressional insiders told him in the days before the bill was passed that he and his forces "were out of their minds" in not opposing the entire upper-river project.[37] In the Senate, key leaders told him they now had the clout, with the Sierra Club's support, to kill the entire project. Senator Paul Douglas of Illinois wanted to know why the Sierra Club had bailed out. On the day the House voted, Brower was sitting in the gallery and was spotted by Congressman Wayne Aspinall. Aspinall, a floor leader, approached Brower and asked about the Sierra Club's plans. Brower told him the club had dropped its opposition, releasing a bloc of more than two hundred votes that had been opposed. Congressman Craig Hosmer, a longtime dam supporter from California, was amazed by the club's new stand allowing Glen Canyon Dam to be built.[38]

Brower, the Sierra Club board, and other environmental organizations *had* discussed the option of killing the Glen Canyon Dam. Brower had met with the board's Executive Committee on December 27, 1955, and again with the full board on January 7, 1956. The price of peace, board members declared, was that Congress needed to agree that no dam would be built in a national park or monument.[39] Robinson recalled Brower arguing that the Glen Canyon Dam should not be built. "You see, he [Brower] was the purist who thought that it was important to take a purist stand even though you went down to defeat," said Robinson. "And we were hinging our whole case on 'go ahead and use Glen Canyon.'" Robinson felt that it was far more important to defend Dinosaur and downstream at the Grand Canyon than it was to protect Glen Canyon.[40] Leonard agreed. "Congress would have been convinced that the preservationists were unreasonable and were urging that the entire Colorado River be unused and just allowed to flood away into the Gulf of Mexico," said Leonard. "That kind of argument would have been so strong that we would have had both Echo Park Dam and Glen Canyon Dam."[41]

Both the Wilderness Society and the National Parks Association, the two other key opponents of the Dinosaur project, also struggled with this issue.

Trustees for the Parks Association concluded that they did not want to be accused of "being excessive in [their] demands." Zahniser at the Wilderness Society concurred.[42]

What is remarkable about these discussions is the volume of warnings that all these organizations had received up to this point. Historical records, a strong organized effort within Utah, and the advice from some of Brower's closest associates extolled the beauty of Glen Canyon and urged that it be preserved. The Sierra Club even ran river trips through Glen Canyon before the dam was authorized, and at least some of those passengers afterward resolved to fight the dam. Alice Joy Keith of San Diego toured Glen Canyon on a river trip in the summer of 1955. "In the name of all that America means to us, I beg you to vote against the Glen Canyon Dam," she wrote to 125 congressional representatives in February 1956. A copy of her letter went to Brower.[43]

Brower would call Glen Canyon "the Place No One Knew," and there was certainly truth to that assessment. It was isolated, hundreds of miles from towns, highways, or railroads. The U.S. Geological Society listed the region as the most isolated in the continental forty-eight states. But John Wesley Powell had described the beauty of Glen Canyon, which he named for the verdant emerald groves that often shaded the river, when he took the first of his famous journeys down the Colorado in 1869.[44] In 1940, Interior Secretary Harold Ickes proposed making Glen Canyon part of the Escalante National Monument, which would have covered 280 miles of the Colorado River canyon lands. Utah politicians and merchants, however, saw this proposal as a land grab by the Park Service, and then the outbreak of World War II killed its prospects.[45]

Wallace Stegner; Charles Eggert, a filmmaker who worked with Brower; and Lewis Clark, a Sierra Club board member, all told Brower about Glen Canyon in the years before the March 1956 vote. At a June 20, 1954, board meeting, Clark described his journey through the canyon and argued that some of the most spectacular features would be destroyed by a dam. Board members agreed that they should try to preserve the region's scenic values, but their effort would have to be coordinated with whatever Congress did.[46]

Stegner's warnings came during and after his work on the Dinosaur book. Stegner had grown up in Salt Lake City; he had explored the Utah wilds as a youth; and in 1948 he had published an account of his journey down the Colorado and through Glen Canyon in the *Atlantic Monthly*. He described his first camp in the canyon as "almost unimaginably beautiful—a sandstone ledge below two arched caves, with clean cliffs soaring up below two arched caves, behind a long green sandbar across the river." Nearby was a canyon that was commonplace in Glen "but that anywhere else would be a wonder."[47]

Eggert wrote to Fred Packard of the National Parks Association in August 1955, exclaiming that the region contained "the most incredible works of Nature I have ever witnessed! I cannot think of the possibility of destroying such magnificent beauty."[48]

Utahans mounted at least two concerted campaigns to oppose the Glen Canyon Dam, urging Congress to make Glen Canyon a national park. One effort was organized by a group of river guides, a second by a group of prominent Utah residents. The Utah Committee for a Glen Canyon National Park spent several years campaigning. William Halliday, who headed the group, did not oppose the Echo Park Dams. This stance may have been designed to curry favor from western politicians, but it likely alienated Brower and other conservation groups.[49]

Brower worried that Echo Park would be put up against Glen Canyon. He told Halliday and an ally, Malcolm B. Ellingson, in a letter on February 7, 1955, that although Glen Canyon sounded beautiful, he did not want to get trapped into comparing it with Echo Park. Conservationists were suspicious of the organizations supporting a national park at Glen Canyon. In December 1955, Congressman William Dawson of Utah mentioned the Glen Canyon national park campaign to Zahniser, who acknowledged that he had been getting letters about it for two years. Zahniser worried that it was part of a disinformation campaign against Dinosaur. "I at first wondered if it might be someone trying to mix me up," Zahniser told Dawson. The Glen Canyon proponents were political neophytes. "We really did not know how to speak out," said Ken Sleight, a river guide and member of one of the groups fighting the Glen Canyon Dam.[50]

Brower's support of a larger dam at Glen Canyon was not an aberration; he did at times support dam projects. Such support included campaigns to build bigger dams on the Snake River at Hells Canyon in Oregon and Idaho and on Flathead River in Montana. The alternative to the large dams at each of these sites would have been a string of smaller structures that Brower and the Sierra Club thought would create even more environmental damage. Brower's willingness to compromise, to allow some dams rather than none, surprised some of his fiercest critics. Ottis Peterson, the public-information officer for the Bureau of Reclamation, was so pleased that he supplied some key details to Brower when he was developing his case for supporting the large Paradise Dam on the Flathead. None of the Flathead dams, including the Paradise, were built. The smaller Hells Canyon dams were constructed, and it appears in hindsight that neither Brower nor the environmental community clearly thought out the ramifications of their Hells Canyon stand. The massive dam they wanted would have presented a far more formidable concrete barrier for the fish that spawn in the Snake. It also might have encouraged federal dam builders to erect even more dams on the rich fishing grounds of the nearby Salmon River.[51]

Brower stressed to friends and colleagues that he wanted to be a pragmatist, not an extremist, and that small victories and compromises were worth the price. About the Paradise Dam, he wrote in October 1956, "Merely to say 'we're against all dams until . . .' does nothing but lump us as aginners [sic] with no specific alternative. It's a sure formula for being ignored by all hands, including other conservationists." McCloskey, whom Brower hired in 1961, said Brower taught him about compromise. "He emphasized when he first oriented me that the park system had been built incrementally, that it did not happen overnight with one big victory, but that it was systematic." Brower told him he believed in incremental victories. That belief would later change.[52]

Brower finally got to the Colorado and into Glen Canyon, but not until after Congress had authorized the huge dam. It was an unforgettable trip, one he repeated for several years until the reservoir finally submerged most of the canyon's beauty. "The river itself was a spectacular sight," he said. "But the side canyons are beyond belief."[53]

This was slick rock country, at times Navajo Sandstone, stained in vertical stripes. Elsewhere it was a vivid pink, the same rock, according to Stegner, as the intricate formations at nearby Zion and Capitol Reef. "It is surely the handsomest of all the rock strata in the country," Stegner continued. "The pockets and alcoves and glens and caves which irregular erosion has worn in the walls are lined with incredible greenery, redbud and tamarisk and willow and the hanging delicacy of maidenhair around springs and seeps."[54] Brower visited side canyons that carried names such as "Music Temple," "Hidden Passage," "Mystery Canyon," "Twilight Canyon," "Forbidden Canyon," and "Labyrinth Canyon." Water had undercut Twilight Canyon and created a cave so large, according to various reports, that it could accommodate either Madison Square Garden or the Hollywood Bowl. Mystery Canyon ended in another domed cavern, with a very large pool. The narrow walls of Labyrinth Canyon were so high that it was not possible to see the sky. Visitors seemed to favor Music Temple more than any other place. It was yet another domed chamber, 500 feet long and 200 feet high, of Navajo Sandstone. A break in the roof allowed a spool of creek water to pour down into a clear pool. The roof fissure served as a skylight. Stegner wrote, "The shadows in a chamber like this, the patterns of light and shadow, are miraculous and utterly unphotographable, and the walls re-echo with the slightest sound." Alice Joy Keith, the San Diego woman who had visited Glen Canyon, said her party arrived "on a Sunday morning, as did John Wesley Powell, and like him, we held an impromptu service of song and praise to God."[55]

Downstream there was one last spectacular rock formation seven miles east of the river, Rainbow Bridge. It was a sandstone arch 309 feet high with a span of 278 feet. The Navajos considered the natural rock bridge sacred, and a 160-acre parcel containing it had been declared a national monument in 1910. It was so isolated that only a few hundred visitors took the trouble each year to hike in to see it, but the reservoir meant that water would potentially be lapping at its footings.[56] Brower and his allies made sure that the legislation authorizing the entire Colorado River Storage Project, including Glen Canyon, would specifically indicate that no dam or reservoir could intrude into a national park or monument, including

Rainbow Bridge. Perhaps, thought Brower, he could use Rainbow Bridge to stop the water from rising too high. Perhaps he could even use it to stop the Glen Canyon Dam altogether.

It has been more than half a century since Echo Park Dam was stopped and Glen Canyon Dam was authorized. Yet the victory at Dinosaur has been obscured by the loss of Glen Canyon, and Brower is more responsible for that loss than anyone else. Yet his martyrdom, his continued lament about his failure, obscures a central point; the clash over the Echo Park Dam was a historic milestone that could not have happened without his daring leadership. He succeeded where John Muir had failed. Never again would the federal government build a dam or other major invasive structure in a national park. Never again would the conservation community be as powerless as it had been in the past.

A major reason opponents of the Dinosaur dams organized was that many remembered how Muir had lost the fight at Hetch Hetchy. He had been unable to convince the federal government of the travesty of building a dam in such a beautiful sanctuary as Yosemite National Park. Dinosaur opponents began their campaign to protect the sanctity of national parks and to uphold the protections provided for in the National Park Service Act of 1916. Protecting those provisions became, in early 1956, more important than Glen Canyon, which did not have national park recognition. Historians have called this recognition that national parks can no longer be violated a watershed in both the history of the parks and the conservation movement. Bureau of Reclamation officials would attempt later to build more dams, including two at the Grand Canyon, but both of those projects as well as the others were outside of what was then the established park boundaries, so they required a different kind of fight. For environmentalists, the two proposed Dinosaur dams crossed a line, a national park and monument boundary line. The tragedy was that there was no such line around Glen Canyon, although there should have been.

Another difference from the Hetch Hetchy campaign was in the scale of the campaign mounted by conservationists. Muir had likewise run a nationwide campaign, but only a handful of conservation organizations joined him, and even the Sierra Club was divided. In contrast, historians have count at least seventy-eight organizations in the Dinosaur campaign. Most joined because they wanted to protect a national monument.[57] Would they have fought to stop the plan to enlarge and build Glen Canyon, the alternative that Brower himself had pushed to replace the Dinosaur projects? Brower maintained later that he did have the votes to end the entire Colorado River Storage Project. Yet the man who was developing a reputation for taking rash and reckless gambles fatally paused in the spring of 1956 when he had the opportunity to prevent Glen Canyon. We have only his regrets as an explanation of why he did not act.

5. Wilderness

This generation is speedily using up, beyond recall, a very important right that belongs to future generations—the right to have wilderness in their civilization, even as we have it in ours; the right to find solitude somewhere; the right to see, and enjoy, and be inspired and renewed, somewhere, by those places where the hand of God has not been obscured by the industry of man.

DAVID BROWER,
"WILDERNESS—CONFLICT
AND CONSCIENCE"

San Gorgonio is a mountainous wilderness 50 miles east of the sprawling urban tentacles of Los Angeles. It is dominated by the 11,499-foot Mount Gorgonio, and there are ten other peaks nearby, each 10,000 feet or higher. They run along an east–west ridge that swings in a semicircle southward, creating a magnificent alpine basin. On one side, Mount Gorgonio, sometimes called Grayback, thrusts a gray granite wall that sweeps up thousands of feet from the valley floor. The winter snowfall and spring avalanches feed two small lakes, several streams, and lush meadows and wildflowers. Ponderosa pine, white fir, and cedar grow in thick stands on some slopes before giving way at higher elevations to lodgepole pine. Higher yet, limper pine struggles in stunted heights of only three or four feet. Above the timberline, there is only the granite and snow.

Writer William Heald called this mountain southern California's rooftop, and visitors have been climbing these mountains since 1872. By the 1890s, resorts surrounded the peaks, and guided tours delivered more visitors by wagon. But human impacts were minimized until winter skiers in the 1930s discovered that the northern slope of Grayback was covered with snow from November to June. The slope was mostly bare of vegetation, making it the best run in a southern California where skiing opportunities were scant. Developers in the mid-1940s proposed building a ski resort. The resort was to be limited to 3,500 acres, a tenth of the wilderness area. But the site was within a national forest, and hikers and wilderness proponents urged the U.S. Forest Service to deny the developers' request. The problem, Heald wrote, was that the ski resort "would include the scenic heart and leave the remainder an empty shell." The Sierra Club was initially divided on the plans for San Gorgonio. The club was dominated by skiers, including David Brower, and many in southern California wanted another ski resort. But the club's and Brower's views on wilderness were evolving, and the organization eventually fought the proposal. In June 1947, the Forest Service rejected the project.[1]

The conflict at San Gorgonio was but one of many beginning in the late 1940s for Brower, the Sierra Club, and conservationists. The federal government owned hundreds of millions of acres of untrammeled land, rich forest, open prairies, wind-swept shorelines, and high mountains. In the years after World War II, the demand for these natural resources soared. Loggers needed the timber for new homes and buildings; miners wanted to tap minerals needed for the growing nation; ranchers wanted rangeland for their livestock; and millions of campers, hikers, skiers, and other outdoor enthusiasts simply desired a weekend or vacation refuge.

Norman B. "Ike" Livermore had been working summers as a packer responsible for bringing the mules loaded with supplies and camping gear up into the Sierra each year as part of the club's outing program. A Sierra Club member, Livermore would go on to become a timber executive and to work in California governor Ronald Reagan's administration. But in the late 1940s Livermore was concerned about how mules and people were affecting the pristine, isolated backcountry. He proposed a meeting to discuss what was happening to the wilderness of the Sierra Nevada. That first wilderness conference convened in Berkeley in 1949. Out of that first meeting came the biennial conferences that the Sierra Club ran until 1975. The conferences grew increasingly bigger, as did the stakes. They became part of a movement to permanently protect the nation's wilderness, whether the threat came from loggers or from those who wanted to build many more San Gorgonio–type playgrounds. Brower helped lead the legislative fight to establish a landmark wilderness bill, an effort that stretched from 1956 to 1964 against the bureaucrats at the Forest Service and the Park Service; against private lumber, mining, livestock, and other financial interests; against Congress; and even against the American public. The Sierra Club Wilderness Conferences became an important forum for those in the fight to gather, recount what had happened, learn where they stood, and decide how to proceed.

Michael McCloskey of the Sierra Club saw the evolution in both the conferences and the gathering consensus that the wilderness had to be preserved. In the 1950s, the conference sessions were similar to a board of directors getting together, in which the activists and leaders most concerned about wilderness would report what was taking place and discuss options. "[The conferences] were the key to catalyzing ideas," he said. In the 1960s, the wilderness activists and leaders stepped back and invited others to give the keynote talks and participate in the panel discussions. Sometimes these speakers had very public positions in the Forest Service, Park Service, or Congress. Other times they were less associated with wilderness. They may have even been ambivalent about the issue, but by being invited to come and speak, they had to work to understand it and take a position. "I think this was a clever strategy that made dividends," said McCloskey. "At first I wondered, Why are we inviting people who do not know very much about the idea? Then I realized it was not to have them develop it as much as it was to listen."[2] Brower organized and produced the Sierra Club's bound volumes of each proceeding. Some of the material he included was dry and dull, as conferences tend to be. Some of it was informative and thought provoking as speakers debated the definition of wilderness, what wilderness meant to most people, the importance of wilderness to science, and how to preserve wilderness for future generations. Some of what Brower chose to include in these volumes was beautiful, especially the glossy photographs by some of the nation's greatest nature and landscape photographers—Cedric Wright, Philip Hyde, Eliot Porter, and Ansel Adams.[3]

Besides the books and the conferences, Brower played another key role in the success of this campaign, second only to that of Howard Zahniser of the Wilderness Society. Beginning with the early conferences, Zahniser was convinced that legal protection was the only way to save up to 100 million acres of the nation's wilderness. Brower was his key ally, counseling him and playing the role of public scold, critical about how the nation's resources were being managed. The movement would eventually grow, and new recruits would include such luminaries as associate U.S. Supreme Court justice William O. Douglas and Pulitzer Prize–winning author Wallace Stegner.

Wilderness can be a thorny idea to grasp. As Brower once pointed out in a speech at one of the Wilderness Conferences, humans have always struggled to understand and define wilderness. As an example he cited the biblical prophet Isaiah. "Woe unto them that join house to house, that lay field to field, till there be no place that they may be placed alone in the midst of the earth!" declared Isaiah.[4] For the many centuries after Isaiah, civilizations considered wilderness only a nuisance. American settlers brought progress to the new country by chopping down forests, damming rivers, and eventually polluting waterways and skies. By the nineteenth century, however, there were exceptions to this sense of what progress meant. Portraits were produced of the American wilderness and its meaning by artists such as Thomas Cole, Frederic Church, and Albert Bierstadt; adventurers such as Meriwether Lewis and William Clark; writers such as Henry David Thoreau and Ralph Waldo Emerson; and activists such as John Muir and George Perkins Marsh. This alternative view led to the protection of Yosemite, Yellowstone, and other areas. The early twentieth century began with Theodore Roosevelt, Gifford Pinchot, and the establishment of the Forest Service as well as with Stephen Mather and the creation of the National Park Service.[5]

Muir for many years believed that the new national parks being developed early in the twentieth century were the best hope for preserving wilderness. But early park administrators in the Interior Department were less interested in wilderness preservation than in developing playgrounds for tourists. The Park Service and its concessionaires built roads where there had been trails and small towns of hotels and cabins where there had been isolated pastures of wildflowers.[6] The lead in preserving the wilderness fell to the Forest Service, a branch of the Agriculture Department created by Pinchot, but Pinchot was Muir's rival. These two conservation pioneers held fundamentally contradictory views. Pinchot believed strongly in overt management of public lands, whereas Muir wanted government lands preserved as wilderness, not actively managed. Muir approached nature as if he were a monk, but Pinchot viewed nature the way an engineer would, according to one comparison.[7] And yet Pinchot's management style allowed a young Forest Service employee, Aldo Leopold, to get the agency to set aside 700,000 acres of the Gila National Forest in New Mexico as the agency's first wilderness area in 1924. Although there was nothing permanent about the declaration, so that timber harvesting or other development might be allowed at any time, support for the idea grew. By 1929, the Forest Service had established the L-20 regulation, which declared areas such as the Gila Forest "primitive," allowing them to remain in their natural state. San Gorgonio also obtained that classification. Leopold would go on to be hailed as a prophet, his books on wilderness held up as spiritual icons, and by 1933 the Forest Service had established sixty-three so-called wilderness or primitive areas. But the dissonance between Leopold's vision and the Forest Service's practice was wide. Local rangers had significant discretion in how they managed these areas. Logging was occurring in twenty-three of them, and grazing was taking place in fifty-three.[8]

Leopold, the scholar and philosopher, was succeeded by Robert "Bob" Marshall, who was more of an enthusiastic apostle. Marshall also had the wealth and the connections to make things happen. In 1935, he helped establish the Wilderness Society, which quickly began battling in behalf of the Olympic Mountains in Washington and the Everglades in Florida. Born into old money, Marshall poured $400,000 of his own wealth into this new rabble-rousing upstart of an organization. He pestered Franklin Roosevelt and his administration to honor the concept of wilderness. Working first for the Interior Department and later for the Forest Service, he achieved his greatest triumph in 1939 with the establishment of a new policy that stripped regional foresters of some of their powers and made it more difficult to change these wilderness classifications. Marshall would not live to see the results of those decisions. He died two months later of a heart attack. He was thirty-eight.[9]

Howard Zahniser was perhaps a cross between Leopold and Marshall. As a longtime writer and editor, he had Leopold's academic and bookish leanings. Although not as extrovertly aggressive as Marshall, he nevertheless utterly embraced the wilderness ethos and understood the politics necessary to achieve the next step. By the late 1940s, it was obvious what that step was—the nation's wilderness needed statutory protection because an administrative fiat could always be broken.

The son of an itinerant preacher, Zahniser as a child moved from town to town in Pennsylvania, primarily in the Allegheny River valley. He retained his Christian faith and often tied his spiritual beliefs to his love for nature and wildlife. After graduation from a small Christian college in Illinois, he headed to Washington, D.C., where he worked for several federal agencies, including the U.S. Fish and Wildlife Service, writing, editing, and doing publicity. In 1945, upon the death of Robert Sterling Yard as executive secretary, the Wilderness Society asked Zahniser to be his replacement. The offer was both flattering and risky. The Zahnisers were expecting their fourth child, and he would have to leave the security of federal employment, but the offer was just too good to resist.[10]

Brower was a Sierra Club board member when he first met Zahniser. In the next few years, he would spend countless hours with the man in Zahniser's home and office, on the telephone with him, at lunches and dinners, in the wild, and in the plush offices of bureaucrats and congressmen.[11]

They became a team. When Brower was in Washington, Zahniser was the key friend he would go to. Brower once estimated that between 1954 and 1964 he was in Washington so much that he must have stayed at a room in the private Cosmos Club for six months altogether. The strategy sessions were often held at lunch at the Cosmos, abetted by Zahniser's Rob Roy and Brower's bourbon and soda. "We were seldom alone in our corner," said Brower. "Zahnie knew well which people to include in our plotting." When Brower was not in Washington, he was on the phone, often nightly with Zahniser. Brower felt that the phone calls were critical, both in the Dinosaur campaign and in getting the wilderness bill passed so that wilderness would have the statutory protection it needed. "One thing I learned early was that it costs a lot less to telephone than to write letters," said Brower.[12] McCloskey said that Zahniser needed Brower's assistance because the Wilderness Society did not have the resources or the presence in the West that the Sierra Club had. Zahniser "very much took the lead, but he and Brower were partners," said McCloskey.[13]

The two were an odd mix. Brower was the risk taker, the upfront gambler willing to compromise sometimes and enrage opponents other times. Zahniser was the back-room handler, who worked out of the limelight patiently and persistently to press or wait, depending on the political circumstances. In negotiating, he could be flexible up to a point, but he was absolutely unyielding on the principle that some federal land needed to be statutorily protected. Sometimes these characteristics backfired. Richard McArdle, who headed the Forest Service from 1952 to 1962, remembered that "Zahnie was about as thorough a gentleman as I ever knew. I used to disagree with him, but I never lost my respect for him." McArdle once got a call once from Senator Clinton Anderson, a wilderness bill supporter. Anderson had just met with Zahniser and was upset. "He [the senator] told me that if Zahniser came up to beat on him just one more time, he was going to drop the whole effort," said McArdle. "He said he was working as hard and as skillfully as he knew how to work, and he was just fed up with constant yapping at him."[14]

At the Wilderness Conference in 1951, Zahniser had spelled out in some detail what a new wilderness law would entail. But the first draft had flaws, and Dinosaur and other issues delayed a new draft until 1955, when Zahniser unveiled it at that year's conference. The legislation would permanently classify many of the Forest Service's wilderness and primitive areas, along with nineteen wildlife refuges and game ranges as well as fifteen roadless and wild areas on Indian reservations. The National Park Service would have ten years to designate which lands within its parks should be classified as nonwilderness because they would be needed for roads or other development. Each federal agency would continue to manage the wilderness areas, but any changes in status would require an act of Congress.[15]

Even before the bill was introduced in June 1956, Brower and Zahniser received a rude lesson in how arduous their task would be. They took a draft of the proposal to John Sieker, chief of the Forest Service's Office of Recreation and Lands. They were there only to seek friendly suggestions, but Sieker copied the draft and sent it to every possible wilderness opponent that the Forest Service had ties to. Brower and Zahniser were stunned; they thought they were dealing with a friend, not an enemy. "That's what the Forest Service did. And they tried to trip [the wilderness bill] up time and time again," Brower commented later. "I kept trying to like them, and they kept making it very difficult."[16]

The movement could not afford to lose friends, especially early on. Timber companies, miners, and livestock farmers and ranchers were opposed, and the bill remained locked in Congress for years. Most speakers at hearings would begin by saying that they believed in wilderness, that they wanted to see wilderness protected, but then their problems with the legislation would begin. The bill was too far reaching. It was aimed at land in the West, which would harm local economies and deprive many of a job. It would restrict future generations from using the land. Fires and insect infestations could not be controlled in roadless areas. The bill would prohibit dam construction in areas where local communities needed new water resources. Finally, it would just lock up too much land from ever being used. An editorial in the *Salt Lake Tribune* in March 1960 called it a "silly" proposal." The bill should be defeated, it said, to prevent "a costly boondoggle [that] could do irreparable injury in the public mind to the whole wilderness philosophy."[17]

The legislation called for setting aside approximately 50 million acres of land, which sounded significant until it was put in perspective: the federal government owned 477 million acres, and the nation contained more than 2 billion acres of land. Much of the protected property would be in the West because it was the only region that had not already been overrun by development. The vast majority of federal land would still be open for mining, timber, grazing, and other uses, and Congress could always reverse its decision if problems arose. Wilderness fires could be controlled by smokejumpers and modern fire-fighting techniques. As the arguments raged for years, Brower could be blunt in responding to the critics. Special interests have been exploiting the people's lands in the West for years, he said. These special interests "seem determined to use diamonds for common abrasives, to dull and destroy rare and irreplaceable wilderness by using it for utilitarian purposes that could just as well be served elsewhere." He called the conflict a case of "local greed versus national need."[18]

The opposition from the businesses that made their money off the land was expected. More surprising was the hostility from the Forest and Park Services. Both agencies felt threatened because some of their discretion in land management was being removed. McArdle argued that the legislation was in direct contradiction to Forest Service practices, that a forest offered a multitude of uses. He also focused on the elitist argument. Speaking at the Wilderness Conference in 1957, McArdle declared that "ninety-nine percent of the people who hunt, fish, camp, picnic or just ride around enjoying the scenery in the national forests don't use our wilderness areas."[19]

At least with McArdle and the Forest Service, Brower knew where he stood. Conrad Wirth, the Park Service director at the time, would never give Brower a straight answer. When Wirth was given a confidential draft of the legislation, he promptly gave a copy to a mining organization. He also tried to convince an advisory council to the Park Service to oppose the bill. Brower recounted three meetings where he asked Wirth for an opinion. Each time Wirth said that he still had not read the draft legislation. Yet Brower learned from other sources that Wirth knew enough about the bill to influence the Department of the Interior to oppose the bill. When Wirth finally did comment on the bill, he said he did not think it should apply to the national parks because they were already recognized as wilderness areas. This argument rankled because conservationists had just finished a major campaign to protect the national parks by opposing the Dinosaur dams.[20]

In the conservationists' dealings with bureaucrats such as McArdle and Wirth, Justice Douglas was an invaluable asset because he understood Washington so well. Douglas had been on the high court since 1939, when Roosevelt plucked him from his teaching post at Yale. He understood·power in a city that knew no other calling, and he helped guide Brower. And although Douglas had lived for many years in the East, his heart was still in the West, especially the mountains. He had been raised in Yakima, Washington, and remained an avid conservationist and outdoorsman, enthusiastic in fishing, hiking, and eating game (his best meal, he claimed in his autobiography, was a blue jay).[21]

Brower first introduced himself to Douglas in 1957. By the next year, they were planning a hike. They ultimately went on several together. Brower spent a week on the trail in the wilderness of the Sierra and North Cascades with Douglas and his wife, Mercedes—the second of four wives—and they got along well. Douglas was extremely interested in the North Cascades, whose mountains he had hiked as a boy. He had never been to the Sierra, and his trip there in 1959 involved more research because he was writing a book about wilderness areas of the West. He carried a looseleaf notebook, and every time they stopped, he would scribble notes about what they had seen and talked about since the previous stop. These notes were eventually translated into *My Wilderness: The Pacific West*.[22]

The wilderness bill was introduced in 1956, and over the next three years it did generate a series of congressional hearings both in Washington and in field sessions across the West. In Washington, Zahniser was often in the hallways of Congress pitching the bill's merits. He hired a Georgetown tailor to create two large overcoats with numerous voluminous pockets. It was a virtual file cabinet with copies of editorial reprints, speeches, and drafts of the wilderness bill that he could distribute. By 1958, public support seemed to be growing, with editorials supporting the wilderness appearing not only in eastern newspapers but also in western newspapers such as the *San Francisco Chronicle*, the *Oregon Journal*, and the *Bend Bulletin*.[23]

Yet even as positive public sentiment grew, the bill remained locked in committee. Frustrated, Brower searched for alternatives. The Forest

Service had begun what it called a Timber Resources Review, a study to determine how trees were available in its forests and what the nation's need for wood products would be for the next twenty-five years. So in response Brower suggested in December 1956 a Scenic Resources Review that would examine local, state, and national parks as well as forests and other natural areas. Its aim, Brower said, would be to identify needed recreational, wildlife, and scenic areas before development threatened them.[24]

Brower spent two years promoting the study plan. It encountered little opposition, which should have been a clue to him that there might be a problem. The amended legislation created an independent presidential panel, the Outdoor Recreational Resources Review Commission (ORRRC), composed of fifteen members. Eight came from Congress, and few were friends of the environmental community. Seven were to be citizen members, with credentials in resource-conservation planning. McArdle and Interior Secretary Fred Seaton handpicked those nominees. An advisory council was added, and most members included representatives of virtually every commercial interest.[25] Brower and his conservation allies even lost the one stipulation that might have made a difference. They wanted a moratorium on any development on public lands while the study was under way. "The reasoning was obvious," said Brower. "It would be useless to survey some national forest wild area and find [what] it needed if, when the survey is over, that area had already been turned over to lumbering or some other non-recreational use." The proposal was rejected, so Brower appealed directly to the committee members. The appeal failed. Laurence S. Rockefeller, the chairman of the ORRRC, pointed out that if the commission granted a suspension of some projects, other petitioners would quickly come forward with similar demands.[26] The Scenic Resources Review ploy backfired in other ways, too. Opponents of the wilderness bill now had a new reason to delay it, arguing that the legislation should be postponed until the ORRRC completed its work. Until the nation's recreation needs were examined, such a comprehensive wilderness bill was premature and possibly even unnecessary. The ORRRC stalled and would not complete its work until 1962.[27]

More than just politics was in play here. Wilderness proponents were speaking about far more than recreation; it had been arguments against recreational pursuits that had created conflicts in places such as San Gorgonio. Rather, Brower argued that preservation of remote lands was a form of stewardship for the future. The wilderness was being set aside not just for the current generation of visitors but also for countless visitors of tomorrow. Zahniser once tried to liken wilderness to art. "None of us feel[s] that the National Gallery of Art is there for just the few people that happen to be in the gallery at a particular time," he said. "We maintain it for everybody; and sooner or later anyone who is concerned can visit it."[28] The misunderstanding was so great that Grant McConnell, a Sierra Club member who was working with Brower on wilderness issues, wondered if it could ever be bridged. "We are not talking about recreation, as they [the bill's opponents] really seem to believe," McConnell told Brower. "What we are trying to fight for is something a good deal closer to religion. It would be good if they tried to understand this."[29]

In frustration, Brower turned to Stegner. Among Stegner's many gifts as a writer was his ability to transform political fodder into literature. In December 1960, Stegner wrote a letter to David Pesonen of the Wildlands Research Center in Berkeley, which was preparing one of several commissioned reports for the ORRRC. Brower had prodded Stegner to write the letter, and when he received a copy, he was thrilled. Stegner's letter addressed the values of the wilderness, and, according to T. H. Watkins, it "was a manifesto, a central document, what Stegner himself would come to call a coda."[30] He wrote:

Something will have gone out of us as a people if we ever let the remaining wilderness be destroyed, if we permit the last virgin forests to be converted into comic books and plastic cigarette cases, if we drive the few remaining members of the wild species into zoos or to extinction; if we pollute the last clear air and dirty the last clean streams and push our paved roads through the last of the silence, so that never again will Americans be free in their own country from the noise, the exhausts, the stinks of human and automotive waste. And so that never again can we have the chance to see ourselves single, separate, vertical and individual in the world, part of the environment of trees, and rocks and soil, brothers to the other animals, part of the natural world and competent to belong in it. Without any remaining wilderness we are committed wholly, without chance for even momentarily reflection and rest, to a headlong drive into our technological termite-life, the Brave New World of a completely man-controlled environment. We need wilderness preserved—as much of it as is still left, and as many kinds—because it was the challenge against which our character as a people was formed. The reminder and reassurance that it is still left, is good for our spiritual health even if we never once in ten years set foot in it. It is good for us when we are young—because of the incomparable sanity it can bring briefly, as vacation and rest, into our insane lives. It is important to us when we are old simply because it is there—important that is, simply as idea.[31]

The letter remains a powerful, romantic, and inspiring testament. Stegner said he wrote the letter in an afternoon. It struck a chord, drawing attention and praise. Portions of the essay were reprinted in a number of publications at the time. Since then, it has been widely disseminated. Stegner said that a friend saw it on the wall of a Kenya game park, and it was reproduced on posters in South Africa, Canada, and Australia. "Altogether, this letter, the labor of an afternoon, has gone farther around the world than other writings on which I have spent years," wrote Stegner.[32]

It was one thing to compare wilderness with art, literature, or religion. It was another to find the tools to preserve it. So far Brower and his allies had not found such tools, and his clashes with the Forest Service and the Park Service had begun to escalate.

6. Forest

Any interest group that we start to oppose immediately goes to work to try to make us out [as] *extremists* or kooks. One group consistently upset by us is what I call the saw-log foresters, whose one interest in forests seems to be getting the board feet of timber out. This type thinks a forest that has been there for millions of years is wasteful.

DAVID BROWER, IN HAL WINGO, "CLOSE-UP, CALIFORNIA'S DAVID BROWER"

David Brower first noticed the logging on U.S. Highway 395 in July 1952 on his way to and from a high trip in the Sierra Nevada. Loggers were felling Jeffrey pine to the edge of the road, and Brower wondered why. He had long thought that this forest, with trees that dated to before the American Revolution, was magnificent.

Just as concerned were John and Barbara Haddaway, who followed up and asked officials in the U.S. Forest Service why the logging was necessary. The resulting tissue of misunderstandings and mistruths from the Forest Service would lead to mutual hostility.

"That's the place my disillusionment with the Forest Service began," Brower said years later.[1] Over time, his disenchantment only ripened, which was evident when he made such statements as "I have called many of the actions of the Forest Service vindictive. They will go way into the middle of a forest that is still virgin and cut right in the middle of it" and "It would be nice if we had a Forest Service instead we have a timber service" and "The Forest Service very richly deserves every bit of criticism that I throw at it. There are still too many saw-log-first foresters at the helm." To the Forest Service, what "is supposed to happen to a tree is that it is supposed to fall."[2]

The logging was at Deadman Summit, more than a mile above sea level on the eastern Sierra on an almost imperceptible rise within the watersheds of the small, meandering Deadman, Glass, and Dry Creeks. The trees in this forest of Jeffrey pine were spaced almost as if they were in a park. They flourished because of a gap to the west in the Sierra range that allowed moisture to dampen the high-desert terrain. The pines, rooted in fine-grained pumice that spewed from the nearby craters and volcanic peaks—cones of volcanic dust and obsidian—rose up to 2,000 feet above the landscape.[3]

The Haddaways ran a successful manufacturing plant in Bishop, California, that produced small pumps for aquariums and other uses. They owned a second home 40 miles away in Mammoth Lakes, a ski resort bordering the Inyo National Forest and the logging area. They would spend $60,000 of their own money fighting the Forest Service and its cutting of the Jeffrey pine.[4] On August 1, 1952, in a telegram to the Forest Service's regional office in San Francisco, they complained about the logging along Glass Creek. George James, a forestry official, responded by return telegram that the cutting was needed to control rust (disease) and beetle infestation. He added that recreation, not timber harvest, was the Forest Service's ultimate aim in the area.

So the Haddaways hired a private entomologist, who analyzed the trees and found no evidence of disease or insects. He reported that it appeared to be one of the healthiest forests he had ever seen. The Haddaways also consulted with entomologists in the Forest Service itself, who concurred that even the likelihood of rust or beetle outbreak was unlikely in the Deadman Summit area.[5]

When the Haddaways confronted them with these findings, forest officials began to retreat. First, they said they were building new roads in the area. When that explanation did not work, they conceded the logging was a straight timber sale. The nearby privately owned Inyo Logging Company had become dependent on timber from the government to survive.

The extent of the logging escalated in 1953. So in March 1953 at Mammoth Lakes, the Haddaways met with James, the San Francisco forest official, and John Haddaway presented his findings from the entomologists. James was no longer interested in talking about insects. He denied that was the problem and said the Forest Service was removing high-risk trees and making the area safer for recreation. Haddaway, who had already found that 90 percent of the trees were healthy, asked James how he defined a high-risk tree. James replied that they were willing to remove any tree that might have a problem in fifty to one hundred years. Barbara Haddaway and her husband began laughing. Sheepishly, the Forest Service official joined them.[6]

John Haddaway learned to take photographs and eventually had thousands of them, which he condensed to a slide show that lasted seven minutes. One of the Haddaways' biggest concerns was the excessive piles of debris called "slash" left by the loggers. Although the cutting had been selective throughout the forest, the slash piles were big and unsightly. Forest officials told them that the area was not a national park; it was a national forest, and as such the trees needed to be harvested. In response, the Haddaways prepared an ambitious plan to declare the area the Inyo Craters National Monument.[7]

The Haddaways and the Sierra Club had begun conferring in 1952, and at one meeting Brower noticed by looking at a map that a buffer zone between the highway and the logging had been violated.[8] By August 1953, Brower also was hearing conflicting stories from top forest officials about Deadman. One said the cutting was to remove high-risk trees, but a second reported it was to prevent disease and insects. "I began to see for the first time what Haddaway had been up against," said Brower.[9] But in 1953 he was still relatively new as the Sierra Club's executive director, so he moved cautiously. As Dick Leonard, Sierra Club president in the early 1950s, told one forest official, the Sierra Club did not take sides in disputes involving the Forest Service. That policy was about to change.[10]

Brower decided to drive out to Deadman Summit and see for himself what was happening. He spent August 27 with several forest officials driving through large sections of the forest but seeing very little in the way of timber harvests or debris left behind. At one place, they did stop and peer at what Brower would describe as modest little piles. He said they told him that this was the type of problem that the Haddaways were complaining about.

Only it wasn't. Brower had decided to tour much of the forest with Haddaway the day before, and he now realized that the forest officials stopped far short of where most of the damage had taken place. He was also told during the tour that the timber sale had generated so little money that after administrative costs were deducted from the revenue, there was nothing left. Jeffrey pine had very little commercial value, the Forest Service officials commented, and most of the wood was going to be used as packing crates for fruit and produce. Brower told them it would be tragic to continue harvesting the Jeffrey pines.[11]

The controversy continued for another three years. It eventually made its way to Washington, D.C. In April 1954, the Haddaways appealed to officials at the Department of the Interior to get the president to declare the craters a national monument. Interior Secretary Douglas McKay would not meet with them. Undersecretary Ralph Tudor gave them fifteen minutes and concluded, "The nation has too many parks and monuments now." They did meet with Richard McArdle, chief of the Forest Service, who bristled angrily after watching Haddaways color slides of the debris left behind by the loggers. McArdle told them that such piles should have never happened and that, from what he could see, the logging had not helped the forest's recreational aspects. But two of McArdle's aides strongly defended the local foresters, and when the Haddaways returned home, a letter was waiting for them from the forestry chief. They were dismayed to read what amounted to a defense of the forest officials in charge of that area.[12]

Richard McArdle was a professional forester who had spent his entire career alternating between the Forest Service and academic forestry and research posts. He had been a dean at the University of Idaho School of Forestry, had headed a research facility in Colorado, had run a forest station in North Carolina, and in 1952 had been appointed the Forest Service chief, whose job was to manage 180 million acres of federal land. Unlike many of his predecessors, he really did intend to manage that land, with an emphasis on long-term planning. He wanted to curb the Forest Service's excessive timber, mining, and grazing abuses. Yet he could also be very "old school." The nation after World War II was discovering the forests as a playground that should be set aside and left untouched. To McArdle, a forest was for both play and work, recreation and timber. Such a stand ran counter to Brower's developing ideas of forest and wilderness, but Brower initially could not help but like McArdle. The forestry boss could be brisk and yet disarming and informal when dealing with subordinates. He was great at recalling names; he avoided formal trappings; and he was well liked within the Forest Service. Everyone, it seemed, called him "Mac" or "Chief."[13]

Brower asked McArdle to conduct a more thorough review of the Deadman Summit logging. In December 1954, McArdle agreed that some of the cutting may have been too extensive. He would not stop all logging, he said, because it would be too harmful to the local economy and the single sawmill in the area. However, he did order Forest Service officials to take recreational concerns into account when planning future timber sales. He also proposed that representatives of the Sierra Club or other citizens should have the right in the future to review and offer advice on such timber sales. The immediate reaction from the Sierra Club board was that these steps were a significant victory. The Deadman controversy, declared Ed Wayburn to the board, "had been a blessing in disguise."[14]

But McArdle's decision was in fact a classic forester's ruling, dictating that recreational concerns had to be balanced by logging interests. Edgar Wayburn had a chance to see what had been cut at Deadman, and he began to change his mind about McArdle's decision. "I went back, and I was saddened," he said later. "The forest shouldn't have been logged. I think it would have been of more value staying in its natural condition."[15]

No one—not Brower or the Sierra Club or the Haddaways—succeeded in stopping the Forest Service from cutting in the Deadman Summit area. The Haddaways became increasingly bitter as they saw more and more trees felled in what they believed were virgin forests, some of those trees dating back centuries. Each harvest degraded the once pristine value of the land, making it ever more difficult to mount an argument for a crater national monument. "Forest supervisors?" Haddaway scoffed years later to a reporter. "They're supervisors of destruction, not forests." The Inyo Forest had some old-growth trees before the harvesting began, but some logging had been going on in the area since the nineteenth century.[16] Although the forest clearly had value, it also had some blemishes from that previous logging, which complicated the question of whether it should be protected. But Brower maintained that there were many forests like this around the nation, where the lumber value was minimal and the recreational or wilderness values far higher. Brower told the Haddaways in September 1956 that he had lost sleep worrying about what had happened at Deadman. So far, he said, neither he nor others had figured out how to make a case to protect forests such as the one at Deadman Summit. And until they did, the Forest Service would only continue to cut trees that deserved to remain standing.[17]

The conflict at Deadman was only the first of many for Brower, the Sierra Club, and other conservationists. Begun in the early years of the twentieth century, the fissures between John Muir's philosophy of forest preservation and Gifford Pinchot's preference to manage and use forest resources would widen significantly in later years, especially after World War II. Pinchot had had strong conservation reasons for convincing President Theodore Roosevelt in 1905 to appoint him as chief of the new U.S. Forest Service and to transfer the national forests from the Department of the Interior over to the Department of Agriculture. Both Roosevelt and Pinchot were alarmed at what they called the timber barons' rapacious practices in the wholesale cutting of trees on private lands. They wanted to protect and manage the nation's public lands to avoid that kind of mismanagement and especially the wholesale destruction of forests. The new management style Pinchot proposed, which allowed for a variety of uses ranging from limited lumber sales to establishing the primitive or wilderness areas, caused little controversy for the first forty years of the Forest Service. Until then, logging in the national forests had been benign and produced only 5 percent of all timber production. But logging companies were beginning to exhaust many of their private resources, and, as Pinchot had feared, they had not been diligent in any long-term planning. Plus, the demand for lumber products soared after World War II.[18]

Although timber harvesting was a high priority among the professional loggers of the Forest Service, the service also faced other demands for mining, grazing, exploiting the forest water resources, establishing recreational areas, and, finally, protecting wilderness. The pivotal player whom Brower had to deal with in these clashes was McArdle, the academic and planner, who viewed his role as that of a forest policeman. "Those conflicting pressures for use of the resources on the national forests helped to keep me standing up straight," McArdle said later. "If one group was silent, it made me feel that perhaps they were getting more than they should get."[19]

The Forest Service had adopted a multiple-use policy, according to which it worked to balance competing interests. But McArdle was under extreme pressure in the 1950s to choose logging over other uses, and there is strong evidence that he and his lieutenants did just that. Between 1940 and 1971, lumber consumption rose 49 percent, pulp-product consumption by 235 percent, and veneer and plywood consumption by 475 percent. Fueling much of that increase was the annual timber production from the national forests, which increased from 3.5 billion board feet to 9.3 billion board feet by the end of the 1950s before leveling off at between 10 and 12 billion. Regional foresters stated that their task was to "get out the cut" under the pro-business Eisenhower administration, which wanted not only to meet builders' demand but also to add income to the federal treasury.[20] McArdle maintained that timber sales were always made within the context of proper planning, yet the pressure was relentless. "There was never a week that I wasn't being beat upon to raise the amount of timber," said McArdle. "We called it the allowable cut and a more disastrous term was never invented."[21]

Brower rejected such reasoning, repeatedly reminding McArdle that there were other problems with the multiple-use concept. Most of these various uses were categorized by their commercial value. Unspoiled wilderness could not be classified that way. It was impossible for conservationists to make a case for wilderness based on dollar values, Brower told McArdle. The highest bid would always win for something that could make a profit.[22] Privately, Brower was even more critical of the policy, calling it nothing more than a game of musical chairs. He said it would be used to pit one user against all of the others. "Whatever user got out of line was to be put upon by the other four or five," he said.[23]

By the mid-1950s, Brower was tracking numerous conflicts with the Forest Service throughout the West. Not all were logging issues. Sometimes the Forest Service opposed giving up land it managed to establish new national parks, such as in the Cascade Mountains of the Pacific Northwest and Nevada's Great Basin. Road construction in pristine areas was another problem. In the fifteen years since World War II, the Forest Service had bulldozed 65,000 miles of new roads, nearly all in isolated and uninhabited national forests. In contrast, the new interstate highway system consisted of only 50,000 miles.[24] The Forest Service characterized these new roads as improvements, but wildlife advocates pointed out that the roads irrevocably altered a forest. One major conflict was at Waldo Lake in the Oregon Cascades, where the Forest Service had improved the road into the alpine lake to open it to camping and fishing. Before the road improvement, the only access in had been on an extremely long and bumpy dirt path that was closed by snow except for two months in the summer. Waldo was spectacular and clear to a depth of 180 feet. Robert Aufderheide, forest supervisor for the Willamette National Forest in the Oregon Cascades, rejected the wilderness arguments because he said that only one-tenth of 1 percent of all visitors had used the area when it was wilderness. "Where is a good balance?" he asked. "How much wilderness should we set aside for one-tenth of one percent?"[25]

The greatest conflicts were over logging, and, according to a list Brower drew up, such conflicts ranged across the West from Alamo Mountain and the Kern Plateau in California to Flat Top in Colorado, Tracy Arm in Alaska, and elsewhere. Nowhere was there more controversy and alarm for conservation than in Oregon's Three Sisters Wilderness and Washington's North Cascades.[26] Deadman contained forests that had never been officially labeled wilderness by the Forest Service, but that was not the case in the Three Sisters and the North Cascades, where the Forest Service was attempting to open up large tracts of officially designated wilderness to logging. Deadman had been primarily a local issue; the Three Sisters and North Cascades became national campaigns that lasted for years.

The Sisters were three spectacularly beautiful volcanic peaks in the Oregon Cascades. David Simons, a young college student hired by the Sierra Club to survey threatened wilderness areas in the Oregon and Washington Cascades, called the Sisters "the shining mountains." "Glacier-sheathed, they dominate a living wilderness of near-rain forests, volcanic wonders, calm lakes, rushing streams and flashing waterfalls," he wrote.[27] The Three Sisters peaks and surrounding forest had been identified as a wilderness area under the original Forest Service regulations. The new rules established in 1939 were designed to provide further protection for these L-20 primitive areas. They were called U-1 and U-2 wilderness areas, depending on their size, and could not be reclassified without going through a lengthy, administrative process.[28] However, the new rules also said that the Forest Service should reassess these areas in the future, which it did. In 1955, the Forest Service announced it wanted to reclassify 53,000 acres in the Three Sisters and make them available for timber sales. Loggers and conservationists fought each other bitterly over the proposal. Brower was opposed to any portion of the Three Sisters being lost to logging, but he was especially upset about the new lines for the protected area: "Rather than put boundaries logically on ridges, where at least you would have skylines that were free of the scars of logging if you were within a wilderness area, they brought the boundaries right down Horse Creek so that one side of the stream would be wilderness, the other side logged."[29]

The North Cascades was even more beautiful and more controversial than the Three Sisters. Brower first learned about the magnificence of the area in 1955, and he first journeyed into the mountains in the summer of 1956. It was wet. When the rain did not pelt down, it arrived as a mist. Moisture overruled sunshine. Between the rain drops, he found a pristine wilderness of white-capped soaring peaks nestled by ancient icy glaciers, which fed waterfalls that plunged down steep canyons into verdant rain forests of rushing rivers, bubbling creeks, towering evergreens, tangled underbrush, meadows carpeted in a profusion of multihued wildflowers, and deep-blue finger lakes. This wilderness region northeast of Seattle was characterized by its highest peak, the 10,700-foot-high volcanic cone of Mount Baker. The Cascades feature numerous other peaks—Bonanza, Goode, Eldorado, and Forbidden, to name just a few—which in the 1950s could not be seen until one ventured into the region by foot or air. There were other features that prompted some to call the Cascades "among the prime scenic areas of North America." When Brower visited the mountain region, the area between Snoqualmie Pass to the south and Canada contained 519 glaciers: three times more living ice than what remained in the other forty-eight contiguous states.[30] A centerpiece feature was Lake Chelan, a 55-mile-long finger lake that snaked between high peaks. The difference in height between the tallest peaks that flanked the lake and its bottom was 9,000 feet. The water from the falls and creeks fed fifteen sizeable rivers, including the Columbia. The rain forests were ancient and cathedral-like. As writer Weldon Heald wrote for a collected volume on the Cascades published in 1949, "Hidden away among these twisted, convoluted mountains are enough lakes, meadows, waterfalls, alpine basins and sweeping panoramas to keep the lover of the outdoors busy for a lifetime."[31]

The North Cascades wilderness covered close to 1 million acres, nearly all of it federally owned and managed by the Forest Service. For decades, its isolation was its staunchest protector. Federal and state officials repeatedly studied it both as a possible national park and as a primitive or wilderness area. In 1940, under the new wilderness rules just put into place, the Forest Service agreed to protect 352,000 acres as the Glacier Peak Limited Area but to leave the balance open to logging. By the mid-1950s, logging in many of the alpine valleys of the North Cascades was upsetting both visitors and conservations. Also by then, the Forest Service had been working for several years on a plan to open yet more of the Glacier Peak wilderness to timber companies. Opponents called the proposal it unveiled in February 1957 the "starfish plan." Just as at Three Sisters, the new boundaries of the 434,000 wilderness area were highly subjective, mostly edging out from Glacier Peak as tentacles that primarily protected high-altitude rock and snow, and so arbitrarily drawn that they would again border logged regions.[32]

Grant McConnell was instrumental in getting Brower and the Sierra Club to wage what became a fifteen-year campaign to protect the North Cascades. McConnell was a political science professor from the University of Chicago who owned a summer cabin in Stehekin, a small village at the head of Lake Chelan and one of the few gateways into the wilderness.[33] McConnell and his wife, Jane, proposed that the Sierra Club make a film about the North Cascades. Brower agreed and spent $1,000 on a film-maker. His work was disappointing. Jane McConnell then asked a former college roommate, Abigail Avery, for help. Avery had inherited $2,500 from an elderly relative, and she wrote to Brower that she would give it to the club for the film if he could promise "an A-1 job." Brower accepted the challenge.

This time Brower told the McConnells, "I'm going to do it myself." Grant McConnell could not accompany Brower, but Jane and their two children went on the trip. Brower brought his two older boys, and a packer, Ray Courtney, led them. That was the trip where it rained. "Dave would get pictures dripping drops from the leaves, cloud mists swirling around," said Grant McConnell. "It was a miserable occasion, and it was generally regarded as a failure."

McConnell was now free from his other responsibilities, and Brower agreed to a second journey. They went up Park Creek Pass, and the weather was better. Brower filmed an avalanche on Booker Mountain and scenes of the Brower and McConnell youngsters picking blueberries. In the fall, Brower invited McConnell to see the raw footage. He had run off 5,000 feet of film. McConnell was disgusted: "It looked like utter junk."

"Dave, this is terrible," he said. "How can we ever face up to what we did with Abby Avery's money?"

"You just wait," Brower replied. "I'm going to edit this."[34]

The resulting thirty-one-minute film, *The Wilderness Alps of Stehekin*, is charming. The soundtrack of inspiration music by an organ and choir is a decided liability, and over time the original color has faded. Yet the film still vividly portrays the great assets of the North Cascades. There is a waterfall that cascades down hundreds of feet into a gorge as well as rushing torrents, meadows of wildflowers, the Booker Mountain avalanche, and an ancient glacier. The film's magic is conveyed in Brower's script, both his lyrical descriptions and his plot line. Aerial views of the mountain give way to a boat ride up Lake Chelan to Stehekin. Brower drives an old Jeep with his two boys, and then they take to the trails. They meet McConnell and his daughter, Ann. The kids, all preadolescents, struggle interacting. More hiking, a meeting up with a Sierra Club outing, and finally the Brower crew finds itself going back down the river and out of Stehekin. Brower closes by exclaiming that he hopes that these mountains will always remain wilderness: "Other people will want to be walking our trails, up where the tree reaches high for the clouds, where the flower takes the summer wind with beauty. And the summer rain too. They will want to discover for themselves the wildness that the ages have made perfect."[35]

The film was completed in 1957, more than fifty copies were made, and it was shown across the country. "It came out beautiful, absolutely beautiful, and it played a very large part in the campaign," said McConnell.[36]

Yet it was going to take more than a film of a beautiful, unknown region to make changes in the Forest Service. The growing conflicts between conservationists and professional foresters in the 1950s produced both attempts at reconciliation and, when those failed, greater polarization.

Brower tried to broker peace or at least calm in a talk he gave in Portland, Oregon, in April 1957 to a crowd of Forest Service regional supervisors. At the very least, he implored them, plan long term, not short term. "This generation is speedily using up, beyond recall, a very important right that belongs to future generations—the right to have wilderness in their civilization," he said.[37]

Some foresters did recognize that certain logging practices were causing them to lose the public's support. "Freshly clear-cut areas look like devastation," conceded Henry Harrison, assistant director of recreation for the Forest Service, in a talk in 1958. Added Ed Cliff, an assistant director to McArdle, "People became enraged when there were changes in scenic values due to logging, road building and other development."[38]

Brower and the Sierra Club produced a plan that called for modest changes in Forest Service practices. The club's so-called Forest Policy took years to complete because of internal disagreements on how far the Sierra Club should go in urging changes. Brower had a role in fashioning the final document, which endorsed the concept of multiple use with the understanding that some areas should still be set aside exclusively for wilderness preservation. The document also appealed to the Forest Service to become more open, to hold hearings, and to allow for public comment.[39] Today such administrative procedures are common, but that suggestion especially rankled McArdle. These types of hearings, he told the Sierra Club in 1961, were often "used by groups seeking special purposes." Besides, he said, the Forest Service knew best how to plan and manage its lands, and its planning tools were already better than what the Sierra Club had offered.[40] McArdle also still had strong reservations about the wilderness bill or attempts to permanently set aside national forest land as wilderness.

Over time, it seemed to become increasingly difficult to find common ground. The Pinchot philosophy was deeply imbedded in the forestry culture, not only in industry but in academia. By the late 1950s, loggers, government forestry officials, and representatives of forestry schools had begun using stronger language in attacking the other side, calling their opponents ignorant, short-sighted, and selfish. W. F. McCulloch, dean of the School of Forestry at what was then Oregon State College, labeled conservationists "urban bird-watchers, the daffodil wing of nature lovers." Nature lovers did not understand resource management or the limits to how much land could be taken out of productive use. "The theory is that if you shout loud enough, you don't need to know anything about the subject," McCulloch said.[41]

Brower's language also over time became increasingly strident. The Forest Service's multiple-use plan, he said, was not just wrong; it was corrupt. Multiple use, he said, meant "that everything in the national forest is for sale . . . that the primary purpose of the Forest Service is the management of timber and to let nothing get in the way of that." Professional foresters would never change. "The Forest Service is still imprinted with the Pinchotism that everything in the national forests is for sale," Brower commented on one occasion. "Forest Service thinking, in my prejudiced mind, epitomizes the mentality that measures usefulness preponderantly as what is salable now to *Homo sapiens*."[42]

Three years after the Forest Service proposed some logging in Three Sisters, an administrator hearing officer settled the dispute by issuing a ruling that satisfied neither conservationists nor loggers nor the Forest Service.

He decided that the area could be reclassified for logging but that additional land-use studies would have to be conducted before any portions were opened to logging sales. As a result of these studies, much of the timber in the area was classified as immature, unsuitable for logging.[43] The greatest blessing for conservationists was that the ruling strengthened their resolve toward getting a new wilderness-protection law. "This decision shows that a wilderness can be made or unmade by simple administrative decision," observed Karl Onthank, who headed both the Federation of Western Outdoor Clubs and Friends of the Three Sisters Wilderness. "Are we to expect that with every change of administration new pressure will be brought to bear upon the Department of Agriculture?"[44]

The starfish plan in the North Cascades was subject to a similar round of hearings and criticism. By the fall of 1960, popular sentiment to protect the area seemed to be growing, as indicated by the number of speakers at a hearing in Bellingham, Washington. In September, the Agriculture Department overruled the Forest Service, sharply limiting where logging could take place, even in some of the corridors that had not been protected. Although it did not expand the wilderness area, it set aside large portions of the forest for "recreation and scenic preservation."[45]

By this point, Brower was so fed up with the Forest Service that in August 1960 he published a newsletter designed to embarrass McArdle and his Forest Service. The World Forestry Congress met that month in Seattle. McArdle had a significant role in planning the meeting, which would draw international foresters, and he wanted to showcase what he and the U.S. Forest Service had accomplished so far. Brower published and delivered four thousand copies of a newspaper-style publication titled *Sierra Club Outdoor Newsletter Number Six*. The message in this publication—in numerous stories, photos, drawings, maps, and graphs—was clear: American forest officials were not to be trusted. There were stories about how multiple-use policies had turned North Africa into a desert and destroyed the forests of ancient Greece. There were striking photos of a woman and child viewing a forest that had been devastated by clearcut logging, with the satirical caption, "Relax in a state of excitement, Oregon."[46] Brower admitted later that he meant to attack the Forest Service.

He would call the newsletter "a major presentation of the mistakes in forest practices and the brainwashing tactics of the Forest Service."[47]

McArdle was furious. He had copies of the newsletter removed from the meeting hall and dumped in the basement. He sent a blistering letter to the Sierra Club complaining about this clear attempt to embarrass him. He and all of his top aides refused to meet with Brower when he came to Washington, D.C. A trade group, the Council of American Foresters, agreed that the newsletter had "shed a poor light on a domestic issue" and was in bad taste.[48]

Most board members supported Brower, especially the club president at the time, Lewis Clark, who had never liked the Forest Service. "The Forest Service, at least in modern years, is very heavily committed to clear-cutting, and I don't like clear-cutting," he once said. In reply to McArdle's letter of complaint to the club, Clark wrote that if "some minor transposition of material was offensive to foresters, we regret it." But he also added that the Sierra Club had been concerned for several years about how the Forest Service was treating many of the nation's forests.[49]

Dick Leonard, who had played such a central role in hiring Brower, repeatedly came to his defense in these years when he used tactics like this. Once a forestry official approached Leonard and exclaimed, "Dave Brower and the Sierra Club are just plain libelous. They are unfair. Look at this picture." He held up a photo in the *Sierra Club Bulletin* depicting Mount Rainier and a clear-cut forest in the foreground. "This is not Forest Service land, it is private land," the forester said. Leonard knew that the property in that area had a checkerboard ownership, every other property either privately owned or owned by the Forest Service, but they all had been clear-cut. "What difference does it make?" responded Leonard. "I'm not going to blame Dave."[50]

On another occasion, Brower published a photo of two streams. One was very clean, and according to the caption it was emerging from Sequoia National Park. The second was dirty and was described as coming from a national forest.

"This is outrageous," Charles Connaughton, a regional forester for the service, told Leonard. "That is private land that is causing the erosion."

"Well, don't you cut your forests the same way?" Leonard asked.

"Yes," said Connaughton.

"The photograph shows the kind of damage that is coming from logging the forest there," responded Leonard. "It doesn't matter whether it is private land or Forest Service land. If it is the same type of logging, it is going to produce the same results."[51]

But there were occasions when even Leonard questioned Brower's judgment.

The Deadman Summit controversy was winding down when Brower received an article about it. The story suggested that the actions of the Forest Service were so egregious that bribes must have been offered and accepted by various Forest Service employees. Brower showed the article to Leonard, a lawyer, who told him that unless what it stated were true, it was libelous. If it offered the truth, Leonard said, Brower should take the information to the U.S. Attorney's Office in Los Angeles.

Leonard was never sure if Brower agreed that bribes had been taken, but it was clear that Brower believed that corruption was one of the few logical answers to what had happened at Deadman. Leonard, quoting Brower, said, "'It was so illogical on environmental grounds, and even on the grounds of forestry, that they felt that it had to be bribery in order to explain it.'"

Leonard thought Brower had killed the article until one day when Brower showed Leonard how the story had been set in type, with a warning that it was being "distributed by the Sierra Club." Normally the wording was "published by the Sierra Club." If anything, Leonard commented later, the change in wording increased the libel danger because the Sierra Club was signaling that there might be a problem with the article.

Leonard felt the decision to publish the article was reckless and stupid, and it upset him. He asked Brower if he had ever forwarded the accusations to the U.S. Attorney's Office. Brower had not. Added Leonard, "He just finally said, 'Well, they wanted to go ahead with it, and so I went ahead.'"[52]

Brower was once asked if he had ever impugned Forest Service officials' motives. He hedged at first, and he was asked if he thought they were dishonest. He recalled that some had been dishonest when they had taken him on the tour of Deadman, Dry, and Glass Creeks, purposely avoiding the problem areas that John Haddaway had already shown him. "To claim that I had said that I didn't want a single tree cut was dishonesty," he said. "So I guess I thought some of them—not all of them—were dishonest."[53]

He did not like Forest Service officials, and he was convinced that they did not like him. Shortly after the clash with McArdle over the *Outdoor Newsletter*, Brower was in Washington, D.C. In a memo to the Sierra Club board on October 12, 1960, he reported that he had learned from friends that "the Forest Service was out to get Brower by any means necessary."[54] Two years later the number of times Forest Service officials had attempted to get him fired was up to at least three, he wrote in a letter that was circulated to a number of people in the conservation field.[55]

Brower was done working with the Forest Service. If he could not protect pristine areas by law or through the Forest Service, he would attempt to convert them into national parks. The Forest Service was second only to the H-bomb in posing a danger to wilderness, he declared in a letter to a friend.[56]

7. Parks

Tenaya Lake was originally called the Lake of the Shining Rocks, and it was described as "one of the most beautiful, glaciated lakes in the High Sierra." A few miles to the west, a sheet of granite sloped gently down for hundreds of feet. What made this sheet distinctive was that its surface had been polished by glaciers thousands of years ago. From the crest of this slope, the gleaming blue glint of Lake Tenaya appeared to the east and Yosemite Valley's Half Dome to the west. Both the lake and the granite wall were spectacular—until they were violated in 1958.

Ansel Adams, David Brower, and others of the Sierra Club accused the National Park Service of destroying these two scenic icons when it rebuilt a 21-mile leg of the Tioga Road. Adams would call the destruction "the tragedy at Tenaya."[1]

Brower and the National Park Service should have shared similar philosophies about protecting the country's most priceless wilderness sanctuaries, but Brower was well ahead of the federal bureaucrats who ran the park system in determining the value for these national refuges. Because of that significant distance between them, he would often clash with the Park Service.

The Tioga Road is north of Yosemite Valley but still well within the national park. It runs from the western side of the park along high ridges and passes to Tuolumme Meadow and then down the eastern flank of the Sierra. The first road was laid out in 1883 by a mining company, and despite some major repairs in 1915 it remained notoriously slow, meandering, and for some downright scary. Drivers sometimes became so frightened by the steep grades that they froze in their cars and refused to move on. Grades could be 18 percent, with lanes so narrow that turnouts were provided for oncoming cars. The road often clung to the edge of a high, steep cliff.[2] Beginning in the 1930s, the Park Service began discussing upgrading the road. The Sierra Club initially supported those plans, but by the 1950s club policy was changing. A more modern Tioga Road would offer a new route across the Sierra, bridge Yosemite Valley with the unspoiled Tuolumme Meadow, and bring new and unwanted traffic to the park.

In the face of those qualms, construction began in 1957, and the first reports reaching the Sierra Club were not positive. The new road was so wide it resembled a highway. Club officials appealed to Conrad Wirth, the Park Service director, who agreed to reduce the road width by two feet. Work resumed again in the late spring of 1958, and on July 4 the club's board of directors met at Tuolumme and got a good glimpse of the new road. Again, most were not happy. The highway had been flattened and the curves eased so that vehicles could speed through the park. Even more alarming for Adams was that the highway, which once skirted only one end of Tenaya, now traversed the entire shore, at times coming within feet of the lake. During a subsequent inspection on July 13, he discovered that gravel and other rock at the bottom of a rocky granite slide, deposited there during the most recent ice age, had been harvested as fill material for the road. Workers had blasted through forest and canyon elsewhere, destroying graceful ridges and swells. To Adams, Lake Tenaya had been desecrated. "Only another glacier age can heal the scars of Tenaya," Adams said. He telegraphed protests to Wirth and Wirth's boss, Interior Secretary Fred Seaton.[3] He also wrote to Harold Bradley, Sierra Club president, announcing that he was resigning from the Sierra Club board. Adams declared that the organization was too prone to compromise and not militant enough in protecting Yosemite.[4]

The route west of Tenaya had been logged, but no blasting had yet begun. On August 3, Brower walked that section of highway, taking numerous photos. He was so alarmed that he sent a telegram to Wirth, and the Park Service chief reluctantly stopped construction. Brower discovered that the highway's new proposed route was directly over the sheet of polished granite and, in Brower's opinion, would needlessly damage it. On August 19, Wirth came to Yosemite and walked portions of the controversial route with his aides, Brower, and two other Sierra Club officials. At the site, Wirth was conciliatory, even though he would state later in his memoirs that he found Brower and Adams to be "violent protesters." Brower thought he had won a concession from Wirth and that the road would be rerouted around the polished granite. Wirth then left for Greece for five weeks. In his absence, workers began preparing a new path that Brower soon discovered would cause even more damage to the polished granite face.[5]

By October 3, news stories reported on the controversy and how the Park Service was ignoring protests that the new highway was destroying the granite facing and Tenaya. Bradley stopped at the work site the same day. Boxes of dynamite were being readied so that blasting could begin later that day. The Park Service had reported that the first shutdown of work had cost $77,000 in penalty fees. Bradley did not identify himself while talking to a construction foreman. He asked if there had been any work stoppages. "'We're laughing about that one,'" Bradley quoted the supervisor. "'The papers have been talking about stoppages. We've never stopped once. We've never lost a day.'"[6]

No one was happy, although Wirth, Brower, and other Sierra Club members met and tried to patch up their differences in November. Wirth denied ever agreeing to the Sierra Club's preferred route around the granite slope. He said he had only agreed to consider alternatives, which his aides did. He also blamed the Sierra Club for any lack of communication. The Tioga Road, he said, had been studied for twenty-five years, and he regretted agreeing to Brower's first request to stop work. "I can discover no valid reason why the matter had to be considered under conditions of high emergency," he told the club. In his autobiography, Wirth would single out this episode as an example of how Brower had turned what had once been a very good conservation organization against the Park Service.[7] Brower always believed that Wirth had misled him that day up in the mountains. As for the road, Brower said, "It was a great mistake, and I'm still very sad about the Sierra Club's weakness. I think if the club had gone all out, it could have reversed what happened." Brower widely disseminated a story that ran in the *Los Angeles Times* about a couple who drove to Lake Tenaya on the new Tioga Road. On previous trips, they had loved the tranquility of the lake, and now they reported they found noise, people, and too much "hurly burly." When they viewed Lake Tenaya, the woman cried. Brower added his own postscript when he distributed the column to a wide and varied audience. He said that his wife had cried too.[8]

Brower wanted to prefer the Park Service to the Forest Service. The Park Service's mission, by law, was to protect the great scenic resources of the nation. To Brower, that meant to preserve them as wilderness, not to chop down trees and build roads. However, too often Brower found the Park Service doing the latter. Since the Park Service's creation in 1916, it had to balance its role to preserve existing natural resources with making those sites available to visitors by building roads, hotels, and other developments, but it had concentrated far more on the latter mission. Stephen Mather, the first park director, once wrote that the Park Service policy should be the "greatest good for the greatest number." That meant hiring engineers and landscape architects to provide the roads, housing, and visitor facilities that tourists would need. Later, as the Park Service expanded and hired wildlife biologists and other scientists, these new personnel would remind park administrators of their mission to preserve. Yet this remained a minority viewpoint, and Wirth was only the latest in a line of builders to run the agency.[9] On the Tioga Road, this viewpoint meant that he preferred a sixty-five-car parking lot providing a 180-degree view that took in everything from Tenaya to Half Dome over leaving untouched a unique granite wall that nature had polished and left on display for eons. To Brower, this approach was maddening. He had offered a plan for an alternative road route at Tioga that would have avoided the granite cliff and included instead a short walk featuring a

360-degree view, but he had been rebuffed. "The glacier-polished slope would be unspoiled and spectacular from the alternative viewpoint, instead of being a shatter foreground of cut and fill," Brower had argued in defense of his alternative plan.[10]

When the Park Service was not being too bold, it was being too timid. Too often the Park Service was too shy to reach out and protect land that was not under its jurisdiction but nevertheless needed help, especially if it was managed by the Forest Service. Gifford Pinchot had been opposed to the creation of the national park system because he said it would be a needless duplication of the protection provided by the Forest Service. His successors grudgingly dropped their opposition to the system as long as no attempts were made to "dismember the national forests."[11] Yet over the years, national parks were carved out of forests in Washington's Olympic Peninsula, Sequoia and Kings Canyon in California, the Grand Tetons in Wyoming, and elsewhere. Few of these transfers went smoothly, and the Forest Service was often one of the most aggrieved parties because it lost land. By the 1950s Brower, the Sierra Club and other environmental organizations were so frustrated with Congress and the Forest Service that they saw the creation of new parks as their next best alternative in saving threatened wilderness areas. They began calling for new parks in the California coastal redwood groves, Three Sisters in Oregon, Washington's North Cascades, and beyond. Establishing these parks would not be easy. The Eisenhower administration had little interest in creating new parks, so no new national parks and only a handful of national monuments were established in the 1950s. The prime opponents to such parks and monuments were usually private logging companies and the Forest Service, while the Park Service stayed on the sidelines, often mute, especially when another voice was needed.

For decades, the federal government did nothing to protect the magnificent redwoods on the California coast. *Sequoia sempervirens*, or coastal redwoods, once stretched along the shoreline from Monterey to the Oregon border. Two million acres of majestic sentinels rose up to 350 feet high, and some of them were two thousand years old. Then the loggers arrived. By 1954, about 50 percent of the old-growth forest was gone.

Only a few small groves had been saved and donated to the state by private benefactors and a conservation group, Save the Redwoods League. The realization that more needed to be done dawned in the mid-1950s at Humboldt Redwoods State Park. Established in 1931, the park included 9,400 acres of redwoods within the Bull Creek watershed. Loggers had continued to harvest redwoods in the greater Bull Creek basin, and in the winter of 1955/1956 a wave of storms struck the California coast. The rain slid down the barren slopes, and the runoff created a wild and raging Bull Creek that grew to 300 feet in width. Erosion and flooding that winter felled 420 giant redwoods in the Bull Creek basin and at the state park. This human-enhanced natural disaster was a wakeup call for preservationists, who began to question the strategy of protecting only relatively small groves of the most priceless trees. The ecosystem was more complex than that; a better solution would be to halt logging throughout the watershed.[12]

The lessons learned at Bull Creek triggered a fifteen-year campaign to create a redwoods national park that would encompass not just groves but also forests. Timber companies fiercely opposed giving up their redwood crops—even if the government paid top dollar for their property. The redwood logging industry had changed by then. The band of small local loggers that had owned the private lands where the redwoods grew had been replaced by huge, diversified logging corporations based in distant cities. Demand for redwood was high after World War II, and to loggers the old-growth redwoods drew the greatest amount of revenue. Paychecks in local California coastal towns depended on logging, which translated into political power. To many locals, the park campaign was a choice between saving a tree or putting food on the table.[13] Brower would play a distant but strategic role in the redwoods park campaign. The Sierra Club was a key player in the campaign, and Brower was involved, but he rarely visited the redwoods, and sometimes he may have been given more credit than he deserved.

Brower was far more involved in the campaigns to create parks in the Pacific Northwest because of David Simons. Simons was a student at the University of California in the spring of 1956 when he talked Brower into giving him a summer job photographing and surveying the still fairly remote valleys of the North Cascades. Throughout his career, Brower would take gambles on young men. Sometimes they worked, sometimes they failed, and either way the results could be spectacularly good or bad. With Simons, Brower's judgment paid off. Simons worked for several summers for Brower and the Sierra Club, eventually writing a series of reports about the Three Sisters and the North Cascades. They were surprisingly articulate, bold, and direct for a man in his early twenties, and Brower may have had a strong editorial hand in them.

Simons did not want to protect just the 247,000 acres of the Three Sisters Wilderness. He recommended instead the creation of the Cascade Volcanic National Park, which would cover 970,000 acres in central Oregon. Besides the Three Sisters, it would protect other high peaks, including the 10,470-foot Mount Jefferson; fragile alpine lakes; and valleys such as Waldo Lake. It would also encompass the volcanic fields at McKenzie Pass, which Simons described as "a great black wilderness of basalt—nearly one hundred square miles," spotted with small islands of tree stands that had been spared by the molten lava when it had poured into the pass. Brower published Simons report on the Cascade Volcanic Park in the *Sierra Club Bulletin*. Simons was not only a promoter but also a skilled writer, and he described the Cascades wilderness as a "fragile film of life. Susceptible to even the slightest cosmic tremor, and subject to many whims, life clings, ever-changing but luxuriant, to the surface of the Cascades. It pulses in the swift motion of wild creatures through grove and upland. . . . Life—free, untouched life—is the essence of wilderness."[14]

The proposed North Cascades National Park was just as ambitious. Simons wrote that the North Cascades was of such national significance that more than 1 million acres should be set aside. It was a bold proposal, but it was bolstered by both history and other supporters. Between 1906 and 1929, several planning groups had studied and considering making the North Cascades into a national park. In 1937, a committee created by the National Park Service had reported that the North Cascades "is unquestionably of national park caliber." It said that if the park were created, it would "outrank in its scenic, recreational, and wildlife values, any existing national park and any other possibility for such a park within the United States." And more recently, in 1963 a local organization, the North Cascades Conservation Council, advanced yet another plan to incorporate more than 1.3 million acres for a new Cascades park.[15]

Simons was so enthusiastic about his work with the Sierra Club that he began to neglect his studies. His grades fell, he lost his student deferment, and he was drafted. But even while based at Fort Bragg, North Carolina, Simons kept up his conservation activities, taking a cross-country bus trip to attend a meeting of the Oregon Cascade Conservation Committee. And then, just as suddenly as he arrived on the conservation scene, he was gone. In December 1960, Simons contracted a rare form of hepatitis and died two days later in an army field hospital. He was twenty-five years old. He death was hard on everyone, especially Brower. As he looked around the Sierra Club offices, Brower could see tokens that Simons had left behind, drafts of letters and other documents, photographs, field notes, maps. Brower had last seen him in Washington, D.C., when Simons had traveled up from North Carolina on a weekend pass to discuss with Brower the need to save the wild places of the Northwest. "One of the sad things is that so few people know how very much David did," Brower wrote in a tribute published in the *Sierra Club Bulletin.* "It means that much history which should have been written now cannot be. But perhaps we can improvise, put most of the pieces together in some semblance of the order he would have worked out."[16]

Brower was committed to winning the battle for both the Oregon Volcanic and North Cascades Parks as a tribute and memorial to Simons. But when Mike McCloskey took over as the club's first regional representative in the Northwest, he found Brower's resolve to be a problem. Besides the push for the two Cascades parks, a third effort was under way to create the Oregon Dunes National Park on the shoreline of southern Oregon. Although McCloskey agreed that all three ventures were worthy, he felt that the Sierra Club "simply did not have the resources . . . to mount credible campaigns on behalf of all of these proposals simultaneously." The Oregon Dunes had the least support, the North Cascades the strongest, but Brower had difficulty giving up on the Oregon Volcanic Park.

McCloskey finally convinced him to concentrate on North Cascades, primarily because the opposition from the Forest Service was just too strong to take on both of Simons's proposals. The Oregon dunes shoreline was eventually declared a national seashore, and the Three Sisters retained its wilderness protection, but the Oregon Volcanic National Park idea quietly died.[17]

Brower and the Sierra Club concentrated its efforts on the North Cascades in a campaign that took as much time and effort as creating the California redwoods park. Brower never has received the credit he is due for his efforts to protect the North Cascades. One reason is that the area itself is remote, not just geographically but also physically, and remains not very well known. Also, much of Brower's work on the North Cascades film and other ventures was done early in his Sierra Club career. By the 1960s, he had moved to other projects.

As expected, the Forest Service for a long time opposed any transfer of its lands for the North Cascades park. McArdle argued that the Forest Service was a better steward because it opened up the land for multiple uses, as opposed to the Park Service, which had a single mission—recreation. Few bought such self-serving arguments, however. "The greatest enemy to the creation of new national parks in the west is the United States Forest Service," John Oakes of the *New York Times* baldly proclaimed in a column published in October 1960.[18] Meanwhile, Park Service interest was low, and for environmentalists that was disappointing. Wirth preferred not to tangle in a fight with McArdle and the Forest Service. That sort of political maneuvering exasperated park supporters. They should have known better—Wirth was a career bureaucrat.

On one occasion, Brower said he thought he understood why Wirth could be so obstinate: like Brower himself, Wirth was Dutch. A colleague had described Wirth as such a "stubborn Dutchman, [that] if he can walk through a door instead of opening it, he'll walk right through it." However, the analogy had a flaw: Wirth's ancestry was Swiss. Wirth was born in Elizabeth Park in Hartford, Connecticut, where his father was the park superintendent, and he was reared in a municipal park in Minneapolis, where his father headed the city park system. He joined the National Park

Service in 1931, eight years after graduating from college. At the Park Service, he supervised projects undertaken by the Civilian Conservation Corps, which meant building things in the parks. Such a perspective fit perfectly with the Park Service preference for promoting tourism over preserving wilderness. It took Wirth twenty years to rise through the ranks and take control of America's crown jewels.[19]

Brower learned not to trust Wirth, and one of the best lessons came during a meeting in 1955 at Seattle–Tacoma International Airport over a controversy at Olympic National Park. The Olympic Peninsula is a richly endowed ecosystem with forests ranging from high-altitude alpine to coastal rain forests. At the time, it sported some of the largest stands of Douglas fir, red cedar, and western hemlock as well as the unusual and iconic Sitka spruce. The Mount Olympus National Monument was created in 1909 to protect those stands, a bitter loss for the Forest Service and loggers who craved the opportunity to cut such rich and varied stands of timber. Even the Park Service opposed converting the monument into the Olympic National Park, which occurred only because Franklin Roosevelt played a personal role in the effort in 1938.

Researcher Carsten Lien has documented in great detail what happened next. Local park supervisors allowed logging to begin in Olympic National Park during World War II, and by the mid-1950s logging there had become blatant. Logging is by law prohibited in parks, so several of the park's seasonal naturalists vowed to stop it. They gathered evidence, showing photos of stumps that had once held Douglas firs five feet in diameter, and fields that had been clear-cut. The photos clearly trumped Olympic Park superintendent Frederick J. Overly's claims that the only logging done was to remove scrap and bug-infested trees. The evidence was turned over to local groups, which gave it to Brower and other national leaders. They then confronted Wirth, who was Overly's boss, and Wirth reluctantly agreed to the airport meeting. Much of the evidence had been gathered by Paul Shepard, who had been a summer ranger and naturalist at Olympic. "His eyewitness story was devastating," said Brower, who was at the meeting. "Wirth saw the threat." At the meeting, he promised to stop the logging, but he began to hedge afterward, until he was confronted with a transcript of the meeting. Wirth had not known that he was being recorded, and he had clearly implicated himself by acknowledging that he had long been aware of the logging and Overly's role in allowing it.[20]

The national parks have changed significantly since the 1950s, when they hosted hotels, tourist gimmicks, and even logging. One reason for this change is the pressure that Brower and others in the conservation movement applied against the Park Service in the 1950s and 1960s. For instance, Brower fought against a hotel at Wonder Lake in Alaska's Denali National Park. Wonder Lake is at the foot of Mount Denali, so remote that it can be reached only by foot or on an 85-mile journey by shuttle bus. Today, Denali has no hotels. Brower traveled great distances to oppose unacceptable developments in national parks and preserves. He opposed a freeway through Rock Creek Park in Washington, D.C., a tram and a hotel at Mount Rainier, and many other building projects in the park system.[21]

As organizations such as the Sierra Club grew under Brower's tutelage, they became stronger in opposing the Park Service's building aims. The climax of this conflict was over a project called Mission 66, which Wirth conceived in February 1954. It was a massive public-works project designed to accommodate more visitors in the parks. It came at a time when even some of the park's biggest boosters by the mid-1950s had been complaining for years about deteriorating conditions, overcrowded facilities, and inadequate staffing. Bernard DeVoto wrote in *Harper's* in October 1953, "Let us, as a beginning, close Yellowstone, Yosemite, Rocky Mountain and Grand Canyon National Parks—close them and seal them, assign the Army to patrol them, and so hold them secure till they can be reopened." Automobile traffic had overrun the parks, and these natural sanctuaries had to be refashioned to accommodate these new demands.[22] By 1954, the national parks hosted 47.8 million visitors, double the number in the year before World War II started.

Mission 66, unveiled in February 1956, increased the Park Service budget by 50 percent and proposed doubling accommodations inside national parks. Other priorities—eight were listed—included additional campsites, parking and improved roads, more staff, and more equipment. "This is a realistic program to remedy the deplorable conditions created in many of our park areas by years of necessary war-imposed economies, followed by an overwhelming postwar increase in park travel," Wirth said in the announcement.[23] The Park Service would spend about $1 billion on Mission 66 over the next ten years. From the beginning, Brower and many other conservations did not like the plans. Wirth, according to Brower, "saw that power would come from money in the Park Service budget. His way to get a big budget was to get a lot of construction going. That's what we were up against."[24]

Wirth created some of his own problems at the beginning of Mission 66 by not including conservationists in on the planning. Brower and Howard Zahniser of the Wilderness Society first met with Wirth to review the plans in December 1955, only two months before the plans were publicly announced. They noticed that there was no mention of the need to preserve the park's wilderness, and they asked that it be included. It eventually was, although Brower would later complain that the inclusion took months of prodding. Brower and Zahniser also protested that conservationists had been kept out of the planning, whereas both the American Automobile Association and road construction companies were given far greater opportunity to offer suggestions. According to Brower's version of what transpired at this meeting in December 1955, Wirth became upset and told Brower and Zahniser that he had expected them to praise the plan, not criticize it.[25]

But Wirth's irritation with Brower over Mission 66 was just beginning and would continue for years. In a letter to Wirth dated April 11, 1957, Brower complained in more detail about the plans and the Park Service's failure to back causes that the conservationists felt were important. At the time, Wirth was opposing both the wilderness bill and Brower's proposal to create a Scenic Resources Review. Brower maintained that a review of the most important natural resources within the parks was far more important than a new road or campground implemented by Mission 66. It was better to leave a road in poor condition or not to upgrade a campground than to make so many changes that they altered an area for generations, he wrote. Besides, he noted, the Park Service alone did not have the clout to protect

Dinosaur National Monument from being flooded by two dams or to get the Forest Service to release land in the North Cascades for a new park.[26]

Wirth became more upset when he learned that Brower had sent a copy of this letter to representatives of various environmental groups and several members of Congress. On May 2, Wirth sent a telegram to Alexander Hildebrand, the Sierra Club president, exclaiming that he was "stunned" that the letter had gone to a wide audience. "I consider such conduct unfair, unfriendly, and unproductive of any kind," he said. Hildebrand defended Brower and said that all Brower was trying to do was point out how a national Scenic Resources Review was needed because its powers would be far greater than that of a single federal agency, such as the Park Service. In addition, some members of Congress were concerned about the Park Service's position, and they wanted to know how the Sierra Club felt about it, Hildebrand told Wirth.[27]

One board member, Ansel Adams, was only disturbed that Brower was being "too gentlemanly" in dealing with Wirth and his staff. Mission 66, he said, was harming Yosemite. In a letter to Bradley, Richard Leonard, and Brower dated July 27, 1957, he wrote that the Park Service was "guilty of criminal negligence" if it destroyed a national park's "basic values." He believed that Mission 66 was doing that and argued that it was time to reorganize the Park Service and fire those responsible. "Heads must roll," he declared.[28]

Meanwhile, Brower was escalating his criticism. In an article in the January–March 1958 issue of *National Parks Magazine*, he poked fun at a recently published Mission 66 brochure. He recounted the story about meeting with Wirth in December 1955 and having to push the Park Service to include any mention of wilderness in the plan. He had an original draft of Mission 66 that suggested that wilderness could be found along a roadside. That line was edited out of the final version, which also bragged that intrusive developments such as aerial tramways, dance pavilions, airports, and summer amphitheaters had been eliminated. And yet, said Brower, the Park Service was planning summer theater at Glacier, a theater and dance pavilion at Yosemite, a tramway at Rocky Mountain, a golf course in Yosemite, and airports in Death Valley, Grand Tetons, and

Katmai. He worried that the national parks and their wilderness features were being "Babbitized, reduced to their lowest common denominator that can produce maximum visitation."[29]

Wirth was furious. He designated Harold Crowe, a former Sierra Club president and a board member, to represent him at the Sierra Club board of directors meeting in January 1958. Crowe read a long statement from Wirth defending himself and Mission 66. Afterward, at dinner, board members regaled Crowe with stories of the many "bloopers" they had witnessed in the park-construction program. They also wondered why Wirth was devoting so much time to rebutting a single article in an obscure magazine. "I don't think I will ever forget it," Wirth wrote two months after that meeting. "I am not used to being called, or having it implied, that I am dishonest." He added: "I can't imagine a more unfortunate outburst coming at a more unfortunate time than this one."[30]

Although Wirth said that he would not write a reply to Brower's article in *National Parks Magazine*, he authorized Horace Albright, a retired Park Service director and former superintendent at Yellowstone, to draft one on his behalf. Albright wrote a long and rambling defense, calling Brower's article a "bitter indictment." He said that he could not remember anything "published or said that has caused such widespread anger and resentment through the Park Service as this attack." However, by the time Albright's letter was received and Brower was deciding whether to publish a response, the magazine editors decided not to publishing anything else on the controversy. [31]

On the one side, Wirth was being overly sensitive to criticism, but perhaps he remembered the old Sierra Club, where gentlemen disagreed in private, never in public. He seemed surprised that so much had changed. On the other side, although Brower was extremely critical of Mission 66, and his comments painted a very negative picture of it, the program did have its benefits. For one thing, it allowed the Park Service to expand its staff, including naturalists who worked in the park. And for another, roads may have opened unwelcome development, but the plan also produced new visitor centers that went a long way toward educating the public about what was valuable in the parks.

Crowe, the former Sierra Club president who had stood up for Wirth, was sorry to see the old days go. "The Park Service got so that they could hardly even talk to [Brower], and when he would show up in Washington, they would turn their backs," said Crowe. "I kept insisting to Dave and the board of directors of the club—if you are going to work with somebody who is keeping care of the parks, you must be able to talk to them. You can't work with someone you can't speak to."[32]

Wirth did not know it at first, but his power as parks director began to wane in 1961, when the new Kennedy administration took over. So would the idea that tourism was more important than wilderness preservation. One of the first signals of this shift came on a sunny day in June 1961 when the new Tioga Road was dedicated at what was called Olmsted Point (in a tribute to the landscape architect), in the very parking lot that had been at the center of the dispute between the Sierra Club and Wirth. Ansel Adams, whose resignation the board had not accepted, believed that no one from the Sierra Club should attend. Wirth, who spoke first, defended the road and minimized the controversy. John Carver Jr., the newly appointed assistant secretary of the interior, acknowledged the disagreement, conceded that some believed the road had been a "grievous error," and pledged to try harder in the future. In the months after the dedication, Wirth often found himself at odds with the Kennedy administration, although he lingered as park director until October 1963, when he was forced out.[33]

In subsequent decades, the Park Service would begin to come around more to Brower's vision, although never as far as Brower wanted it to come. Today Olmsted Point is a popular stop on the Tioga Road, and the parking lot is usually jammed on summer days. But in the late spring, the scene is far different. Tioga is closed in the winter, and Olmsted Point is often the last section the plows can clear. Avalanches often keep rolling snow down the slick granite slopes, piling up huge drifts on the road. Some places are so great, Brower would often argue, that the best solution is to leave them alone. Although the Park Service may not agree, Mother Nature does each spring.

8. Glen Canyon

The ceremony inaugurating the construction of the Glen Canyon Dam was scheduled for October 15, 1956. The plans were elaborate. Explosives had been set near river level, tied to a wire connected to a plunger at the canyon rim. In the White House, President Dwight Eisenhower would touch a telegraph key, which would relay a signal to Kanab, Utah, and then by radio to a flagman at the dam site, who would lower the plunger. But nothing happened when Eisenhower pressed the key. He pressed it a second time. Interior Secretary Fred Seaton, standing next to the president, spoke by telephone to engineers in Kanab, telling them that Eisenhower was waiting. The message was frantically relayed to the dam site, where it was misunderstood. The flagger called back, and this time the message was explicit and emphatic. The flagger signaled; the plunger dropped. The canyon roared; boulders arched; dust and grit billowed upward. Eisenhower shouted in glee and then, slightly confused, said, "I guess it takes electricity a long time to travel out West."[1]

Work commenced on a massive engineering challenge. Just to prepare to build the dam, nearly a hundred miles of new highway had to be built to reach the dam site near the Utah–Arizona border. A town of five thousand, Page, Arizona, had to be established to house the workers and their families. A bridge spanning the 1,300-foot chasm had to be erected 700 feet above the Colorado River.

It would take six years to build the diversion tunnels and complete the 700-foot-high arch dam and another seventeen years to fill the 186-mile-long reservoir. The numbers for the construction materials were staggering: 5 million barrels of cement, 10 million cubic yards of aggregate, 3 million board feet of lumber, 130,000 tons of steel, 20,000 tons of aluminum, and 5,000 tons of copper. The dam was 300 feet thick at its base and tapered to 35 feet at its crest. The cost was $275 million and the death of 18 workers out of 2,500 who were on the job at its peak.[2]

While the work was under way, David Brower mounted arguments on why it should stop. He was cautiously optimistic that he could succeed. Engineers at the Bureau of Reclamation had indicated during the Dinosaur fight that they had concerns about the quality of the sandstone at the Glen Canyon Dam site. They worried that it might be too porous, and for a while opponents hoped that these concerns might scuttle the project. However, engineers and geologists tested the site and found the sandstone soft but with a high compression strength that brooked no apparent weaknesses.[3] Brower also suspected that the costs to build and operate Glen Canyon would be staggering, making it an expensive embarrassment. However, the initial construction contract came in well below what federal engineers had forecast. Electricity sales were designed to pay off the cost of the project and its maintenance. Brower forecast that with nuclear power coming, the electricity the dam produced would never be competitive. Hydropower turned out to be more economical than nuclear, although a combination of drought and concerns about the impacts to the downstream environment eventually did diminish discharges and power sales. Critics such as the Utah Committee for a Glen Canyon National Park even suggested that the resulting reservoir would have no recreational benefits, but each year millions of visitors would come to play on what would be called Lake Powell.[4]

Brower's most powerful argument was the one he had begun advancing even before the dam was authorized: that it was not needed because the river was already overdeveloped and supplied enough water for what was required. Glen Canyon Dam, he would say later, "was just an engineering, job-making project and a money-making machine."[5] There was truth to this argument, but it was for these very reasons that so many of the political forces of the Rocky Mountains had joined to support the dam. They deserved this dam, they said, and all of the side job and moneymaking benefits that came with it. After years of watching California swill so much of the Colorado River, the states upstream wanted an insurance policy, a pool of water they could hoard for the future. The continued damming of the Colorado River meant that the Reclamation Bureau and local construction companies would remain at full strength. That mixture was a powerful elixir for western politicians.[6]

Brower was thus left with only Rainbow Bridge as an argument. The sandstone arch was protected as a national monument, and it would be within reach of the reservoir. The legislation authorizing Glen Canyon and the dams being built as part of the Colorado River Storage Project prohibited a reservoir inside Park Service territory. Bureau of Reclamation engineers had indicated they could find ways to protect Rainbow Bridge. Brower did not care what solution the engineers came up with as long as they protected the national monument. And he wanted a guarantee before the water began to rise.[7]

To prevail, Brower would have to convince Floyd Dominy, who had been working his way up the federal bureaucracy and in 1959 became chief of the Bureau of Reclamation (figure 8). Dominy, who thrived on power, quickly realized that the massive Glen Canyon construction project would be the hallmark of his administration. He became its biggest booster, and that meant attacking any critics: "I have told Dave Brower that I'm a bigger environmentalist than he is. I've changed the environment, yes. But I've changed it for the benefit of man. We haven't destroyed the world; we've made it habitable for a lot more people."[8]

Dominy versus Brower, one unmovable force striking another, their debate a clash that would reverberate for years.

No one tangled with Dominy unscathed. He was born and raised far from California, by warring parents in the hardscrabble landscape of

FIGURE 8 Floyd Dominy

Floyd Dominy served as the flamboyant director of the U.S. Bureau of Reclamation from 1959 to 1969, and he often clashed with Brower, especially over Glen Canyon Dam. He once bragged that the only changes he made to the environment benefitted humans. (Courtesy of the American Heritage Center, University of Wyoming)

Hastings, Nebraska. He graduated from the University of Wyoming in 1932 in the Great Depression. He became a county agricultural agent in Wyoming's desolate Campbell County, where farmers were failing even with 640 acres for their cattle because the average rainfall was 12 inches a year. Dominy learned that the federal government was willing to pay ranchers 15 cents a cubic foot to move dirt to create dams as holding pools for their livestock. To stretch the budget, he took the cash and paid the ranchers only what it cost each of them to move the dirt, which was considerably less money. He had no use for many of the program's rules. Everyone griped, but he was saving cattle and ranches. Dominy told one writer that he built three hundred dams in a year, another that he had built a thousand. "When I was twenty-four years old, I was king of the god-damn county," he bragged.[9]

He attracted attention, moved on to other federal posts, and realized that what he liked best was building dams. In 1946, he joined the Bureau of Reclamation. He became adept at making friends in high places (especially Congress); he easily outmaneuvered the engineers he despised who ran the bureau; and in an amazing thirteen years he was commissioner. As historian Marc Reisner writes, "Most Commissioners of Reclamation were dull, pious Mormons—or if not Mormons and pious, then at least dull. Floyd Dominy was a two-fisted drinker, a gambler, he had a scabrous vocabulary and a prodigious sex drive."[10] Dominy favored Stetsons, string ties, and gaudy silver belt buckles. In some respects, he was the antithesis of David Brower, his despised nemesis. Anyone who objected to the bureau and its aims, according to Dominy, was "a David Brower type."[11] "I like people. I like taxi drivers and pimps," said Dominy. "They have their purposes. I like Dave Brower, but I don't think he's the sanctified conservationist that so many people think he is. I think he's a selfish preservationist for the few. Dave Brower hates my guts. Why? Because I've got guts."[12]

And yet they shared similarities. Both men had a phenomenal work ethic and were able to marshal great energy resources to accomplish their goals. Dominy once told a story about quitting college and moving to a farm in western Nebraska with his young wife. He grew disgusted with how unkempt the farm was run, so the owner challenged Dominy to demonstrate how he would change it. Dominy climbed onto a tractor, ran it until ten that night, went to bed, got back on it at three in the morning, and finished that afternoon. All was now in order. The farmer told him, "With that kind of drive, you're wasting yourself. You ought to go back to college."[13] Dominy did, and the quest for power eventually led him to try to walk over anyone who got in his way. His adulterous affairs with so many women, however, would outrage the self-described puritanical Nixon administration, which would force Dominy out of office. Brower, too, stayed night after night until a job was done, traveled days on end with little rest, and wrote prodigiously. His quest for power was still growing, although his downfall would eerily resemble Dominy's.

Glen Canyon was Dominy's highest priority when he became the Bureau of Reclamation chief. Virtually everyone was certain that the reservoir would seep onto the 160-acre Rainbow Bridge National Monument unless it was blocked by a dam. There were several options, but the most favored was to erect two dams. One would be a 200-foot-high barrier to prevent the reservoir from encroaching on the monument. A second would be upstream to divert floodwater that otherwise would get trapped by the downstream structure, creating a secondary reservoir. Pumps would also be needed.[14]

Dominy immediately wondered about the feasibility of that plan, and he trekked to the site by horseback to investigate. He joked later that the journey was so arduous that he "walked out and carried the horse." Dominy was flabbergasted by what he discovered. The cost and the degree of construction and upheaval to such a remote area would be staggering. "This is nonsense, absolute nonsense," he declared. "The cure is far worse than the disease, and it will look much more like a bridge with a little water under it." He told his many friends in Congress not to appropriate any money for the Rainbow Bridge dams.[15] The cost of building the dams escalated from $3.5 million to as much as $25 million, and Dominy and his staff called the project a "boondoggle."[16]

Brower was alarmed. A stream of communication, stories in the *Sierra Club Bulletin*, and then resolutions passed by the Sierra Club board were sent to officials in the Eisenhower administration, which was about to leave office. At one point, Brower thought he had an understanding with Interior Secretary Seaton, who was Dominy's boss, that the barrier dams would be built. But then a new administration arrived.[17]

The odds looked better with Seaton's successor, Stewart Udall. After all, Brower had helped Udall get his job. Udall, an Arizona congressman, had approached Brower after John Kennedy's November 1960 election. He wanted the post at Interior. So did the Colorado congressman Wayne Aspinall, who had far more experience than Udall's six years in Congress. Udall asked Brower to help him. "I searched my soul for a while before I decided to do something about it," recalled Brower. "Then I thought, well, the guy does have some good things that he wants to do."[18] Brower's role in helping Udall was limited, although he did convince John Oakes at the *New York Times* to write an editorial in favor of Udall's nomination. Udall got the job primarily because in the Democratic Convention of 1960 he swung Arizona's votes to the Massachusetts senator, which was pivotal to Kennedy's nomination.[19]

Udall was grateful, and for a while he welcomed Brower into his inner circle. Udall spoke at the Sierra Club Wilderness Conference in 1961 and asked Brower and others to give him some elbow room. Recalled Brower, "Later on we were asking, 'What's Udall got going on in his elbow room these days?' The elbow room became a legendary spot."[20]

Yet Udall could never quite be counted on. On issues such as wilderness, Brower had a loyal ally. On others, such as dams on the Colorado River, Udall would range from a reluctant enemy to an adamant foe. He was multidimensional. Sired from a western Mormon family that helped lead territorial Arizona, Udall was a westerner who appreciated the region for its open lands while acknowledging its need for life-giving water. Yet even with his western roots he was still comfortable in the eastern, cosmopolitan Kennedy court of Camelot. It was Udall who talked the new president into inviting Robert Frost to recite a poem at his inauguration. One of Udall's first aides was Wallace Stegner, an author and fellow westerner. Udall could bridge the frontier West to the posh East because he was a skilled politician; the Udall clan could rival the Kennedy's for political lineage, at least in the West. Udall's grandfather served in the Arizona territorial legislature, his father was the state's Supreme Court chief justice, his brother succeeded him in Congress and ran for president, and both a son and a nephew would serve in the U.S. Senate. Finally, Udall was ambitious, and no sentimentality about the wilderness West was going to block him from the developments his constituents favored. Publisher Alfred Knopf told Brower not to trust Udall. He was a young man too anxious, too much in a hurry. He compared Udall with the Catholic who ordered ham on Friday and then remarked, "God knows I asked for fish."[21]

Even before Udall arrived at Interior, there were warning signs that he could not be counted on. When he was a congressional member of Aspinall's House Committee on Interior and Insular Affairs, Udall decided to investigate Rainbow Bridge on his own. On August 8, 1960, when the desert heat was at its peak, he arrived at the small community of Hite, Utah. Udall had been an all-conference guard on the University of Arizona basketball team, and he kept in shape. He rode downstream to Forbidden Canyon and then hiked seven miles to the monument. He spent three hours there, immensely impressed. He found the red sandstone of Rainbow Bridge, with the 10,000-foot-high Navajo Mountain in the background, "unquestionably the most awe-inspiring work of natural sculpture anywhere in the United States." In his report to Aspinall, he wrote that the area had "a rugged beauty comparable only to that of the Grand Canyon itself," but the problem confronting the Congress was "delicate." Rather than building dams, Udall preferred doing nothing at the site, the same conclusion that Dominy had reached. He thought that the boundaries of the national monument should be pulled back from the reservoir and expanded to include natural bridges to the east and a large white-domed throne at the foot of Navajo Mountain. Udall's letter was very influential with the committee and Congress.[22]

Once Udall was interior secretary, he attempted to tread a more middle ground about Rainbow Bridge at a January meeting with conservation groups (Brower did not attend). In a statement given after the meeting, he reported that both the Glen Canyon reservoir and the construction of the proposed barrier dams would be "disfiguring" to the monument. The statement added that "permitting water to enter Rainbow Bridge would not be a precedent in the national park system as long as he was secretary." Udall proposed further studies.[23]

Dominy initiated the next step. "Let's take them in there, Stewart," he recalled telling Udall. "We can get a helicopter in there now. . . . No problem. So let's take Dave Brower, the director of the Park Service, and all these guys. Let's take them in there and rub their nose in it. They've got to agree that the cure is worse than the disease."[24]

Udall agreed, although Dominy was not prepared for the details necessary to carry out the one-day junket that Udall planned for April 29, 1961. "Well, that crazy Udall, when he got through, I had sixty-six people," exclaimed Dominy. "Good God almighty. I had to hire Navy helicopters. They were so big they couldn't land in the canyon." Instead, a ferry system with smaller choppers became necessary. "It was a real logistics undertaking, and I was scared to death we'd lose somebody."[25] As one writer described the expedition, the doors were taken off the glass bubbles of the smaller craft, and, "thus, as the little machines whirled their passengers off the mesa and over the canyons, in some cases the last few inches of a congressional rump or a conservationist elbow hung over empty space."[26]

Those invited included writers for *Life, National Geographic, Sports Illustrated, American Heritage*, the Associated Press, *Deseret News, Denver Post*, and *Los Angeles Times*. Author Russell Martin described how Udall, in jeans and hiking boots, standing on the rocky terrain before the gathering, began an impromptu debate with Brower. He told the assembled crowd that both the Bureau of Reclamation and the U.S. Geological Survey had concluded that even if the water from the reservoir reached to the narrow streambed below Rainbow Bridge, it would not undermine the structure or cause it to fall.

Brower, who had been quiet, said that other geologists did not agree.

Udall said that if the water did infiltrate the boundaries of the national monument, it would not set a precedent for allowing similar encroachment on other national monuments.

Brower responded that if the government allowed it once, it would be that much easier for officials to permit it a second time.

Dominy joined in the conversation, saying that the downstream dam that Brower and others favored would require 5 million cubic yards of earthen fill, and that much soil was not available in this rocky landscape.

Brower shook his head. He was disgusted.

Udall attempted to clinch his argument by pointing out that it was now too late to build the downstream dam, anyway; the water would rise above that dam site before it could be built. He moved on to his previous idea about enlarging the national monument from land that was held mostly by the Navajos.

Even though this idea was a diversionary tactic, it excited Brower, who called the countryside there "extraordinary" and unique. "We are moving now to make a reservation for all the world, and we hope for all time," he said. "And I hope that the Navajos will be willing to help this be set aside as one of the great national parks."

Paul Jones, a former college professor who headed the Navajo tribal leadership, listened and said very little. As Martin would write, "Jones knew that Nonnezoshi itself would have to rise into the sky before the tribe would trade away Navajo Mountain, Head of Earth Woman, Naatsis'ana, but he saw no reason to say so at the moment, letting Udall and these park people speak their minds."

Later, Udall and Brower scrambled to the top of the natural bridge, and with Brower's help Udall rappelled down.[27]

From Brower's standpoint, the trip had accomplished nothing except for a few headlines, and the political process remained unyielding. In the *Sierra Club Bulletin* for June 1961, he said that promises were being broken and that unless positions changed, "the greatest welsh in the history of conservation will have been accomplished."[28]

Udall asked for money for the Rainbow Bridge dams in 1961, 1962, and 1963, but the administration made very little effort to convince the key congressional committees to act, and officials at the Bureau of Reclamation and the U.S. Geological Survey openly urged the committees to stay the course. Meanwhile, the Navajo tribe indicated it was adamantly opposed to giving up any of its lands, including the sacred Navajo Mountain, as part of a new national park. As the months slipped away, Brower watched, increasingly frustrated and bitter.[29]

On March 18, 1962, Brower wrote a personal note to Udall urging him to act on Rainbow. He told Udall to call him collect; he would fly to Washington; he would even change his political registration from Republican to Democrat. He wanted Udall to ignore Congress, to order Dominy, and to begin construction of the barrier dams. If Dominy said he did not have the money or the time, Udall should announce that he would not fill the reservoir. Then, Brower said, the money would become available for the barrier dams. At the bottom of his note, he scribbled that he had shown the letter to his wife, Anne, who believed that it had an "ultimatum-ish sound to it." He told Udall not to read it that way.

Udall did not respond, so Brower published his letter in the *Sierra Club Bulletin* and released it to be published in newspapers across the country.[30]

While Brower's tactics and rhetoric to get the dams at Rainbow approved were escalating, friendly sources were telling him privately that the dams might be a problem. The Sierra Club had retained an engineer who had initially been in favor of the downstream barrier dam. In May 1962, however, he changed his position. He told Brower that he never should have taken that position without first visiting the site. His report was kept confidential for a while.[31]

By the end of 1962, conservationists were desperate to protect Rainbow. A coalition of organizations agreed to force a resolution by going to court. Brower warned the Sierra Club board that the case was a long shot, but the board agreed to join the coalition.[32] The suit was filed December 14, 1962, asking that Udall be ordered to prevent the impoundment of the reservoir until Rainbow Bridge was protected. Less than two weeks later a federal judge ruled that the environmentalists had no legal standing to sue.

On January 8, 1963, Brower made one final appeal to Udall, releasing a letter designed to publicly embarrass the interior secretary. The letter said, "If Rainbow is not protected, it is not your subordinates who will be held responsible. It is you. Don't let yourself down. Nor us."[33]

The Sierra Club labeled 1962 the "year of the last look," and it scheduled several trips down the Colorado. Brower went more than once. In

June 1962, he and his youngest son, John, were on the river, which was riding high. They waded in water neck deep to reach several side canyons. On the last day of the trip, the river was falling. They could see a 15-foot band, the former high-water mark, across the cliffs. The rock below the mark had a bleached look, and much of the grass and shrubs had been washed away. This, Brower said, was a preview of the new reservoir.[34]

He turned his attention to recording what the dam would ruin. On each trip down Glen beginning in 1960, he filmed images with a motion-picture camera. The camera worked fine on the open river, but he could not capture the magic of the side canyons because the light was poor, and on film they were often too dark. He produced more than 10,000 feet of film but gave up, throwing the film into storage, never using it. He later asked Phil Pennington and Chuck Washburn to do something with the film, and they produced a forty-five-minute slideshow that captured both the canyon's beauty and its transformation when the waters began to rise in January 1963. Brower worked with a professional filmmaker, Larry Dawson, to convert the slide show into a twenty-six minute film. It was first released in 1964.[35]

Brower also produced the book *The Place No One Knew: Glen Canyon on the Colorado*. The photos were by Eliot Porter, the editing by Brower. Porter, the one-time East Coast physician turned photographer, was a master at both taking pictures in color and capturing nature and extraordinary landscapes. Some have called *The Place No One Knew* Brower and Porter's greatest book. It has also been labeled "elegant propaganda."[36]

The text was brief but powerful. In the foreword, Brower lamented his role in killing Glen Canyon. The book featured passages from great writers and thinkers including Owen Wister, John Wesley Powell, Stegner, Joseph Wood Krutch, William O. Douglas, Aldo Leopold, Henry David Thoreau, Alexander Pope, and G. M. Trevelyan. But the printed word was no match for the eighty photographs depicting river and canyon, geology and botany, light and shadow, with slender and delicate maidenhair fern, waving yellowing canes, shoots of tamarisk and grass set against backdrops of a cool, glistening river or baked, brittle red sandstone, and more than anything, the rocks in round cobbles, razor-sharp cliffs, cascading terraces, uplifting arches, rippled sheets and twisted strata dressed in black desert varnish, gray Wingate Sandstone, red Navajo Sandstone, and the blues, yellows, oranges, and purples that constituted a geological impossibility. The colors were rich, stunning, and beautiful, so sharp that they seemed ready to crackle and snap. In all, the work was a powerful testament to the strength of photography.

The first copies of *The Place No One Knew* came off the press in June 1963. Brower sent copies to the president, members of Congress, and other conservation organizations, accompanying each with copies of two letters he addressed to Udall. The letters were partially a lament of what had been lost at Glen Canyon and a plea for a new great national monument or park, as Udall had originally conceived. Both were also a blunt warning from Brower that further damming of the Colorado, especially downstream in the Grand Canyon, would provoke a far greater resistance.[37]

Harold Gilliam, an aide to Udall who would go on to cover Brower and environmental issues for the *San Francisco Chronicle*, was working at the Department of the Interior when the book arrived. He recalled several aides crowded around Udall while the pages turned. They were awed by the photographs.

Udall was sad. He shook his head.

"I was a member of the House when we voted for this dam," he said. "We had never seen the canyon and had no idea what was there."[38]

But Udall was also pragmatic enough not to be budged. Wilderness, no matter how valuable and necessary to humankind, had to be offset by water, mineral, and agricultural development. He simply did not feel the same as Brower.

Brower was soon receiving letters in response to the book. Congressional letter writers praised its beauty while generally ignoring its political message.

Senator Barry Goldwater of Arizona had first marveled at the beauty of Glen Canyon during a trip in 1940.[39] He now wrote to Brower, "Eliot Porter's collection of photographs of Glen Canyon is the finest that I have ever seen."

Aspinall wrote simply, "The artistry is magnificent."

One of the few to take issue was Alaska's senator Ernest Gruening, who wrote, "My information, however, is that the calamity is not as serious as the book indicates" because the new reservoir would make the canyon accessible to so many more people.[40]

A separate letter went with the book to newspaper and magazine editors around the nation—including editorial page editors—pleading for publicity about the impending loss. However, if Brower was hoping that this public-relations effort would awaken the public, he was wrong. Editors and reviewers were more interested in the book's artistry than in its politics. The *Atlantic Monthly* noted that the publication was "not a fighting book but a requiem for the Glen Canyon."[41]

The dam gates closed on January 21, 1963, and the 186-mile long reservoir, at the time the longest in the world, began to fill. For Brower and others who had run the river, that year was even more difficult than the farewell trips of the year before. Looking back, Brower commented, "The tragedy became greater and greater year by year."[42] On April 11, he was at the Music Temple. The water was creeping over the sand bar that John Wesley Powell nearly a century earlier had used to beach his boats to explore the side canyon. There was a new waterline every day on the canyon. At least six pairs of blue heron could be seen attempting to tend nests under an overhanging wall. Brower guessed that generations of herons had been raised there. He estimated that by late May the rookery would be under water and predicted that the dead water would stretch to the projected end of the reservoir, at Hite, by early July. Only a motorboat would be able to get up Hidden Passage to the waterfall that "Eliot Porter thought should be a national monument just for itself—provided you get there before it goes under," Brower told his readers in the *Sierra Club Bulletin*. He was angry, and it showed in that report.[43]

And yet once the water began to rise, new visitors arrived. A huge marina and plush resort were planned for the southern end of Lake Powell, at what once had been an isolated Navajo homestead called Wahweap. Motorboats, houseboats, tour boats, and skiers came to this blue oasis in the desert. Soon more than 2 million were vacationing there each year.[44] Stegner returned to Glen Canyon in the spring of 1966 and wrote a mixed review in *Holiday* magazine. Although fretting over what was being drowned by the dam, he found a region that was diminished but not ruined. "Lake Powell is beautiful," he wrote. "It is not Glen Canyon, but it is something in itself." He marveled at the vistas that had opened up now that visitors were higher up the canyon and how the contact between the blue water and the canyon walls was "bizarre and somehow exciting." Yet, in opening up the lake to more visitors, "in gaining the lovely and the usable, we have given up the incomparable," he said. History, archaeology, and wildlife had vanished, and although some upper canyons were still mostly untouched, he feared for their future.[45]

To Dominy, Lake Powell was unparalleled beauty. He said that an area where no one lived, where no railroads or highways or towns existed, had been transformed. "We flooded out the rattlesnakes and the prairie dogs and a few deer and a beaver or two," he said. "That's all that was flooded out."[46]

Fifty years earlier, John Muir had lost Hetch Hetchy and went to his grave regretting it. Glen Canyon now became Brower's Hetch Hetchy. He rarely recalled his victory at Dinosaur National Monument, how he had stopped the Echo Park and Split Mountain Dams, and what a momentous boost that victory had given to the conservation movement. He instead dwelled on what had been lost at Glen Canyon. He could not get over it, and he always kept close track of the dam and reservoir. In 1972, Brower returned to court and the Rainbow dam issue as president of Friends of the Earth. This time a federal judge agreed and ordered the government to build the dams. But the ruling was overturned by an appellate court, and the Supreme Court refused to intervene.[47] Time will tell whether Rainbow Bridge is in geological danger of collapsing. So far the reservoir has surged and ebbed while never actually reaching the footings of the sandstone bridge. Meanwhile, motorboats now bring thousands of visitors on a half-day journey to hike and explore Rainbow Bridge.

In the 1990s, Brower became part of a quixotic movement calling for Glen Canyon Dam to be torn down, the reservoir drained. Although this proposal gained some traction with other environmental organizations, western politicians called it everything from "nonsense" to "silly," "absurd," "myopic, selfish, and impractical." The government was not going to dismantle a dam that had cost $275 million or pull the plug on a reservoir that drew up to 2 million visits a year.[48]

In August 1999, shortly before Brower died, an old friend, Polly Dyer, wrote to him suggesting that he had to get over the loss. He had to stop blaming himself. She did not blame him; "blame" was not the right word.[49]

What Dyer did not understand was that Brower had channeled his bitterness and disappointment into a new resolve never to allow another Glen Canyon Dam. "I made mistakes in giving away things I had not seen; that was my mistake in Glen Canyon," he said one day during a talk at the Grand Canyon in 1977. "To think that I was arguing for a higher Glen Canyon Dam really shudders me. But I did. But I did learn from that never to give anything away that I had not seen."[50] In an interview in the 1990s, he was asked about his willingness to compromise, how flexible he could be. "I try to be as inflexible as possible in saving resources," he replied. "And I see no reason to be flexible about that. And I think that inflexibility has some values that, in certain ways, it builds character, it comes out of boldness."[51]

Public-policy advocates always face the same challenge: To what degree do they push their cause? Extremism has its attributes, but it also has it faults. Go too far, and radicalism will so antagonize not just decision makers but the public that nothing can be accomplished. For decades, the conservation community and especially the Sierra Club had very little public support, and in its weakness the club had no choice but to employ moderation and compromise. Brower's skillful use of public relations in his books, films, and political efforts helped build public support, allowing him and others in the movement to be more demanding. In addition, Brower's growing extremism came at an opportune moment. American society was becoming more demanding in the 1960s; activists in numerous fields were becoming more bold, their demands more extreme. Brower and his anger over Glen Canyon would eventually establish a new philosophy for the environmental movement. Not all individuals or organizations would embrace it; some would not only reject it but also attack Brower. Nevertheless, although Brower antagonized some, his idealism inspired generations of young people and helped create new environmental organizations that espoused a radical philosophy. It was the anger over Glen Canyon that prompted Edward Abbey's novel *The Monkey Wrench Gang*, about a group of environmental saboteurs who want to destroy Glen Canyon Dam. Members of Earth First!—for a time the most radical clan of environmentalists—credited Brower with the group's formation. David Foreman, who founded Earth First!, dedicated his book *Confessions of an Eco-Warrior* "to Dave Brower and thanks for showing me the promise of our species." In 1981, representatives of Earth First! unfurled 300 feet of black plastic from the top of Glen Canyon Dam to symbolize a large crack.[52]

However, even as Brower's philosophy was evolving, he began to understand how it would be tested. The date January 21, 1963, was scheduled as the first day that the diversion tunnels would be closed and the reservoir would begin to fill. Brower flew to Washington, D.C., that day to see Udall, without an appointment, and urge him not to close the tunnels. Brower told Gilliam, Udall's aide, of his mission. Gilliam took the back halls and private passages, found Udall working alone on a stack of papers, and told him why Brower was there.

Udall frowned. "Well, I'll see him," he said. "But Dave just doesn't understand Washington. There's nothing I or anybody else could do to keep those gates open."[53]

That same day Udall held a press conference. Brower stood in the back as Udall announced the creation of the Pacific Southwest Water Plan, an ambitious dam, pump, and piping plan to feed additional water to Arizona and California. A key component was to build two dams within the Grand Canyon. Brower had heard reports of the dam project. He had hoped that this day would never arrive. At the top of his journal for that day, he wrote: "BAD DAY: Udall at crossroads; went South."[54]

Brower's convictions would clearly soon be put to the test, not just by his opponents but also by those he had considered his allies and friends.

9. Progress

I hope the American people remember well who the groups are who think so much of their present, and so little of everyone's future, that they fight and stall this bill—the dammers, the sawlog foresters, the graziers, the miners, and strangely the oil men, who above all should know the importance of keeping America full to beautiful places to drive or to drive near.

DAVID BROWER,
TESTIMONY IN U.S. HOUSE
OF REPRESENTATIVES,
SUBCOMMITTEE ON PUBLIC
LANDS, *NATIONAL WILDERNESS
PRESERVATION ACT*

By the early 1960s, David Brower and his allies believed that they were close to getting Congress to pass a wilderness law. The new Kennedy administration strongly supported the bill; events were forcing the Forest Service and the Park Service to end their opposition; the Senate leadership was backing the bill; and even the Outdoor Recreation Resource Review Commission was no longer an impediment.

There was one problem: the congressman from the Western Slope of Colorado, Wayne Aspinall.

Interior Secretary Stewart Udall would call Aspinall "a very crotchety, difficult person to work with."[1] Throughout the 1950s and 1960s, not only in the wilderness bill debate but also on a range of other issues, Brower and Aspinall would clash. After Aspinall lost the Dinosaur fight, he retaliated by refusing to release a bill making Dinosaur a national park. He was bitter about losing the Dinosaur dams and wary that the wilderness bill was another attempt to rob resources that he felt belonged to the West.[2] So for four years he blocked passage of the bill.

John Kennedy endorsed the legislation in 1960 while running for president; it was part of the conservation plank of the president's New Frontier agenda; and the Democrats controlled both houses of Congress. Agriculture Secretary Orville Freeman and Interior Secretary Udall were in charge of getting the bill passed, and Udall emphasized the need for the law. "Wilderness preservation is the first element in a sound national conservation policy," he said. Even after Kennedy's death in November 1963, conservation and wilderness had a strong supporter in Lady Bird Johnson.[3]

That level of support made it difficult for the Forest Service and Park Service to oppose the law. But Brower's increasing outspokenness and criticism of each agency did create some problems for a bill that was going to require compromises to get passed.

The Park Service's position began to change as it moved away from its construction agenda and more toward the acquisition of new parks, thus earning greater support from most environmental advocates. Unlike the 1950s, the 1960s would be a time of major expansion of the national park system. Mission 66 had set the foundation for some of this change by surveying potential recreational land that might be threatened by development. That survey led to the creation by 1962 of three national lakeshores and three national seashores, including Point Reyes, California (figure 9). Those parks contained more than 700,000 acres and 718 miles of shoreline.[4]

With Point Reyes so close to Berkeley, Brower worked hard for the new shoreline preserve. He knew it well, having taken his family to the peninsula many times. Much of Point Reyes was cut off from the mainland by a rift in the San Andreas Fault, which ran undersea beneath the narrow Tomales Bay. The beach topography varied from sandy beaches and coastal bluffs to a three-mile long sand spit where herds of elk fed on the high grass that waved in the wind that swept across open meadows and knolls. Despite the Kennedy administration's support, the campaign for the preserve was not easy; many of the landowners were opposed because property values were rising, and the patches of Douglas fir and pine had value on the timber market. Brower countered by commissioning a twenty-eight-minute film, *An Island in Time*; sending out a flurry of press releases

FIGURE 9 John F. Kennedy signing the Point Reyes National Seashore law
President John F. Kennedy signs the bill creating the Point Reyes National Seashore on September 13, 1962. Those looking on include Congressman Wayne Aspinall (*far left*), Interior Secretary Stewart Udall (*third from left*), and Brower (*far right*). (Courtesy of the Bancroft Library, University of California, Berkeley)

and brochures; and organizing a lobbying campaign in Washington, D.C.[5] It helped that the new president had a summer home on Cape Cod near the boundaries of one of the other protected seashores. Other park campaigns were less successful. Local opposition killed the Oregon Dunes and Oregon Cascades national park proposals. But more frequent success was obvious by 1972; by then the government had created eighty-seven parks and protected areas, including historical sites, which consisted of 3.7 million acres.[6]

Those efforts helped ease Conrad Wirth's opposition to the wilderness bill. Yet Wirth remained angry at Brower and other conservationists, who he felt were too extreme, especially after Brower's published critique of

Mission 66 came out in 1958. In his autobiography, Wirth denied that the Park Service opposed the wilderness bill, but he also admitted that park support was withheld until Howard Zahniser redrafted some of the language to Wirth's liking.[7]

The Forest Service reluctantly agreed to drop its opposition after most conservation organizations supported a separate bill sought by Director Richard McArdle. Brower was virtually the only environmentalist opposed to McArdle's legislation, and his opposition nearly ended his friendship and alliance with Zahniser.

The legislation in question, the Multiple Use–Sustained Yield Act, had first been introduced in 1953, and it lingered for years until the Forest Service rewrote it in 1958. It was designed to ratify the type of planning that McArdle had already instituted. By 1960, the only foe to it besides Brower was the timber industry, which feared that the bill would dilute logging's historic priority over other uses in the national forests.[8] Brower worried that this Forest Service legislation would kill the wilderness bill. On June 3, 1960, in a memo addressed to twenty-nine friends and colleagues, including Sierra Club leaders and representatives of other conservation groups, Brower spewed out his frustration with the Forest Service. He said efforts to work with the Forest Service and to protect wilderness were not working. The Forest Service wanted to cut trees; most of its revenue (95 percent) came from timber sales, and the multiple-use act was simply a guise that would allow logging everywhere, including the wilderness. Brower also believed that the bill locked the National Park Service out of acquiring forestland. The bill, he said, threatened the prospects for passing the wilderness bill and could spell further trouble for rounding out the wilderness system in parks and national forests.[9]

Although the memo was an excellent summary of Brower's frustration, it came too late to stop the bill. McArdle had sent one of his assistant chiefs to the Sierra Club board meeting on May 7, and there was a long and cordial discussion about both bills. Two motions were officially approved, one urging that the multiple-use legislation allow some single uses in the forests and the other calling on the bill sponsors to ease the transfer of Forest Service land to the national parks. What came out of the meeting unofficially was an understanding that the Sierra Club would not oppose the multiple-use bill. In return, McArdle would soon drop his opposition to the wilderness bill.[10]

Even more striking than Brower's refusal to make a deal with the Forest Service was that his position exposed a rift with Zahniser, who had supported the forestry bill. The same day Brower wrote the memo, June 3, he sent a separate note to Zahniser. After working with Zahniser for a decade, he said it was disappointing to see their routes diverging. Not only had Zahniser disagreed with Brower's opposition, but he had also criticized it privately and publicly in at least two publications. There had been past disagreements, both with Brower's backing of the North Cascades National Park and with the creation of the Scenic Resources Review. Zahniser had felt both efforts diluted the lobbying needed for the wilderness bill. He quickly responded to Brower's letter. They had drifted apart, he acknowledged. For whatever reason, the phone calls and the meetings in Washington had virtually stopped, and Zahniser did not understand why. He said that he was no longer sure what Brower was doing, so he had felt it best to go ahead on his own in the public statements about the multiple-use bill. Zahniser added that he was pleased that Brower had sent both letters to him; at least they were talking again, and he hoped it would continue. Brower, who in future years would break so many ties with so many once close friends, did not part here with Zahniser, but the relationship had changed.[11]

Following Kennedy's inauguration, the Senate convened, reorganized, and—in a key development for wilderness proponents—named Clinton Anderson of New Mexico the new chairman of the powerful Senate Committee on Interior and Insular Affairs. By September 1961, the Senate passed the wilderness bill overwhelmingly, seventy-eight to eight.[12]

Although support was strong for the bill throughout the nation, Aspinall knew that many in the West resented this intrusion into lands that they had always used. The House Interior Committee scheduled hearings in October and November in McCall, Idaho; Montrose, Colorado; and Sacramento, California. Turnouts were high; nearly five hundred people came to the Colorado hearing, forcing the committee to move into a larger movie theater. For every speaker who favored the legislation, more than two others opposed it. Brower spoke in Sacramento, joking that he had heard people ask what was going to end first, the hearings or the wilderness. The legislative process had already been exhausting, and by the time the bill was approved, it had gone through sixty-five versions and eighteen hearings.[13]

Brower was critical of those who had held up the bill for so long, declaring "it is high time for the public to blow the whistle on rapacity in its last vestige of the American wilderness." He continued: "The important thing today is to stop the compromise, the shilly-shally, and hesitation that selfish interests press upon us and get on with the job. Every week of delay brings another short-term profit to a few and forecloses forever on something irreplaceable in the national estate."[14]

In February 1962, the long awaited report was issued by the ORRRC, the outdoor commission that Brower had helped create. The ORRRC's work had taken three and a half years, cost more than $2.5 million, and ultimately was a disappointment to Brower. The ORRRC supported establishing a wilderness bill, agreed that demands on parks and recreation would increase, and called for the creation of what would become the U.S. Bureau of Outdoor Recreation, which would supply federal funds for new parks. The bureau would go on to pave the way for park planning and unleash significant money for recreation in local, state, and national parks and recreation areas. But it also endorsed the multiple-use concept in many areas. "More than recreation is at stake," Brower would complain later. "What is needed is a broad public understanding of the meaning of wilderness to civilization." A friend in the Seattle chapter of the Sierra Club, John Warth, was so disappointed that he suggested the best tactic might be to call "the report a sham and use that to arouse public opinion."[15]

However, Brower may have been too pessimistic because the report also strongly endorsed the need to establish wilderness areas in federal areas as well as the prohibition of mining, dams and water development, and other uses in these designated areas. To reinforce these endorsements, wilderness bill supporters took to the floor of Congress and read those portions of the report.[16]

Despite those pleas, the legislation remained locked in a House sub-committee, and by March 1962 many were very worried about its prospects of getting through the second chamber. "I have very grave doubts from the latest reports whether the wilderness bill will pass the House," Supreme Court justice William Douglas, now a Sierra Club board member, wrote to Brower. "Every special interest seems to be against it. These are indeed very discouraging days."[17]

The chief problem now was Aspinall. The Democratic congressman from Colorado had enormous power (figure 10). Udall would call Aspinall "autocratic," with the "good and bad traits of an industrious hedgehog." Aspinall, Udall said, was "one of the last in this century's chairmen to run his committee as though his vote was the only one that counted."

FIGURE 10 Representative Wayne Aspinall
Wayne Aspinall served as a Democratic congressman from the western portion of Colorado from 1949 to 1973, and during much of that time he chaired the key House Committee on Interior and Insular Affairs, which oversaw dams, national parks, forests, and wilderness. He was often an irascible foe to Brower. (Courtesy of the Penrose Special Collection, University of Denver)

Aspinall did not like the bill, anymore than he liked the aims of most con-servationists. If Brower and Wallace Stegner wanted to talk about spiritu-ality, Aspinall would talk religion. God made the land to be used, Aspinall declared in one newspaper interview. The environmentalists, Aspinall said, were seeking a version of an earthly heaven, "and they expect every-one else to conform to their views."[18]

The western slope in Colorado never generated the political power, the money, and the prestige of Denver and the other big cities on the opposite side of the range. Aspinall would carry that chip on his shoulder, a belief that he was the underdog. He needed somehow to overcome the odds, to show "them"—be they front-range politicians, California environmental-ists, or East Coast elitists. He grew up on a ten-acre peach orchard that was nourished by water irrigated from the Colorado. He understood the needs of the ranchers and farmers and how precious and necessary water was in the West. After law school, he ran for the state legislature and won in 1930, rose through the ranks, and wanted to run for governor. In 1948, he had to settle for the Fourth Congressional District. With the aid of Harry Truman's coattails, he made it. His district covered all of west-ern Colorado, a rugged, high-mountain land good for grazing and mining, with valuable lead, zinc, and silver deposits. He sought out the commit-tee assignments that would protect his district's water and mineral inter-ests. Then Dinosaur made him a player on the national scene, the go-to authority in the House when it came to dams and water in the West. As he became increasingly powerful, he also grew more conservative. In the late 1960s, Brower and his new employee, Jeff Ingram, encountered Aspinall in a Washington restaurant. After introductions, Aspinall exclaimed to Ingram, "You have more hair on your face than I have on the top of my head." Ingram knew it was not a compliment. He soon shaved the beard. In the 1960s, after race riots broke out in major cities, Aspinall declared it was time for the white majority to stand up for its rights. "Those who dissent, must dissent within well-defined limits," he said, "and must not be led by shysters."[19]

From 1961 to 1964, Aspinall controlled the fate of the wilderness bill and came under heavy attack from environmentalists and the press.[20]

In *Harper's*, Paul Brooks accused Aspinall of singlehandedly thwarting the overwhelming popularity of protecting the wilderness in Congress and the nation. "Aspinall's chief ally is not so much public apathy as it is the apparent complexity of the problem," wrote Brooks. "His master plan for 1963 is beautifully designed to confuse the issue to the point where clear judgment becomes impossible."[21]

The lead role in lobbying for the bill remained with the gentlemanly but persistent Zahniser. Brower was just too caustic at times to work the deals that needed to happen to get the bill through Congress. Yet Brower could be surprisingly cooperative at times. In a House Subcommittee on Public Lands hearing in May 1962, Brower toned down his criticism of his opponents while urging passage. The wilderness, he warned, was in danger. "It is not reassuring enough to know that excessive damage has not yet been done," he said. "You can't claim a dropping egg is safe merely because it hasn't broken yet. We can measure the need for the wilderness bill by the very intensity of the effort to defeat it." He was even more conciliatory under questioning from Congresswoman Grace Pfost from Idaho. Roads into wilderness areas should be closed, and new transmission lines rerouted, he said. But livestock ranchers could still graze in the designated wilderness areas, and even dams and reservoirs might be permitted under some circumstances. "We might wish to argue a bit about that," he said, "but we want to be reasonable."[22]

In the spring of 1963, the Senate passed yet another version of the bill and sent it to the House. Aspinall remained maddeningly difficult to pin down on his intentions. One day he was conciliatory, the next he was defiant. A writer for the *Washington Post*, Julius Duscha, met Aspinall off the House floor in April 1963 and asked when hearings would occur on the bill. "I don't plan to hold a thing at the present time," Aspinall replied. "If by calling names and putting pressures, they think they can move me any faster than I intend to move, they're mistaken." The interview ended, and Duscha noticed that Aspinall was hurrying back to the floor. A bill affecting silver mines in Aspinall's district was under consideration.[23]

Yet Brower remained committed to the battle. In a letter to an old friend, Ike Livermore, about the need for the bill, he recounted a story. Two weeks earlier, he had been coming back from a ski trip at Squaw Valley with his sons Ken and Bob. They had to slow to pass some road construction. The highway was being widened to four lanes. A moment later, Bob told his father how much better off the older generation had been.

Brower asked his son what he meant.

Bob said his younger generation faced the potential for nuclear war, and now the mountains were going, going, going.[24]

In the fall of 1963, Brower sent staffer Mike McCloskey to fight Aspinall within his own district. When McCloskey arrived in Grand Junction, Colorado, he literally had no plan and no allies to help him. He walked into a sporting goods store and asked a clerk who in the area might care about wilderness. He was told packers and guides cared, and there was a list, produced by the state Game Department, with names, addresses, and phone numbers. For several weeks, McCloskey tracked down each guide, discussed the need for the wilderness bill, and in most cases got the guide to agree to send a telegram to Aspinall. McCloskey would write down what they wanted to tell Aspinall and then find a telegraph office and wire the message. Aspinall could literally track McCloskey's work as the telegrams came in, and he complained about how Brower had sent outside agitators into his district. But McCloskey always felt this telegram campaign was a very effective tool in nudging Aspinall toward passing the bill.[25]

Finally, House hearings were set for December 1963 at the Capitol and in Denver. Brower testified in Washington; he sent McCloskey to Colorado. In the most recent hearings, the number of speakers on each side had come close to splitting even, with opinion clearly tilting toward the wilderness side.[26] It was not yet well known, but Aspinall had had two key meetings and was prepared to stop blocking the bill. One was with Senator Clinton Anderson, the second was with Kennedy, just two days before the president went to Dallas, Texas, and was assassinated. A series of deals were cut; the most important was that Aspinall was able to add a provision allowing mining to continue in the designated wilderness areas for another twenty years.[27]

A final vote was scheduled for the summer of 1964 when Brower received a telegram the morning of May 5. It said that after a normal day at the office, Zahnie had died in his sleep early the following morning.[28] He had a heart condition, it was well known, and yet his death from a massive heart attack at the age of fifty-eight was still shocking. Brower remembered Zahniser and his family meeting up with the Brower family above Stehekin the summer he did the film on the North Cascades. The film shows Zahnie, who by then knew he had a weak heart, plodding at a slow but measured pace up the trail from White Pass toward Glacier Peak. He may have tired on such trails, but it was where he wanted to be, in the wilderness. In some respects, Brower said in his eulogy of Zahniser, it had been political madness to take on so many opponents in the fight for the wilderness bill, but Zahniser had the political fortitude to preserve. "The hardest times were those when good friends tired because the battle was so long," said Brower. "Urging those friends back into action was the most anxious part of Howard Zahniser's work. It succeeded, but it took his last energy."[29]

The bill passed in August, and President Lyndon Johnson signed it into law as the Wilderness Act in September in a ceremony attended by Howard Zahniser's widow, Alice. Days before the bill's passage, Wirth asked Aspinall if there were any way it could be named after Zahniser. Aspinall said he only knew of bills that could carry the names of legis-lators, and sometimes they should not have gotten credit. Zahniser was deserving, but nothing could be done.[30]

Brower and other conservationist were upset with two major amend-ments to the new law. One, noted earlier, gave miners another twenty years to operate in the designated wilderness areas. A second amend-ment removed the administration's authority to create new wilderness areas, leaving it to Congress. "The result was, hardly a compromise—it was, rather, an acceptance of Aspinall's terms," wrote James L. Sundquist, a scholar at the Brookings Institution. Despite the flaws and changes, the structure of the bill still incorporated conservationists' early aim— to protect lands that had been untrammeled. Brower declared that the bill's "passage was the most significant conservation development in this decade and perhaps the most significant since the National Park Act of 1916."[31]

The law designated fifty-four areas in thirteen states as wilderness. They totaled only slightly more than 9 million acres, but the law allowed for other lands to be considered, a provision that resulted in lengthy hearings that went on for years. Ultimately, the provision was a blessing. In the fifty years after the act was adopted, the nation set aside 757 areas in forty-four states, comprising 109,511,966 acres, about 5 percent of the entire United States or an area slightly larger than California. In some respects, the growth in the number of areas set aside reaffirmed the popular support for wilderness preservation in the United States. Fifty years after its passage, the law remains a landmark in conservation and environmental history, establishing statutory protection to wilderness for the first time. It inspired the existing generation of conservationists and launched a new breed who would continue to work for more protection. However, environmentalists and scholars have also criticized it, contending that land undeveloped by roads or other intrusions was far less important than poverty, pollution, and other problems in both rural and urban communities. Others criticized early efforts to remove native peoples from some national parks and wilderness areas and suggested the concept of wilderness preservation was ethnocentric, if not racist, because it benefitted privileged whites primarily. Finally, some ecologists have remarked that land, even when left alone by humans, still undergoes changes over time, which raises the question of what is truly being preserved.[32]

The bill was designed to protect wilderness in the United States permanently. But one of Aspinall's amendments that did not survive when the bill was passed demonstrated how protection is never entirely guaranteed. The amendment called for a ski resort to be built in the middle of what would be the San Gorgonio Wilderness in southern California. Developers had not given up. In 1961, a proposal surfaced for a resort with parking for five thousand cars and fifteen ski lifts. The Forest Service rejected the proposal, relying on the decision it made in 1947 when it listened to Brower and the Sierra Club in rejecting a similar plan.[33] The amendment popped up again in a House version of the wilderness bill in 1962. After the Sierra Club and others objected, the amendment was dropped. New special legislation on San Gorgonio was introduced once more in 1965. Congress held new hearings, opposition remained high, and the ski resort was rejected yet again.

Today, many U.S. wilderness areas have geographic names, such as San Gorgonio, which draws a heavy stream of backpackers and day hikers on foot.[34] Other areas have names that are either more or less descriptive, such as the Three Sisters Wilderness in Oregon, the Glacier Peak Wilderness in Washington, and the Inyo Mountains Wilderness in California. A few are named after individuals, and, as Aspinall told Brower, some individuals are less deserving than others in having a landmark named after them. Areas named after dedicated conservationists include the Ansel Adams, Bob Marshall, John Muir, Theodore Roosevelt, and William O. Douglas Wildernesses.

There is no David Brower Wilderness or Howard Zahniser Wilderness, and certainly none was named in honor of Wayne Aspinall.[35]

10. Books

The importance of America's wilderness can be stressed and protected in various ways, and in the late 1950s David Brower and Ansel Adams stumbled onto a tool that significantly defined Brower's career. It began when Yosemite Park officials wanted to make the LeConte Memorial Lodge into a geology museum, and Adams had an alternative proposal. Completed in 1904 by the Sierra Club in the heart of Yosemite Valley, the small, Tudor-style cottage now under the management of the Park Service had become a dry, dusty, dull library.[1] Adams suggested using LeConte as a forum that would feature a photography exhibit promoting the Sierra Club's point of view about the importance of conservation.

That was the birth in 1955 of *This Is the American Earth*, an exhibit produced by Adams and Nancy Newhall, a museum designer. Adams shot about half of the black-and-white prints, including his great portrait of Mount McKinley at sunrise and a winter storm over Yosemite. Other renowned photographers were also represented, including Margaret Bourke-White, Jacob Riis, and Edward Weston. Adams made all the prints to ensure uniformity in presentation.[2] Newhall supplied the text. One reviewer said the message was "terrifying and beautiful." It was frightening because much of it deplored the human tendency to exploit or conquer nature and the results of that propensity. Yet the exhibit attracted attention because the photographs and the scenes that they depicted were so striking. In 1956, David and Anne went to the exhibit at the LeConte. Anne told David that she thought Adams was the greatest of all photographers. David told Adams that Anne virtually never offered such praise.[3] The exhibit eventually was shown in museums and universities around the country.

By 1957, Brower wanted to convert the exhibit into a book. He also believed that the quality of the book should match the exhibit, which would be costly. Adams had connections with the McGraw Foundation, which agreed to supply a $15,000 grant for the book. Brower and the Sierra Club board still hesitated. Although the club had been in the book-publishing business for years, it had never attempted anything of this scale. In November 1958, Brower tried to persuade Paul Brooks, an editor at Houghton Mifflin, to publish the book. Brooks wanted full control and was willing to sacrifice the printing quality to keep the price down and sales high. Brooks also told Brower if he insisted on using the best reproduction techniques, the book would be too expensive. It was an art book, and art books were not sold in bookstores, Brooks said. Brower, paraphrasing Brooks, reported, "Your art book would only convert the converted. We would much rather see 50,000 or 100,000 copies sold to spread the conservation message more broadly, where it's needed." Customers also would be unable to differentiate between Brower's "lavish" format and Brooks's more modest result, anyway.[4]

The Sierra Club had been publishing books for decades. Most were modest guide books or mountain adventure books, and none was of the scale that Brower now wanted. Nevertheless, after a debate the Sierra Club board approved publishing the book. Brower, Adams, and Newhall toiled in Adams's San Francisco home, in Newhall's residence in Rochester, New York, and in New York City, where the book was published in 1960. The 112 pages of *This Is the American Earth* were an oversize 10.5 inches by 13.5 inches. Because of an error in preparing the printing plates, the photos came out so sharp that they had a three-dimensional quality. An additional $10,000 had to be borrowed to meet expenses; ten thousand copies were ordered and would be sold at $15 each. Sierra Club members got a discount, but at the time there were only fifteen thousand members.[5]

The response went far beyond expectations. William O. Douglas called *This Is the American Earth* "one of the greatest statements in the history of conservation." The *New York Times* published two reviews and one story, exclaiming that the book featured "magnificent photographs and eloquent words." NBC's *Today Show* aired the photos while host Dave Garroway read the text. Newhall won a prestigious publishing award, and Wallace Stegner and Stewart Udall considered converting the book into a film. The first press run of ten thousand sold out, so a second was launched.[6]

Thus was born the Exhibit Format book series. The Exhibit Format books were artistically beautiful while also promoting the Sierra Club's environmental agenda. Membership and revenues for the club increased dramatically, and much of the increase was attributed to these books. The books were one of Brower's great achievements, demonstrating how media and art can not only inspire an appreciation of nature but also be used as an effective tool in garnering political support for environmental and conservation causes. Yet for all the brilliance of the book series, Brower became so caught up in their production that it would lead to a fissure between him and the club leadership. He was a poor financial manager, and the costs of producing such high-quality books eventually threatened the club's finances.

Following the success of *This Is the American Earth*, Brower wrote to the board in September 1960 suggesting a significant increase in the club's publishing. He pointed out that the club was already printing a monthly newsletter, publishing books, and producing films on conservation messages. He acknowledged that there were risks, but there were also opportunities.[7] In response, the board created a five-member Publications Committee and asked August Fruge of the University of California Press to be its chairman.

Fruge was an interesting choice. He may have been Brower's former boss, but in this relationship they had often clashed, not least when Fruge was pressured to fire Brower from the press after the incident in which a student claimed Brower had made sexual advances to him at the University of California in 1952. As Brower saw their relationship, Fruge became a jealous competitor who worried that the Sierra Club publishing program would be more successful than the University of California Press.[8] Fruge maintained that his academic press and its objectives were far different than those for the Sierra Club books and that he was always fair in his role on the Publications Committee. "I don't remember anybody having any problem with it," he said. "I never found any competition."[9]

Fruge was a librarian and editor who never avoided a fight. Over the thirty-five years that he would lead the University of California Press beginning in 1944, he transformed it from a small publishing house where local professors published their monographs to a professional operation producing nearly two hundred titles a year. To make changes, Fruge often had to fight, and he never backed down, either with the printers who resisted changing technology or the writers who quelled at having their work edited. His opinion of Brower the editor evolved from respect to mixtures of deference, skepticism, and hostility. Although Fruge acknowledged Brower's commitment to quality in publishing, he questioned Brower's financial and management acumen.[10] Brower took risks that Fruge did not always admire, yet Brower credited Fruge with encouraging him to take chances. He once recalled an exchange that he had with Fruge at a bar. They were discussing human affairs. Fruge "ran his finger along the edge of the table and said, 'You've got to be out here at the edge. You may fall off, but if you're back towards the middle, it's too safe. Nothing happens.'"[11]

Adams was Fruge's major counterpoint on the Publications Committee and Brower's biggest booster. Adams's name was attached to three of the early Exhibit Format books, including his biography written by Newhall, *Ansel Adams*, volume 1, *The Eloquent Light*. Brower's name is sometimes difficult to find in these books, but Brower was the creative force behind all of them. He chose the author and photographer (unless someone approached him first), and he was responsible for the typeface, the photos, the layout, and the final production and printing. "He liked layout," recalled Mike McCloskey, "he loved getting the aesthetic balance the way he liked it."[12]

The photographer who would raise the stakes for Brower and the Sierra Club, however, was not Adams but Eliot Porter. Adams was a master of black and white; Porter made color photographs. A high-quality book in full color was significantly more complicated and expensive to publish at a time when color reproduction was still developing. Brower took risks that commercial publishers refused to take, and the combination of Porter's photographic genius and mastery of color literally made the books irresistible to buyers despite their hefty price tag.

The pivotal book for Brower and the Sierra Club had begun many years earlier, in 1950, when Porter's wife suggested that Porter do a book that combined text from Henry David Thoreau and photographs from Porter. "Your pictures remind me so much of him," she told Porter. "They show his Walden as it is."[13] No one was more qualified than Porter, who ten years earlier had given up a career in medicine and research at Harvard to take photographs of nature. As a child growing up in Illinois shortly after the dawn of the twentieth century, Porter would spend hours in a bird blind waiting for just that moment to snap his camera shutter. He went on to get a degree in chemical engineering and a medical degree and finally taught at Harvard. This career was not fulfilling. Yearning for more, he picked up a camera again. The great photographer and art promoter Alfred Stieglitz encouraged him. He toiled in the field and the darkroom. Finally, in 1939, Stieglitz was offering more praise than criticism. Porter quit the cloistered halls of academia for the uncertainty of freelance nature photography. He moved his family to New Mexico and began to gain recognition.

What catapulted Porter's reputation, bringing Brower and the Sierra Club books along for the ride, was the color that Porter could achieve. His chemistry background enabled him in his own darkroom to experiment with Eastman Kodak Company's new Kodachrome film at a time when other photographers shunned it. Porter spent years experimenting, but the result was clear, crisp color transparencies that dazzled. They pushed Porter to the forefront of photography as the popularity of color surged and that of black and white waned. It was only years later, when technology and other photographers caught up with Porter, that some reviewers began to reassess Porter and to suggest that his work was only pedestrian.[14]

It took Porter ten years to finish the Thoreau book, *In Wildness Is the Preservation of the World*, and the expense of printing it scared publishers until it got to Brower. The beauty of the proposed book overwhelmed Brower, and he told Porter in a letter in February 1961 that he would be willing to begin a life of crime to pay for its publication.[15] Although he need not rob a bank, he knew he would need lots of cash to undertake the book. He assiduously courted Kenneth Bechtel, whose family-owned engineering and construction company based in San Francisco was best known for building oil refineries, power plants, and facilities that in later years Brower would rail against. Bechtel, through the Belvedere Scientific Fund, gave the Sierra Club a $20,000 grant and a $30,000 loan to subsidize the production of the book.

A second problem was finding a printer capable of reproducing Porter's superb color photographs. It took months before Barnes Press of New York passed muster on the samples it showed to Brower and Porter. Barnes needed to produce ten thousand copies of Porter's seventy-two color prints and to get the four colors to balance and register on the presses. The firm used a sixteen-plate form, with four rows of four, each of a different photographic image. The yellows, reds, blues, and blacks had to be matched perfectly in trial runs, with paper spewing off the presses. These experimental runs took an inordinate amount of time and often forced another trial. Brower recalled one evening when he stayed to supervise, while Porter and the owner of the press, Hugh Barnes, went to dinner around 7:00 P.M. Barnes returned at 9:00. Brower stayed until 11:00 P.M. and then returned to his hotel. Porter, who had slept after dinner, returned at 1:00 A.M. and stayed until dawn.[16] This kind of pattern was not unusual at Barnes, and Brower's journal for years in the 1960s is filled with entries of his returning to the printing company at odd hours of the day or night.[17]

Finally, on a day in August 1962, Brower, Porter, Barnes, and others gathered around press number 3 and watched the first 2,500 sheets roar off the presses. The men examined them at a table, using lenses carefully.

They were excellent, recalled Brower, but they were not perfect.

Hugh Barnes agreed. "How about it, Dave, shall we throw out the first 2,500 sheets, and will you go fifty-fifty with me on the cost of the paper?"

How much would that cost? Barnes said $200 for each of them. For Brower, that was equal to the amount of dues the club got in a year from twenty-five members, but he agreed.

Barnes returned a few minutes later. "You did the right thing," he said. "Now they [the Barnes workers] really know that this is a fussy job."[18]

Even though *In Wildness Is the Preservation of the World* would be sold at an incredibly expensive $25 (the equivalent of nearly $200 fifty years later), the first five thousand copies quickly sold out, as did the next nine reprints. Critics praised the book. "Only a bold photographer could try to capture Thoreau's vision again and again. But Mr. Porter succeeds triumphantly," declared the *Christian Science Monitor*. Added *Natural History*, "If there was ever any doubt that Eliot Porter ranks among the world's great artists with a camera, it is dispelled by this volume."[19]

Brower was now the darling of publishing, the upstart who had proved Brooks and other commercial publishers were wrong. He had created a new genre, an expensive, sprawling book that openly touted an environmental message. As *Newsweek* was to comment, "Each book is a graphic reminder to Americans that their land is slowly being turned into shopping centers, superhighways and crowded recreation areas." [20] The glossy books with their bold, beautiful photographs could have a hypnotic effect. John Mitchell, who would one day become editor of Sierra Club Books, recalled a friend who came to his office and waited while Mitchell finished an annoyingly long telephone call. Bored, the friend began to thumb through Porter's *Wildness* book. Mitchell's call finally ended. He remembered slamming down the phone to get his friend's attention. No response. "It seemed nothing less than a cannon shot could have shaken his concentration," said Mitchell.[21] This was not an unusual response. Time after time, people described how they joined the environmental movement and the Sierra Club by first browsing through an Exhibit Format book in a bookstore or at a friend's house.

The Sierra Club was becoming the most recognized environmental and conservation organization in the nation. Membership doubled from fifteen thousand to thirty thousand between 1960, when the first Exhibit Format book was published, and 1965 and then doubled again in less than three years.[22] But there was no way to prove that those increases were tied to the books, even though Brower tried.

The greatest skepticism regarding this connection between the books and the increased membership came from the Publications Committee, whose relationship with Brower only continued to fester. Brower believed that the Publications Committee had been created not to help advise him but to control him.[23] Perhaps that is why from the committee's inception its members complained with some justification that they were receiving scant information about finances, inventory, and other details relating to the book publications. Brower had friends on the committee, but he also had enemies besides Fruge, old-timers who were not interested in seeing the transformations Brower was making within the club and the conservation movement. Even winning approval for *The Place No One Knew: Glen Canyon on the Colorado* did not come easily. In 1962, Brower still hoped he could stop the dam, and he wanted to get the book into the hands of Congress and the Kennedy administration as quickly as possible. Some on the board believed that Glen Canyon was "a dead issue" and that the proposed book was too expensive. Brower had to work to win committee authorization.[24]

If ever there was a time when Brower's book program nearly died, it was the summer of 1963, when Fruge wrote a memo that firmly recorded his reservations. The Sierra Club had to decide, said Fruge, whether publishing was a tool for conservation or the club would become a large general publisher of conservation books. Fruge said he feared that the club was headed in the latter direction, which he opposed. He conceded that the "nature propaganda" that Porter was producing educated the public and potentially could produce additional revenue for the club. But it could also drain off resources that were needed for conservation battles, so he questioned whether the books were the club's best tool and commented that the club already worried too much about sales figures and book issues. Too many books were being considered solely on the basis that they could produce big sales during the Christmas season. If that trend continued and making money to sustain publishing became paramount, then Fruge worried that the Sierra Club would lose its purpose and character. The organization that he had joined would be gone.[25]

Brower was shocked by Fruge's memo, and his defense was basic: publishing was the Sierra Club's most effective conservation tool. There was no need to choose between conservation and publishing. He argued that one promoted conservation through words, voice, and picture and that books, pamphlets, journals, and films were integral media to transmit those messages. The books had made the Sierra Club a national leader, and funds had increased tremendously to hire more staff and spread the word. Attacking the book program would not only harm the club but also deprive it of money and recognition. Critics, he said, had the power not only to destroy the book program but also to harm the club's conservation program.[26]

The debate raised other concerns. One was how Brower, the club's executive director, could run what was becoming a bigger organization, with more employees, members, and responsibilities, while also devoting so much time to editing books. Brower responded that he was spending about 50 percent of his time with the books, which he thought was about right. Brower's personal journals, however, show that by 1963 he had begun spending significant amounts of time in New York. For instance, in the twelve months leading up to Fruge's August 1963 memo, Brower spent at least two months off and on in New York. There are numerous notations for the time he spent working at Barnes Press or conferring with Porter or others about a book issue. Even when Brower was back in San Francisco or elsewhere, it appears that he was devoting more than 50 percent of his time to books.[27]

It was also clear, though, that the books were changing the club. In 1959, the year before the first Exhibit Format book was published, the club's budget was $120,000. In 1962, after the first black-and-white books were published but before Porter's color book on Thoreau was launched, the budget had doubled. By 1964, the annual budget had ballooned to more than $1 million, and the publication budget alone was half of that amount. One other figure in the ledger demonstrated the risks being taken in the book program: that year, the Sierra Club spent $17,000 more in the program than the program brought in. Those deficits on the publications side of the budget would only increase over time. A side issue was that it took time to sell books. In June 1963, the club had $193,000 invested in inventory that was still sitting in bookstores and warehouses.[28]

For Brower, that inventory merely meant that the club's endowment was being converted from stocks to books. He did not have a problem with that, but some board members did. Richard Leonard commented later, "Dave told the board that the money had always come in and always would and that he didn't see what we were worried about." Edgar Wayburn recalled that an analysis completed in 1965 showed that on average each book publication was losing $50,000. Brower strongly disagreed with those numbers, in part because the club assessed an overhead charge for its administrative expenses and he thought that amount was too high. "The specter in my mind," said Wayburn, "and I think in the minds of other members of the Publications Committee, and reflected to the board, was that if ever Dave published a turkey, we would indeed be in trouble."[29]

Brower could be creative when it came to financing a book. He found the money for *The Place No One Knew*—which was going to be even more expensive than *In Wildness Is the Preservation of the World*—by convincing Barnes Press to delay payment, having Porter agree to limit his royalty payments, and tapping into some leftover funds from the Belvedere grant (figure 11). These maneuvers helped, but he needed yet more money. He had received loans and grants for earlier books, so he decided to put out a wider solicitation to Sierra Club members. He received more than forty responses. Some objected not only to the solicitation but to the way their dues were being spent on such appeals. But the money did come in through advance purchases of the book, contributions, and interest-free loans. In July 1963, Brower reported that the combination of these fund-raising activities had produced $60,000. The downside was that promotion and production costs were now up to $80,000.[30]

Both the Publications Committee and the Executive Committee for the board seriously considered the publication and financing issues. After several months, they told Brower that he could continue publishing, but with some limitations. To control the club's financial exposure, limits were placed on how big the book inventory could be and the level of borrowing.[31]

FIGURE 11 Brower and Eliot Porter studying photos

Eliot Porter and Brower choosing what photos should go into Porter's next book from the many laid out on the floor. Porter first approached Brower in early 1961 about publishing a book based on his photos and the writings of Henry David Thoreau. They forged a partnership that created beautiful books and greater exposure for the Sierra Club, but they also increased the organization's financial risk. (Courtesy of the Bancroft Library, University of California, Berkeley)

No Exhibit Format book was more controversial than *The Last Redwoods* by Philip Hyde and François Leydet.[32] Although Brower's name is almost nonexistent in the book, he oversaw and managed its production. Forestry and lumber representatives resented the title because they felt it suggested that the California coastal redwoods were disappearing. At the time, more than 60,000 acres had been preserved. A review in the *Journal of Forestry* complained that the book "resorts to half-truths, distortions, exaggeration and quoting out of context." But the book drew wide praise elsewhere, ranging from the *Salt Lake Tribune*, which called it a "masterpiece of photographs and poetry," to the *Atlantic Monthly*, which said it was "a poignant, infuriating record of American impatience" in destroying priceless natural resources.[33] The book's first chapters were typical of an Exhibit Format book, depicting soaring redwoods, breakers crashing on the coast, clear blue rivers, and verdant glens of moist moss and fern. But by chapter 5 the book's tone changed as it showed clear-cut fields, polluted streams, and wide freeways that slashed through the forest.

When books went well and Brower was praised, he glowed. When he was blamed for the problems, he glowered. He had never taken criticism well, and criticism was an integral part of the book business. He did not like it from the Publications Committee, the Sierra Club board, the authors and photographers, or even the occasional book critic who published a negative review. He told George Dusheck at the *San Francisco News-Call Bulletin* that his negative review about Porter's Thoreau book "financially impaired and caused anguish" to everyone connected with the book, from Porter to individual book sellers.[34] After an unusually emotional argument with Adams, he apologized and said Anne had told him that lately he had been "pitiful, petulant and pompous."[35]

He did not react well even when he received what were meant to be constructive comments from such friends as Adams and Porter. The degree of his reaction was telling. The morning after a Publications Committee meeting in February 1964, Adams wrote a detailed, single-spaced, four-page letter addressing concerns about the book program. Although he closed with the comment that the books had been "an astonishing event in the publishing world" and "the most effective aspect" of the club's efforts, Brower responded with a bitter defense, concluding that Adams's letter was more discouraging than any previous comment about the Sierra Club publishing program. Adams was astonished by this response; he had only been suggesting ways to improve an already great book program. But he should not have been surprised. A year earlier, he had scolded Brower after another outburst—Brower's paranoia that outsiders were intent on getting him fired seemed to be on the rise at the time—and assured him that no one was out to get him. Everyone truly did appreciate him.[36]

In September 1964, in response to complaints from Brower, Porter was just as blunt. If the situation were truly as serious as Brower made it out to be—Brower had been complaining that his life was so hectic and disorganized that he could not get anything done—there would be nothing left to do but to offer Brower a loaded gun.[37]

By 1967, Brower had produced seventeen Exhibit Format books, and he was continuing to work on several others that the Publications Committee was not as willing to authorize. His staff now included a publications department that took care of the monthly *Sierra Club Bulletin*, Sierra Club films, and various non–Exhibit Format books that the club also continued to produce. The number of books that were published is a testament to Brower's talent, energy, and ability to persuade others, including the members of an often divided Publications Committee, to do as he wished. He did not always win, however.

Brower believed in working on projects despite widespread opposition. He was used to finding ways to overcome such resistance. In an especially revealing comment in his oral history, recorded from 1974 to 1978, Brower talked about the issues he had with the Publications Committee and exclaimed, "I led them on a merry chase. I felt that I had to, that there were a lot of things that needed to be done, and if I took the risk, I would rather do it than not."[38]

As time went on, there was no secret about what Brower was doing. "Dave bragged to the board of directors that the books would never have been published if he had not gone ahead without authorization," said Leonard, who increasingly worried about Brower's independent streak.[39] Such independence worked as long as the successes continued.

From 1964 to 1967, they did. There were books on the Grand Canyon, the Sierra Nevada, California's Big Sur, an American climbing expedition to Mount Everest, Summer Island on the Maine coast, the Southwest's Navajo country, Hawaii, Alaska, and Baja California. The best of them may have been the Grand Canyon book, *Time and the River Flowing*, which drew strong support even from the Publications Committee.[40] It was a campaign book, a direct response to Udall's call for two dams within the Grand Canyon. The book was controversial, depending on how one defined the Grand Canyon. The national park boundaries at the time included only a portion of what geologically constituted the Grand Canyon. Critics were unhappy that the photos in the book displayed areas of the canyon that would be affected by the dams but were not within the park.

Controversy sells books, and the Grand Canyon book was picked up by the Book-of-the-Month Club and was selling well. In April 1965, Brower learned that both *The Last Redwoods* and *Time and the River Flowing* had won a prestigious book award. The Carey–Thomas Award honored what it called a "distinguished project of creating publishing." A surprised Brower said in response to the news, "This is good for us, but more important it is good for the Grand Canyon." Book publishing was helping the conservation movement, "and this award is the greatest boost of all."[41]

Yet, to paraphrase an old maxim, one is only as good as one's next book. The Everest book threatened to become the financial failure that Wayburn and others most feared. The old mountain climber Brower really wanted this book, and he sought out the leaders of an expedition in 1963 that challenged the world's highest peak. The author, Thomas Hornbein, conceded in the preface that he had doubts about his writing ability, and some critics said he should have heeded those fears. As had happened with other Sierra Club books, the photographs of the expedition helped save this book.[42]

More controversial within the Sierra Club was Brower's proposed book featuring the California coast of Big Sur and the poetry of Robinson Jeffers. Brower and Adams, who had by now moved to Carmel on the coast, had been discussing how to protect the Big Sur shoreline of California. The concept of recognizing an area that included developments would later become part of the national park system, but it was premature in the 1960s, and so the book's conservation message was muted. Jeffers himself was also controversial. He was not a conservationist, said Fruge, but more of a misanthrope. Adams thought that his poetry was second rate. Wallace Stegner believed the solution was to not emphasize Jeffers poetry because it would overwhelm the book. Instead of a Sierra Club volume, it would be a Nietzsche–Jeffers book with Sierra Clubbish photography.[43]

By a narrow vote, the board grudgingly allowed Brower to continue with the Jeffers–Big Sur project, which would eventually be titled *Not Man Apart*, but Brower had trouble finding time for it. He eventually turned it over to his twenty-one-year-old son Kenneth. Like his father, Ken had struggled at college, which made him available. Although Brower was listed as editor, he noted in the preface that in reality Ken should have taken that title. Ken's inclusion in the project only complicated things, however, because several members of the Publications Committee were unhappy with the younger Brower's selections. Too many of them, complained Stegner, displayed his immaturity.[44]

Adams had doubts until he finally received his published copy. After spending several hours examining every element of the book, he wrote that it was the most moving, unified, logical, and beautiful book that the Sierra Club had done to this point. He was especially pleased with the editing of Jeffers's prose, a task that must have been truly challenging.[45]

Stegner said Brower seemed to be more intent in publishing *Not Man Apart* to get more cash flowing into the club and less worried about the book's quality. The financial picture was getting more complicated and at times dire. In 1966, the publications program was $119,000 in the red, but the club's overall budget of slightly more than $1.5 million ended with a surplus.[46]

If ever there was a time not to expand the book program, this would seem to be it. So of course Brower expanded it. Ian Ballantine, a paperback publisher, was intrigued by Exhibit Format books and wanted to publish them as small, high-quality paperbacks. At the time, paperbacks were rarely priced at more than 95 cents, but Ballantine agreed to pay more to attain the high quality of the hard covers and charge $3.95. That move was risky, but so was the idea of undercutting the more expensive hard covers with cheap paperbacks.[47] Ballantine paid Brower $22,000 as the first advance for the paperbacks, a deal Brower made and then presented to the Publications Committee. The committee's members and even Brower worried how the paperbacks would affect the hard covers. They worried needlessly. The first paperback was Porter's *In Wildness Is the Preservation of the World* in an abridged version with slightly fewer color plates. Its sales would eventually top 1 million. Brower decided that the paperback not only did not cut into the sales of the hard covers but also created a market for the bigger books.[48]

Ballantine also rescued Brower from the Publications Committee when it balked at publishing a highly unorthodox work, *On the Loose*, by Terry and Renny Russell. It was not an Exhibit Format book, but it would become important to the Sierra Club. The two young men, inspired by the Exhibit Format books, found snapshots of trips they had taken through the American West. The photos were taken by low-end cameras and were processed at drug stores, so they had low fidelity. They were matched to Terry Russell's paeans to the immaturity of youth and quotations from the Bible, actor Steve McQueen, H. G. Wells, and others. It took the Russells about three months to produce a draft of the book, which was bound in green Moroccan leather with gold lettering. They were living in the Bay Area when they knocked on Brower's door one evening, and he let them in. The family's pet monkey snapped at them, and Brower poured a drink and began reviewing their manuscript. Renny Russell had misgivings as he studied Brower's face, which remained impassive. "Then, without finishing, he flashed his sky-blue eyes at us and simply said, 'We have to publish them,'" recalled Russell. "I melted."[49]

The Publications Committee was adamant that this book, which was far more about youth than about conservation, was not for the Sierra Club. Brower took the book to Ballantine, who immediately recognized it as a potential best seller. Ballantine agreed to pay $20,000 for the paperback rights to the book. Faced with the potential arrival of a large cash flow, the Publications Committee relented. The plates in the book would incorporate the Russells' photos as well as Terry Russell's handwritten text. Brower ordered fifty thousand copies, which reduced the cost per unit, but he had only fifteen thousand bound. The first copies sold out rapidly, and Brower was able to get the remaining copies bound quickly to keep up with demand. Over the years, the book would remain in print and sell more than 1 million copies.[50]

Brower had done the seemingly impossible. His books were now on best-seller lists, included in the Book-of-the-Month Club, winning awards, and selling in the millions. And yet Fruge's warning in 1963 now seemed more germane than ever. Was publishing still the servant of conservation, or had Brower turned the Sierra Club into a publishing house? And was such a transformation good or bad?

11. Escalating the Risks

In 1959, the Sierra Club board ordered David Brower to be polite to public officials. Richard Leonard and Ansel Adams had consistently defended Brower's increasingly sharp language, but others on the board were uncomfortable with his tactics. They worried that he was causing more harm than good. The board adopted a resolution that said that employees needed to be cordial in dealing with or discussing public officials and issues.[1]

Brower called the resolution a gag rule. To him, he was being told that he could no longer criticize any government official or agency. His supporters questioned how an environmental activist could be tactful in defending the environment. One could not be both polite and aggressive in making a point, said Hugh Nash, a Brower staffer and editor of the *Sierra Club Bulletin*. There was no way anyone should delude themselves in thinking Brower could do both, he said. If an individual rarely offended anyone, Nash felt that individual should be fired.[2]

For those with a cause, this debate is endless—violence versus pacifism, ultimatums or compromise, hostility as opposed to conciliation. For years, the Sierra Club debated what tactic is best. At one board meeting, Brower argued that open hostility with government officials worked, and being polite did not. Some of this debate was with the old guard, the conservatives of the past, but some was also with the new generation that took over in the 1950s, some of whom did not want to go as far as Brower. Edgar Wayburn responded at that same meeting that scathing criticism was warranted only when "we have reason to be scathing."[3]

It was not just tactics, however, that concerned many of these board members. They worried that Brower plunged into controversy less for the cause and more for personal attention. "He coined the phrase 'No blind opposition to progress but opposition to blind progress,'" said Wayburn. "He was impatient: the earth was in a state of crisis; the time for action was always now. Nor was he above using guilt as a primary motivator for preservation. 'If enough people care' was one of his pet phrases in ads and books. Brower displayed no reservations about saying he was against anything or anyone."[4]

Such resolutions precipitated the beginning of a rift between Brower and the Sierra Club board. He felt that he was always several steps in front of them. "It was such a severe restriction that that was the main thing that drove me to the publications program," he said later. "I couldn't do what I was doing in speeches and articles, and I tried to get this general attitude out in books."[5]

Was Brower tiring of the job and the organization that he loved so much? In January 1961, he responded to a request by the newly appointed secretary of the interior, Stewart Udall, to name potential candidates to become Udall's director of information and associate director of the National Park Service. In an eleven-page letter, he offered a number of names, and the last one for each post was named Brower. Brower said that he was not looking for a new job; he already had one that was increasingly challenging. But the opportunity to work with Udall inside government, instead of outside, was irresistible.[6]

A month earlier, Ansel Adams had also written to Udall urging that he hire Brower. He said Brower was not aware of his proposal, and he praised Brower's "amazing courage and tenacity" in protecting natural resources and his knowledge and experience in dealing with conservation issues. Surely, said Adams, there was a place for Brower, with his extraordinary talents, in the Kennedy administration.[7]

When Udall demurred, Brower went to Europe. In the summer of 1961, the Brower family spent three months touring the continent. "My mother had always wanted to go," recalled Barbara Brower, who was eleven at the time (brother Ken was seventeen; Bob, fifteen; and John, nine). "She did all of the reservations; she planned out every aspect of the trip," said Barbara. "She figured out where we would stay and how we would acquire a car in Germany and renting out our place here." They met David's brother Joe and Joe's wife, Gayle, in Zurich, where Joe was stationed as a commercial airline pilot. Brower and his brother hiked in the Italian Alps, rediscovering some of the places he had been during the war, while Anne and Gayle Brower stayed more in town, including making visits to pastry shops.[8] It was Brower's first real vacation from the Sierra Club in nine years.

When Brower returned to work, he would devote most of his time to the publication program while staying at the Biltmore Hotel in New York. It would become a second home, and when he was not at Barnes Press, he would haunt the jazz clubs. On rare occasions, Anne or one or more of his children would come with him to New York, but he mostly traveled alone. That made liaisons easier, and in New York Brower's sexual choices were "an open secret," according to Gary Soucie, a field representative for the Sierra Club based in New York.[9]

The long absences meant that the four children were growing up without their father. "I have a collection of birthday cards," said Barbara Brower. "He was often not home for my birthday, but he always called. While he was absent a lot, he was a very loving dad." The eldest, Ken, described how other children would be confused when he would tell them that his father worked as a conservationist. No one knew what it meant. "It was embarrassing," said Barbara Brower. "It was hard to explain what he did."[10]

Brower was rarely home for dinner, but when he was, Ken recalled, his father often dominated the conversation, turning it into "a discussion of the beauty of the natural world, or the cleverness of evolution, or the obtuseness of humanity, or the task ahead in salvation of the planet." These intense bursts of energy and domination could be both annoying and exciting. As a teenager, Ken would sometime call his father "the Gee Whiz kid B."[11] Their father was not saddled with the conventional thinking of his generation, and the children found that disconcerting (figure 12). "As a father, he spoiled the '60s and the counter-culture for me," said Barbara. "We were already there, and there wasn't any place to go."[12]

On Sunday mornings when Brower was home, he would host strawberry waffle breakfasts for anywhere from fifteen to forty guests. The crowd would fill the main rooms of the house and spill out into the patio, with Brower in the kitchen, several sets of waffle irons going at once. It was a recipe he perfected, calling for Bisquick, oat bran, milk, eggs, grated zucchini, apple, carrots, and a shot of Worcester sauce. Brower was sometimes asked to write down the recipe. One inquiry came from a book publisher

FIGURE 12 Brower's four children atop a summit, mid-1950s

David and Anne Brower's four children take a rest (*left to right*): Kenneth, John, Barbara, and Robert. The two oldest, Ken and Bob, often accompanied Brower on hiking trips in the 1950s. Barbara and John would join their father on a hiking trip on Nepal's Himalayas many years later, when Brower was sixty-four years old. (Courtesy of the Bancroft Library, University of California, Berkeley, and the Brower family)

who pointed out that Brower's recipe did not include strawberries even though he called the dish strawberry waffles.[13]

If there were rifts in the relationship between Brower and his wife, they kept them to themselves. With four growing children and a monkey for a pet, the house was often less than spotless. Anne Brower had very little use for housecleaning, and Brower was a pack rat, but she was the glue that bound the family.[14] Brower could be trying, such as when he delayed their departure for an engagement by playing the piano. Yet the couple's intellectual bond remained strong. Barbara Brower remembered her parents reading the *New Yorker* series written by Rachel Carson about pesticides to each other in the car in 1962 and talking about it. "They were deeply impressed," she said.[15]

The 1960s introduced new ideas about race, sex, music, lifestyles, and even science. Brower mingled with some of the nation's best and brightest thinkers, who were considering such fundamental issues as how to best protect the earth. Michael McCloskey, who would work closely with Brower and eventually succeed him, could see how this contact was affecting Brower. "It was very clear to me in the '60s that he was spending more and more of his time in the East and around the world and that he was around great thinkers of the environment that did influence him. They kept broadening his horizon and giving him ideas on how to save the environment. When I would meet with him over lunch he would talk for quite awhile about how he spoke to whomever and got a better idea of this."[16]

There was Supreme Court justice William O. Douglas, *New York Times* writer John Oakes, editor and author Paul Brooks, and noted University of Pennsylvania anthropologist Loren Eiseley. Brower was on college campuses for speeches, and he enjoyed mingling with students. But he often drew greater satisfaction and intellectual stimulation from the faculty. Barbara Brower said her father was very good at listening. "He was also a very curious person," she said.[17] In 1962, Douglas forwarded to Brower an inquiry he had received from the Ford Foundation, which was interested in funding work in environmental activism and education. Brower responded by writing a detailed fifteen-page proposal. Although the proposal was never funded, it offered a fascinating blueprint of what was ahead for the American environmental movement. He told the Ford Foundation that two major priorities should be considered: improved environmental education and preserving land of "irreplaceable value." Government and environmental organizations would eventually embrace the latter concept. Examples range from local land trusts and conservation easements to government and nonprofit organizations such as The Nature Conservancy that have preserved literally hundreds of millions of acres both in the United States and around the globe. His environmental educational ideas ranged from producing more television programming to new curricula in higher education. Television was only beginning to discover wildlife shows, and colleges were still years away from developing environmental studies and conservation biology programs.[18] These ideas are characteristic of Brower, who was a visionary. He was sometimes so far ahead of others that he would grow impatient when they could not keep pace with him.

Brower met Rachel Carson at a banquet in New York. Their friendship would be short, Carson died of cancer only two years after her polemic on pesticides was published. In October 1963, when she took a trip to San Francisco, Brower, Anne, and Barbara accompanied her to nearby Muir Woods National Monument to see the redwoods. Barbara recalled that Carson was somber but very impressed by the behemoth redwoods. She was frail and needed a wheelchair, which they pushed along the park's paved paths. After Carson's death, her will provided for a significant financial gift to the Sierra Club.[19]

Brower had voiced concern about pesticides, especially aerial spraying to control lodge pole beetles in Yosemite, since he had become the club's executive director. The club struggled with the issue because many members were upset by how an insect infestation could turn green healthy trees into a forest of brown, dead sticks across thousands of acres. Others worried more about the health and safety risks from poisons sprayed in the air. The club's policy would evolve from conceding to limited spraying in national parks in the 1950s to opposing all pesticide use in 1964.[20]

The Sierra Club was evolving and expanding. New Sierra Club chapters were at work with new local conservation efforts: Great Lakes, Rocky

Mountains, Rio Grande (New Mexico and western Texas), the Lone Star, Grand Canyon, and Mackinac (Michigan). Brower could show up any-where in the nation to express concerns about an environmental issue. It might be a hearing in New York opposing a hydroelectric plant at Storm King Mountain in the Hudson Highlands, in Hawaii to support the cre-ation of Kauai National Park, or in Missoula, Montana, to explain his developing conservation creed.[21]

By now he was a polished speaker who was quite skilled in winning over an audience. Part of the attraction came from his tall stature, his broad shoulders, his melodious alto voice. Part came from the eloquence of his encyclopedic recanting of facts and stories—some of which were embellished or not factual. And part was just an intangible, magical trait. Years later, when Barbara Brower was teaching at the University of Texas, she invited her elderly father to talk to a class. Many of the students were not that interested in the environment, and Brower was ill—he looked gray. The talk began poorly, but the more Brower spoke, the more alive he became, the more his color improved, the greater his pace. The reaction from the students was even more intense as he caught their attention and captivated them. "By the end of the talk, they were on the edge of their seats," Barbara Brower recalled. "That was the night we went out and closed out the bars."[22]

Brower was so busy that he often had to delegate important campaigns, even when they were in the Sierra Club's own backyard. In the 1950s, he had testified against a proposed Trans-Sierra Highway that would have slashed across some of the mountain range's most isolated and scenic wil-derness. But by 1964, he was telling Eunice Elton of San Francisco that local people needed to pick up the fight. Good campaigns win because of individual effort, not because of what the staff at headquarters could do, he said. Ultimately, it was a local coalition that successfully blocked the highway.[23]

After passage of the Wilderness Act in 1964, the club established two top goals: establishing national parks in the North Cascades and the redwoods and blocking dams in the Grand Canyon. Brower assigned McCloskey to head up the North Cascades and redwoods campaigns, although Wayburn eventually took over much of the redwoods project. Brower, even though he was heavily committed to producing more books, assigned himself to lead the Grand Canyon fight.[24] Although those goals were the priorities, other issues were unavoidable, such as the proposed nuclear power plant near San Luis Obispo, California, and a ski resort planned in Mineral King, south of Sequoia National Park.[25]

By now, Brower had far more resources to call on when an unexpected campaign developed. The staff was growing rapidly. In 1959, the club had twelve employees; by 1963, it had thirty-three.[26] Separate departments were responsible for outings, publications, conservation, membership, and finances. Brower was fully involved in hiring employees in the areas he cared the most about, so loyalty to him was especially high in the publications and conservation departments.[27]

Yet administration was never one of Brower's strengths, and the hiring process was confusing. Brower sometimes followed the rules and brought on new employees only with the board's authorization. Other times he hired employees without consulting his bosses. Sometimes the board told him not to fill a position, but he did it anyway.

Brower was especially interested in adding regional representatives in the Pacific Northwest, the Southwest, the East Coast, and Washington, D.C. There was a clash about whether to have a Washington office, with Brower arguing he could no longer do all the lobbying.[28] The costs of some of these positions were shared with other local environmental organizations. But when Brower wanted to hire an East Coast field representative, the board turned him down.

Brower ignored the order and hired Gary Soucie, although he called him "assistant to the executive director" instead of "East Coast representative." It would be weeks before Soucie discovered the ruse. "If I had known it, I would have said absolutely no, absolutely no way," he recalled. But Soucie was talented, especially in dealing with the New York press, so the board eventually relented, and he eventually took the title Brower originally meant him to have.[29]

Brower often undercut the board's orders by using his expense account, which would rise to $30,000 a year, 50 percent higher than his annual salary. Brower called the account his mad money, and he would use it for whatever purpose he wanted, an arrangement the board agreed to for years.[30]

According to Stephanie Mills, his management style was loose, chaotic, and erratic. A former worker at the Sierra Club, she described it as "maddeningly unbureaucratic and refreshingly nonprofessional."[31]

Brock Evans was hired as the new Northwest field representative after McCloskey moved to the San Francisco headquarters. Evans described his first trip to San Francisco and going into Brower's office as if he were a marine private, "reporting for duty as ordered, sir." It was a joke, but Brower did not laugh. Evans, who was eager to work on the North Cascades campaign, asked about a plan. He wanted to know the priorities, the battle map, the design of the attack. "I'm your chosen instrument," said Evans. "I'm your spear point. Tell me what to do, and I will do it, sir."

"What the hell are you talking about Evans?" Brower replied.

Evans persisted, asking about a plan. Finally, Brower told him there was no plan.

"Well, what should I do?"

"I don't know, Evans. What would you like to do?"

Evans described options for the North Cascades. When he concluded, Brower said, "That sounds great; go ahead and do it."[32]

Both Soucie and Evans said they were never quite sure what specific areas their territories entailed. Evans elected to take everything north to the Arctic Circle. Soucie had the entire East Coast, but he was not sure about who was representing places such as Michigan and Kentucky.

Evans remembered that on his first trip to Washington, D.C., Brower told him to do some lobbying. Evans asked how he was supposed to do that. "I don't know," replied Brower, "just do it."[33]

Soucie said he also never quite knew what to expect from Brower. One day he got a call from Brower cryptically asking that he meet him the next day in Washington, D.C., for breakfast.

"Should I bring a toothbrush?" Soucie asked.

"Probably a good idea," Brower responded.

The next morning, Brower told Soucie that the Sierra Club was running a newspaper advertisement in the *New York Times* and the *San Francisco Chronicle* opposing a dam project. Brower wanted a press conference in Washington.

Soucie asked, "We're not running it in the *Washington Post?*"

"No, we don't have the budget for it," replied Brower.

"How is this newsworthy in Washington?"

"You'll figure it out."

Soucie also did not know anyone in the Washington press corps. Brower gave him a name to get started, and Brower was pleased with the resulting press conference.

"He gave subordinates a lot of freedom," added Soucie. "He would let us have full rein, and I don't think any of us took advantage of it. We worked very hard."[34]

Brower's most complicated relationship on the staff was with McCloskey. After McCloskey had worked for three a half years as the Northwest field representative, Sierra Club board president William Siri asked him to come to San Francisco as assistant to the president. Brower had repeatedly praised McCloskey, but he opposed this plan. Siri rejected Brower's objections but wondered why Brower was so opposed. He said that Brower's arguments disturbed him; if McCloskey was not qualified to come to San Francisco, by extension one could argue that McCloskey also was not fit for what he was currently doing in the Northwest.[35]

McCloskey was unaware of Brower's opposition, but when told about it years later, he was not surprised. The board had created the position as an attempt to control Brower. It wanted a full-time employee in the office to gather information and pass it on to the board. From Brower's standpoint, McCloskey had been brought in as a spy. McCloskey attempted to defuse an uncomfortable situation by consulting with Brower. Brower responded well, and the two got along. But McCloskey never liked the situation, and after a year he proposed a new post for himself as conservation director. With Brower spending so much time in New York, McCloskey coordinated all conservation efforts for the club. Both the board and Brower liked the arrangement.[36]

The new position gave McCloskey greater freedom, especially in the North Cascades and redwoods campaigns. Although on paper the field and Washington representatives answered to McCloskey, he often felt that Jeffrey Ingram in the Southwest and Soucie in the East ignored him and talked only to Brower.[37] Both Ingram and Soucie deny this claim, and the Sierra Club records do show a trail of correspondence from Soucie to McCloskey.[38]

McCloskey and Brower worked reasonably well together, considering that McCloskey was the mild-mannered lawyer and Brower the verbal-bomb thrower. "Mike McCloskey and Dave had an odd relationship because their approaches were so different, although they did get along well," said Soucie.[39] Brower would confide potentially damning information to McCloskey, who never passed it on to board members, even when asked. McCloskey was alarmed by how much Brower drank during the day and turned down Brower's martini luncheon offers. "I was very abstemious," McCloskey said. "I did not want to buzz my brain for work."[40]

Brower's sexual relations with men apparently were never raised as a problem in the workplace, but at times it was an issue. Both McCloskey and Ingram said that Brower sexually propositioned them at hotels or resorts where they were attending conferences or other meetings. And McCloskey sometimes worried that Brower was hiring a great number of young men more for their perceived sexual orientation than for their commitment to the environment and that, as a result, their work was not of the highest caliber.[41]

Brower also struggled with the bookkeepers and accountants, who often produced financial reports bewilderingly different from Brower's. In 1963, the board reorganized the club's finances and hired a new financial manager, Clifford Rudden. Brower liked Rudden, and when he was in San Francisco, they would sometimes go out for one of Brower's long and sometimes tipsy lunches. But Rudden also sized up Brower and decided that to protect his own reputation, he would answer to the board as much as he answered to Brower. The situation was combustible.[42]

Sierra Club finances rode a roller coaster that was becoming increasingly difficult to control. It was not just that the club teetered between budget deficits and surpluses, but bank loans and book inventories kept growing. Total revenues would increase from $208,000 in 1960 to more than $3 million a decade later, and when the club missed finishing a year in the black, it sometimes missed badly. The good news was that money coming in from new memberships and from the wilderness outings kept growing. The bad news was that the cost of book publications also kept increasing, and between 1963 and 1969 the publications budget always finished in a deficit.[43]

In budget discussions in the mid-1960s, Brower assured the board that he was deeply concerned. Sometimes he would devote days and weekends struggling through all the numbers. But he rarely wanted to make cuts, no matter how dire the situation. Rampant pessimism, he said, was not wise when the Sierra Club seemed to be achieving so much in the field of conservation.[44]

Longtime board member Lewis Clark said club leaders knew for years that Brower was financially reckless. "We should have blown the whistle earlier," he said. "We kept feeling, 'He is a very valuable person to the club. He's a good spokesman.' We didn't want to stop his good work. So we thought, 'Perhaps we can reform him.'"[45]

Brower's generous spending from his expense account did not reassure his bosses. His salary remained comparatively modest for a national leader of his stature, but his lifestyle and his expense account were far richer. Evans remembered staff meetings in San Francisco. "David liked to do things first class," he said. "Often at the Francis Drake Hotel there were martinis served. He liked good things; that was David's style."[46] It was not the smartest message to send, however, when the board worried about budget deficits. "He was running up a $700-a-month bill at the Alley, the little local restaurant," remembered Adams. "Anybody comes to town, he'd take the whole staff out to lunch, put it on the chit. Dick Leonard was acting treasurer and began to see these things come in."[47]

Virtually every club president struggled working with Brower and his difficult personality. The burden may have been the greatest on Wayburn, a physician who led the club twice, from 1961 to 1964 and again in the tumultuous years from 1967 to 1969. Wayburn was often caught in the middle between appeasing a board that wanted to crack down on Brower and his own concern that such discipline would smother Brower's brilliance. "I would have conflicts with myself as to how this should be handled. In large part, I believed in giving Dave his head and did," he said. "I had to compromise a number of occasions in trying to bring together the wishes of the directors and the actions of the executive director."[48]

Brower first worked with Wayburn when the physician joined the club's Conservation Committee in the early 1950s. Brower shared many of the same values as Wayburn—their love for the outdoors, their prodigious work habits, and their desire for power. And yet Wayburn was also different, with his Deep South Georgia roots and Harvard medical degree. He had stumbled into the Sierra Club when he was recruited to join the high trips because organizers needed a physician. He stayed with the club until his death at the age of 103. Wayburn claimed that he sometimes worked ninety hours a week between his physician calls and the administrative work he did for a medical society and the Sierra Club. His greatest accomplishments were protecting the redwoods and later protecting the Alaska wilderness, campaigns that his wife, Peggy, joined in an equal partnership.[49]

Wayburn liked being in power in the club, and he held that power as long as he could. He was elected to the board in 1957, and after he became president four years later, he would often have lunch with Brower, when Brower would complain about how his enemies in government, industry, and even on the Sierra Club board were attacking him. "Dave was always of the opinion that someone was out to get him," said Wayburn. "Dave had a strong paranoid streak in him, a strong paranoid streak." For years, Brower threatened to resign each time he hit a new obstacle. The threat would dissipate after he calmed down. But in the mid-1960s, Wayburn warned him that board opposition to his tenure was growing. "Dave, if you resign this time, I think your resignation might be accepted by a majority of the board," Wayburn told him. Brower wanted a board that advised only, not one that actually set policy, especially if that policy opposed what he wanted to do.[50]

In the early years of his tenure, Brower had reported on a regular basis to the board and its president. Even then, though, there were lapses, and by the 1960s Brower was seeing very little reason to communicate except when it was in his best interest. "Dave was brilliant but demanded absolute self-rule," Wayburn commented. "One of his most used expressions was 'follow me.'"[51]

Theories varied on why Brower was given so much freedom. Glen Dawson, a longtime Sierra Club member, said Brower had too many friends on the board. "I think they gave him too much leeway," he said. "They had grown up together, known each other, climbed together, and skied together, the whole bunch of them."[52]

That was only part of the problem, said Phillip Berry, who was thirteen when he first met Brower and who would eventually serve on the Sierra Club board. "From Dave's perspective, the expectation was that if you climbed rocks, you loved the mountains, and therefore to save them was good," he said. The board "had expectations too. They knew more about finance and knew how to deal politely with these outside influences, the Forest Service, the corporations, what have you. As board members they thought he at least ought to listen in the selection of methods. There developed out of all that a kind of a mutual disrespect."[53]

To a certain extent, the friction between Brower and his board was mirrored at some other conservation organizations. Historians have pointed out that many conservation organizations, including the Sierra Club, had amateur, volunteer leadership for their first fifty years. By the 1960s, professional staffs were growing not just at the Sierra Club but at other organizations such as the Wilderness Society and the National Parks Association.[54]

One reason why the situation may have been more corrosive at the Sierra Club was that Brower and many of the allies he recruited to the board had a poor opinion of many of the Californians on the board who had been serving for years. Brower said that Justice Douglas called the Sierra Club board "a combination of a mourner's bench and ladies' sewing circle."[55] Brower once provided a remarkably candid assessment of the board to Paul Brooks. In a letter written in February 1965, he urged Brooks to read it and then burn it. He stated that the current board had six leaders and nine followers. On too many issues, the board was parochial, with all but one representative from California. It was also stagnating because four had been on the board for twenty-four to thirty-nine years. He said that only a handful of board members had a sense of the club's national role or were on the correct side on issues such as the dangers of pesticides.[56]

Brower remained protected by many loyal followers on the board, including Martin Litton, Patrick Goldsworthy, Eliot Porter, Ansel Adams, Richard Leonard, Wallace Stegner, and William Douglas.[57] Some would never drop their allegiance. "Brower has great charisma, and he has an army of devotees who think he's just the second coming," said Adams.[58] As late as 1965, the majority of board members continued to rally around Brower when he was attacked by former Sierra Club president Joel Hildebrand. A chemist with a national reputation at the University of California, Hildebrand dramatically resigned in May 1965 from the club he had once led because, he said, Brower had made "ridiculous statements" about the dangers of chemical fertilizers in the December 1964 issue of the *Sierra Club Bulletin*. Further, Brower and his staff refused to publish Hildebrand's rebuttal. "The curious result is that the Sierra Club does not permit a member to criticize the published utterances of an officer, utterances that abound in criticisms," said Hildebrand. Hildebrand said that no one could control Brower and that "the organization that I have served with faith and enthusiasm is operating in ways that I regard as undignified, ignorant, futile, and offensive, ways that may even prove dangerous." Hildebrand's resignation surprised Stegner, who wrote a letter of support to Brower. Adams, who by now had begun to harbor increased doubts about the executive director, also defended Brower, but he also worried that such internal clashes "weakened the public image of the club."[59]

Soon after the Hildebrand episode, Brower began to lose support from two board members who had long meant the most to him: Adams and Leonard. Brower had hiked, skied, and climbed mountains with Leonard; he had often been in Leonard's home; and Leonard's wife, Doris, had been nearly as close to him. The break was not easy.

"We were very close," recalled Leonard. "Dave was extremely generous with praise. He used to tell the board of directors with great praise how much money Dick Leonard had given to the Sierra Club by his free legal work. Dave was very proud of what I was doing and was very generous with his statements, and I was very proud of Dave. We worked together magnificently all the way through the Dinosaur campaign."[60]

But they both slowly changed. The personal friendship drifted, and by the late 1950s the Leonard and Brower families rarely gathered together. Leonard did not always agree with Brower's tactics, but for a very long time he stood by and defended him. Leonard knew that Brower's management skills were not strong. "Dave's weakness was that he was unable to budget either finances or time," said Leonard. "He was always trying to accomplish too much." Leonard's concerns multiplied as the club grew and the financial risks increased. "Dave used to say that the money had always come in and always would and that what we saved today is all that ever will be saved. I used to point out to him yes, but if we had gone bankrupt ten years ago, we wouldn't be here to fight the Grand Canyon and the other battles at the time."[61] Leonard worried that Brower could bankrupt the club.

Brower blamed himself in part for the fissure with Leonard. He agreed that there were too many arguments over money. During one board meeting, long before the situation heated up and the break occurred, he was arguing that membership dues would come in and that the club would be financially healthy.

"All we're trying to do is protect you," Leonard told him.

"You protect me with the back of your hand," Brower responded in what he later said was a very abrasive manner. "It really cut him," he added. "His face just fell. It was quite a shock to him that I would say something like that. It kind of shocked me that I had said it, but I was upset. I would get upset at some of the meetings."[62]

The break with Adams was even more emotional and more wrenching. For years, Adams had served as Brower's confidant. Brower had always believed that he could confide all of his fears and frustrations to Adams. "He had been sort of my father confessor," he said. "I would take all of my troubles to Ansel."[63] Adams in turn had staunchly defended Brower. He privately wondered at times if Brower were mentally stable, but he would tell friends that Brower's only problem was that he was overworked.

He seemed tired and, Adams said in one letter in 1959, too sensitive to criticism.[64]

By late summer in 1966, Adams warned Brower in a letter that he needed to change if he wanted to continue running the Sierra Club. He told him that his extreme reactions to criticism had to stop. Brower was far too paranoid far too often, said Adams. He had a "sense of persecution," an attitude that was totally baseless. Adams said the solution was up to Brower, that he needed to calm down, but if he did not, he might be fired. Adams told Brower that he truly believed in him and that he was writing because he wanted Brower to keep up the good work in the same position as executive director.[65]

Two months later, after Brower had not responded. Adams tried again. He feared that Brower now considered him an enemy, part of the old guard, a conservative. Far from it, Adams told Brower, but he added that Brower was often dictatorial and would bypass the board, the Executive Committee, and the Publications Committee. He pleaded with Brower to be more diligent. He said that he was not exaggerating in warning that on the present course Brower was headed for disaster.[66]

Brower and his supporters would long espouse another reason why Adams "turned on" Brower: jealousy. Brower often described making a remark at a Sierra Club board meeting shortly after two of Porter's books had been published, which might have produced such a feeling in Adams. "I made the tactless remark in the presence of the directors, including both Ansel Adams and Eliot Porter, that Eliot Porter was our most valuable property," he said. "I'm sure that offended Ansel no end."[67] Brower mentioned this comment so often that it gained credence with others.

Clearly, the relationship between Porter and Adams was complex. They were both colleagues in the same field and rivals because they espoused two different approaches. Wayburn believed that Adams never could warm to the color photography in the later books. "He felt that they were too expensive, and from a photographic standpoint he was a black-and-white man," said Wayburn.[68] But Wayburn dodged the question whether Adams was so jealous that he would turn against Brower. It defies common sense, however, to believe that Adams would feel and react this way.

What is clear is that when Adams finally did take a public stand against Brower, it was fueled by vitriolic anger and emotion. It shocked Brower.

In the meantime, Brower was being called on to defend one of the nation's greatest natural shrines, the Grand Canyon, from his greatest enemies, Floyd Dominy and the dam builders. To win, he would have to gather all of his resources and mount the greatest conservation campaign of his career.

12. Grand Canyon

The idea of building not one but two dams in the Grand Canyon seems incomprehensible. It is the social, political equivalent of erecting an amusement park in Yosemite with rides using the great waterfalls or of running a foundry at Yellowstone that would tap into the hot geysers. "Look," David Brower would say, "it's the *Grand* Canyon we're talking about."[1] Brower did not just assemble a campaign to oppose the dams. What he directed was a war, one in which absolutely no compromises within the canyon would be tolerated.

And yet Brower had once supported building a massive dam within the Grand Canyon. He even agreed to surrender territory managed by the National Park Service to make the dam project more digestible.

That position came in 1949, and the transition to Brower's philosophy by the mid-1960s was as vast as the gap between the North and South Rims of the Grand Canyon. Brower called himself a graduate of the University of the Grand Canyon, and he employed every resource he knew—along with some startlingly new tools—in this political and propaganda Armageddon.

On the other side were the dam builders, the engineers, the U.S. Bureau of Reclamation. To them, the Grand Canyon was a big hole that needed to be filled. The canyon walls rose up to 5,000 feet, and the Colorado River fell 1,930 feet through the canyon's 277-mile length. The first engineering surveys by the dam builders began in 1916, even though President Theodore Roosevelt had set aside 100 miles of the canyon as a national monument in 1908. Senator Carl Hayden of Arizona tried to trump that protection by getting Congress to convert the national monument into a national park in 1919, with a provision allowing for dam construction.

In 1930, the Bureau of Reclamation proposed building a 570-foot-high dam at the downstream end of Bridge Canyon. Two years later, the outgoing Hoover administration created a new national monument in land between the dam and the national park. No dam provision was included this time.[2] The dam at Bridge Canyon would be outside park territory, but its reservoir would flood the Colorado the entire length of the monument and slightly into the national park. Surprisingly, National Park officials did not oppose the project. Park director Horace Albright wrote in 1933 that he did not see how the dam would adversely affect the national monument.[3]

Other dams and World War II stalled the Bridge Canyon project until 1946, when it was revived and linked to an even more audacious plan. The bureau proposed a dam upstream of the park boundaries in Marble Canyon, plus a 45-mile underground burrow. The Kanab Tunnel would divert 95 percent of the water in the Colorado. The river and the Grand Canyon at that point formed a large U. The tunnel would create a straight line at the top of that U. It was a plan that only an engineer could love. Bestor Robinson of the Sierra Club, who usually embraced compromise, called it absolutely terrible. No longer would the Grand Canyon be the canyon that displayed the river that created the canyon. Instead, he said, it would be nothing but a canyon that had a creek in a gravel bed. J. F. Carithers, a lobbyist for the conservation forces, complained that the trickle of water could be turned on or off at any time. The very force that had created this great spectacle, he said, would vanish.[4]

After the war, however, Newton Drury was chief of the Park Service, and he fought these power projects. In July 1949, under pressure from Drury, the Bureau of Reclamation backed down on the tunnel but pressed ahead on the two dams.[5] Despite that retreat, the tunnel project was still seriously being considered as late as 1961. Engineers testified at a federal hearing that hydropower from the Kanab Tunnel would be four times greater than what could be generated at the Marble Canyon Dam.[6]

Drury's battle was hampered by a lack of support from conservation organizations. On November 12, 1949, the Sierra Club board, which at that time included Brower, met and actually undercut Drury's position on the Grand Canyon dams. It passed a resolution proposed by Robinson, the club president, assenting to the construction of a dam at Bridge Canyon, under the condition that the national monument boundaries be redrafted so that the reservoir would not be within Park Service territory.[7] This resolution hardly made sense, but Robinson had concluded that the downstream area of the canyon was scenically less important and that dam builders had the political clout to win. "If you had to give up any part of the canyon, that would be the part to give up," he explained. He also thought that there was a recreational value in having boats journeying deep into the Grand Canyon. Despite consenting to the dam, Robinson argued privately that the resolution could be used politically as a holding tactic to delay anything from being built. "This was a question of tactical maneuvering as against the purist position," he explained. "Dave Brower never understood the tactics of it; being a purist, he objected to it."[8]

Brower often disagreed with Robinson, and so did Dick Leonard, but both voted for the resolution. Years later Stewart Udall, his brother Morris, and Floyd Dominy would point out in debates that the Sierra Club—and Brower—had originally supported building Bridge Canyon Dam.

The resolution shocked Drury. Changing national park boundaries was "a very dangerous precedent . . . to accommodate reservoir projects," he wrote Leonard. "If that were undertaken, it would be only a matter of a few years before there wouldn't be any national parks worthy of the name."[9]

Drury's isolation did not last long, though. Across the nation, environmental groups were waking up to the dangers in the national parks, especially at Dinosaur. The Sierra Club board found it impossible to fight dams at Echo Park and Split Mountain while supporting a dam at Bridge Canyon. On September 3, 1950, it thus announced that it was opposing any Grand Canyon dams.[10] A legal water war between California and Arizona held up any serious consideration of Grand Canyon dams until the courts ruled in 1963. Udall changed the stakes on January 22 1963, when he announced his grandiose plans for supplying additional water to the Southwest. The Marble and Bridge Canyon Dams (minus the Kanab Tunnel project) would soon be the plan's most controversial aspect, but they were hardly mentioned in news stories about Udall's announcement. The dams were primarily to provide revenue from electric sales to finance the necessary aqueducts and reservoirs that would supply water across the five southwestern states. Neither dam would be in territory run by the National Park Service, which protected less than half of the geological Grand Canyon. However, the plan ignored the fact that the downstream reservoir would intrude into park territory and that water releases from the upstream dam would affect the river in the canyon. Another key feature of the Udall proposal was a costly and extravagant plan to build a series of saline plants to convert salt water to fresh, but it generated little dispute. "A critical period lies ahead," said Udall at the press conference, "for millions of people who are flocking to the Pacific Southwest to establish permanent homes. This burgeoning population will require vast quantities of additional water for industrial and municipal use; greater quantities of electricity and other basic services; and more irrigated land."[11]

Wallace Stegner, who served as an aide to Udall and remained close to him, maintained afterward that Udall never really wanted the dams. "Stewart was against them and wanted to find some way, but he didn't think he could just nix them out," said Stegner. He did not like the other alternatives. Added Stegner: "There's something wrong with producing massive amounts of power, particularly in a wilderness like that where it has to be transported long distances, or where it's made with coal with all kinds of particular pollution problems, or nuclear with dangers of several kinds, or with dams in the Grand Canyon. Which do you choose? You know, you flip a coin."[12]

Although Brower's views of dams had changed over the years, Robinson's had not. At the May 4, 1963, Sierra Club board meeting, Robinson urged fellow members to compromise and accept at least the Marble Canyon Dam. The impoundment would be upstream of the national park, which meant the reservoir would not intrude into park territory, and the dam could create a resource for trout fishermen, he said. Perhaps, Robinson suggested, the club could ask the Bureau of Reclamation to build public elevators in the dam to make the reservoir more accessible to fishermen.[13]

For a while at the meeting, the board seemed to be swayed by Robinson's influence. Brower attempted to swing votes the other way by describing the Udall press conference and warning that even the Kanab Tunnel might be resurrected. Then he asked Martin Litton to talk. Litton's passion for saving the great natural resources of America may have been even stronger than Brower's. When he was growing up in the Los Angeles area as a teenager, he had attended meetings of the Los Angeles chapter of the Sierra Club. All they did at the meetings was talk about wilderness outings, which disgusted Litton, who wanted to work on conservation campaigns. "That [the discussion of outings] did not make me mad," he said. "I wanted something that made me mad." Brower discovered Litton during the Dinosaur campaign. Even though Litton worked in the circulation department of the *Los Angeles Times*, on his own he had been writing stories about the threats to the West's national parks. Litton's stories on Dinosaur published in the *Times* were among the best Brower had seen. Yet when Brower approached him about helping the Sierra Club, Litton told Brower that he had no use for the organization. "They don't do anything," he said. That was in 1953, and in a tale that Litton told many, many times Brower was supposed to have responded, "Well, they are going to do things now. I'm in charge."[14]

Litton, who knew the Grand Canyon, had first gone down it in the mid-1950s. Years later he created a tour company, and he once estimated that he made at least eighty river trips down the canyon.[15] Brower had been so persuasive back then that Litton was now in May 1963 a board member, and he had come prepared to rebut Robinson. He had fashioned a crude map of the Grand Canyon on cardboard backing. It showed the river, the geologic boundaries of the canyon, the dam sites, and, most important, the boundaries of the national park and national monument (figure 13).

"Here's the Grand Canyon, and here's where they want to put those dams, and what are you going to do about it?" Litton told them. "The

FIGURE 13 Martin Litton

Martin Litton displays a rude map of the Grand Canyon that he drew and showed to the Sierra Club board of directors in 1963 when they were debating whether to take a stand on the proposed government dam projects in the canyon. He convinced the board to oppose the dams. Litton was one of Brower's strongest defenders on the board.

Grand Canyon is not just a place at Angel's Point where you stand on the rim and look over the edge. The Grand Canyon is 277 miles long, and it goes a long way."[16]

Marble and Bridge Canyons are on opposite ends of what is geologically one of the most magnificent gorges on earth. Dam proponents argued that Marble and Bridge were inferior, less glorious than the central Grand Canyon. This was the equivalent of arguing that the Empire State Building or the Statue of Liberty could be sacrificed and demolished because the splendor of the New York skyline would remain.

François Leydet, a Sierra Club member and writer, wrote that Marble Canyon grew more fantastic as the river ventured downstream to the dam site, where engineers were already boring test holes.[17] The river roared with booming rapids that had exotic names such as "Hot Na Na" and "Houserock." The red limestone strata on the canyon's soaring walls gave the illusion that one was plunging deeper into the earth. Red pigments streamed into horizontal bands of light and dark stripes. John Wesley Powell had thought erroneously that the rock was marble, hence the canyon's name. The rock was so implausible, according to Leydet, that "the scene might have been out of a child's fairytale book, with rocks made of peppermint candy." At Vasey's Paradise, a waterfall of two jets gushed out of the mountain wall, cascading onto rocks and a garden of lush greenery. Farther down was Redwall Cavern, a yawning cave 100 yards wide, 50 yards deep, and 30 yards high. Powell called it a "vast semi-circular chamber which, if utilized for a theater, would give sitting room to fifty thousand people."[18] The Marble Canyon Dam would bury all of this in a reservoir that would extend nearly to Lees Ferry, just below the Glen Canyon Dam. On the other end of the Grand Canyon, the Bridge Canyon Dam would create a reservoir equally as long. It would be 93 miles long, including 53 miles within the Lake Mead National Recreation Area, 27 miles inside the Grand Canyon National Monument, and 13 miles within the existing national park. The country here was just as spectacular, the dazzling features too numerous to tally. They ranged from the delicate series of pools and waterfalls ringed by ferns and flowers that climbed the wall at Fern Glen, a 200-yard side canyon, to the thundering cataracts at Lava Falls.

Leydet wrote that during a visit to Fern Glen, he climbed as far as he could until he reached a 20-foot-high tongue of travertine that he realized was a petrified waterfall. Drops of water were continuing to fall from the pool above, slowly adding to the limestone. Lava Falls is the remains of a once mighty natural dam created by the upstream Vulcan's Forge volcano. Lava once cascaded into the canyon. Powell wrote, "What a conflict of water and fire there must have been here. Just imagine a river of molten rock, running down into a river of melted snow. What a seething and boiling of the waters; what clouds of steam rolled into the heavens!" Here the river becomes violent amid the massive chunks of basalt, monuments and pillars of the canyon's past.[19]

These were the types of images that Litton sought to convey at the Sierra Club board meeting, images of what would be lost if the dams were built. As Litton spoke, he became increasingly passionate and emotional. He could be blunt, iconoclastic, and unable to acknowledge the other side. One of his greatest attributes was his encyclopedic knowledge of the geography of the West, especially of its natural resources. Besides getting Litton on the board, Brower helped him become travel editor at *Sunset*, a regional magazine based at the time in Palo Alto, California. L. W. "Bill" Lane Jr., the magazine's publisher, called Litton "a fantastic editor," "a fabulous writer and photographer," and a difficult individual. Actually, said Lane, Litton "could be quite difficult." Others could not supervise or edit Litton, so Lane as the boss tried to do so, with limited success. Litton would inappropriately use *Sunset* stationery to write political polemics to elected officials. He once flew Lane's airplane so low over the forests of northern California to view logging activity that the Federal Aviation Administration warned Lane that it was going to ground him. Litton's articles had such a "blatant, arousal-type lead-in" that Lane often had to tone them down.[20] The zealous Litton was just as passionate at the Sierra Club, and no one at the board meeting wanted to challenge him as he appealed to his colleagues to do the right thing. Finally, he finished. Spontaneous applause erupted. After ninety minutes, the board was ready to vote. It approved a resolution that had two parts. First, in recognition of Litton's geography lesson, the boundaries of the national park and monument needed to be extended to their natural geologic limits. Second, in support of Brower, no dams should be built anywhere in the entire length of the Grand Canyon.[21]

Ed Wayburn, president of the Sierra Club in 1963, placed Brower in charge of the Grand Canyon campaign. "He was the logical person to lead that fight," said Wayburn. "Dave had been working on the Colorado River Project since 1950."[22]

As Brower studied Udall's water plan for the Southwest, he was struck by the enormous volume of money that these hydroelectric facilities, called "cash-register dams," would produce in reservoirs. Not only did he fear their environmental impacts, but he became convinced that they were unnecessary. The only reason that much money was needed, he decided, was to raise enough funds to divert water from the Pacific Northwest down to a very thirsty southern California. "The Grand Canyon project," he said later, "was a major diversion of funds that would otherwise return to the general treasury, to build a bank full of money for a major import of water from the Columbia River."[23]

Brower pitched that argument to any audience, anywhere in the nation, that was willing to listen. "Water is not the issue" in the Grand Canyon, Brower told a National Audubon Society audience in Tucson in November 1964. "There are plenty of ways to finance water other than pouring concrete into canyons or sticking wheels in rivers." Erect coal-fired or nuclear-generated plants instead, he maintained. Nuclear was cheap, and coal was plentiful in the Four Corners region.[24]

Floyd Dominy and Stewart Udall also campaigned. After President John Kennedy was killed, Udall was uncertain of where he stood with Lyndon Johnson, but he knew that he could count on Arizona's two senators: Carl Hayden and Barry Goldwater. The crusty, aging Hayden had made solving the water problems of his desert state his life's aspiration. Goldwater was enormously influential as the Republican nominee in 1964 against Johnson. And Morris K. Udall, Stewart's brother, had also been elected to Congress. In the mid-1960s, "Mo" was still gathering his political wits, but he knew it was political suicide in Arizona to oppose any dam, even in the Grand Canyon.

Dam proponents' basic arguments, hashed out in hearings in 1965, were that both reservoirs would be located in remote sections of the park visited by very few people and that 98 percent of the Grand Canyon would remain unaltered. Conservationists were selfish elitists, they said, intent on keeping the canyon for themselves and excluding those with neither the time nor the means to tour the river. In contrast, the two reservoirs would open up previously inaccessible areas of the canyon for boaters and fishermen. Plus, the reservoir would submerge only about 20 percent of the canyon walls.[25]

Dominy stressed to members of the House Subcommittee on Irrigation and Reclamation that the Udall water plan was the most important legislation of his tenure as Reclamation commissioner. It would complete the task of taming the Colorado, a river that "is completely controlled by the works of man," he bragged.[26]

Brower and his allies were closer to the truth than they may have realized. Secret talks were under way between Udall and California officials on the prospects of shipping Columbia River water south. Senator Henry "Scoop" Jackson of Washington State was already suspicious of Udall's motives. Jackson headed the Senate committee that was considering the Udall plan, and, prodded by Brower and others, his fears about the project would only increase.[27]

Privately, even some dam supporters had their doubts. Wayne Aspinall in the fall of 1964 flew over the disputed areas of the Colorado River. Later that year, he wrote to a Colorado constituent that he was not convinced that a "high" Bridge Canyon Dam was necessary. Although he didn't think the dam would harm the canyon, he felt that a "low" dam would be preferable because it would create a smaller reservoir that would not intrude into the park.[28] Dominy had also been studying the canyon. He took helicopter rides down the river, shooting photographs along its entire length. It was not easy to impress Dominy, but he was taken by the beauty of Marble Canyon. "I was perfectly willing to give up on Marble after my trips down the canyon," he said years later. "I said, 'This Marble Canyon is better than the Grand Canyon itself. This ought to be part of the park.'" Dominy also realized that it would be very difficult to get boats or rafts down to the base of the Marble Canyon Dam so that they could traverse the Grand Canyon. He told Udall that the national park boundary should be extended to Glen Canyon Dam. But Bridge Canyon was different, and Dominy desperately wanted to build a dam down there.[29]

As in previous campaigns, the two sides used a variety of media to make their arguments. Dominy's color slides of the canyon were shown to every member of Congress willing to watch. Other photos went into a brochure touting the boating benefits of the new reservoirs. Dominy hired actor Fredric March to narrate a film touting the glories of hydroelectricity. He found government funds for an elaborate scale model of the Grand Canyon, with dams and reservoirs that could be lifted out. The model demonstrated how little of the park or even the river would be affected.[30] Brower produced an Exhibit Format book: *Time and the River Flowing: Grand Canyon*, which was popular and, more important for Brower's purposes, controversial. Written by Leydet and edited by Brower, it featured spectacular color photographs, a narrative about traveling through the canyon, and one-sided arguments against the dams. It was a very specific, planned campaign book in the spirit of *The Last Redwoods*, and dam opponents called it misleading and dishonest.[31]

The sharpest exchange on that point was between Brower and Morris Udall during the August 1965 House subcommittee hearings. "This book, as well written and pretty as it is, and it touched my heart when I read it, is one of the most misleading things I have ever seen," said Udall. He described how a detailed analysis of the book found that forty-five of the seventy-nine photos in the book were in areas that would not be affected by the reservoirs. Only twelve would be in areas inundated by water, and Udall said that none of the six in the Marble Canyon were within the national park.

Brower said the book tried to cover the geological Grand Canyon.

"Does it not leave a false impression when you are trying to preserve a living river, and 95 percent of what you are showing will be left?" asked Udall.

"Our intention," replied Brower, "and I think it comes off pretty well and fools no one, is that we would lose with those dams the living heart of the river."

Udall questioned whether the river was truly living when it was already tamed by upstream dams. Brower assured him it was still alive, although not for long if dams were erected in the Grand Canyon. "It will be greatly reduced," he said. "It will be an off-again, on-again river."[32]

The August House hearings were the first major face-off between the two sides since May, when the pro-dam forces had sustained a setback. The Bureau of the Budget had announced it would not support the construction of the downriver Bridge Canyon Dam. It was something of a split verdict, however, because although the Johnson administration was opposing Bridge Canyon, the president had signed on to allowing the second upstream facility. Brower was convinced that Dominy and the Bureau of Reclamation would try to bring back the Bridge Canyon Dam. In a cryptic note about the decision, he wrote that Ottis Peterson of the Bureau of Reclamation had told a reporter for the *New York Times* that Bridge Canyon Dam had been "tabled and not deleted."[33]

Stewart Udall in his testimony confirmed that he did indeed hope to resurrect Bridge Canyon Dam in the future. He also indicated that building a coal or nuclear generating facility instead of the dams was a possibility, but a remote one. The problem with producing electricity from coal or nuclear fuel, Udall said, is that it was "a completely new departure which obviously would be highly controversial and which obviously would in itself be something that would be debated at great length." Proponents worried that building federally owned generating plants, traditionally the exclusive domain of privately owned utilities, was a form of "creeping socialism." It was not clear that the votes would ever be there for such an option.[34]

Twenty-five opposition witnesses testified, including the seventeen-year-old granddaughter of Aldo Leopold. Brower was articulate and impassioned, but he simply had no tour de force argument as he had ten years earlier in battling Echo Park. Even in his exchange with Morris Udall, he was on the defensive.

Dam supporters may have had the political clout in Congress and the White House, but Brower could always count on grassroots support. By now, he and the Sierra Club had become adept at organizing letter-writing campaigns, and in 1964 he created the Grand Canyon Task Force. Thousands of names of those who had written Brower and the club or had made donations were put on a mailing list. The list permitted Brower to communicate with a wide swatch of voters across the nation when he needed them to write a letter or send in a small donation. Soon there were thousands on the task force.[35] Letter writing had been extremely effective during the Dinosaur campaign, when congressional representatives were startled by the large volume of letters from angry constituents. But that was ten years ago. There were diverse political messages not only for the Grand Canyon but also for the redwoods and other issues. Many of the letters parroted the same message, making it easy for representatives to conclude that these well-meaning letter writers were part of a "disinformation campaign" waged by Brower and others.[36]

National publicity was another tool, and Brower got the *New York Times*, *Fortune*, and *Life* magazine to write important articles about the Grand Canyon dams. *Life* editorialized against the dams, asking why Arizona could not pay for the water it needed as opposed to insisting it be financed from hydroelectric plants in the Grand Canyon. "Besides," the editorial ended, "we haven't any Grand Canyons to spare."[37] But overall news coverage was faltering. Richard Bradley, the Sierra Club member who had helped Brower on the evaporation issue during the Dinosaur campaign, was convinced that dam proponents were successfully keeping the name "Grand Canyon" out of the debate, which confused the public and diminished the threat. Even the legislation itself did not include the name "Grand Canyon" in the title, and pro-dam organizations such as the Central Arizona Project Association went out of their way to avoid using the name. Bradley and his brother David began talking to the editors of the *Atlantic Monthly* about writing a story that would prominently feature the Grand Canyon.[38]

Brower determined he needed more help. During the Dinosaur hearings, he had relied only on his math because the technical experts were too afraid to confront Bureau of Reclamation engineers. Ten years later he was finding it easier to find willing experts, and the three he chose became his "MIT Trio." Jeffrey Ingram, Alan Carlin, and Laurence Moss, all graduates of the Massachusetts Institute of Technology, arrived on the scene in different ways.

Carlin wrote and volunteered his services. He was an economist at the RAND Corporation with a specialty in understanding cost–benefit analysis. The Bureau of Reclamation had been using similar analysis for years, and Carlin was adept at finding the weaknesses in the material and exploiting them. He told Brower that RAND had published a book seven years earlier showing that the type of cost–benefit analysis used by the Bureau of Reclamation was theoretically invalid.[39]

Moss was a nuclear engineer who could argue with some authority that nuclear energy was at least comparable to the cost of generating electricity from hydropower. Nuclear energy did not yet have the opposition that would develop in the 1970s, but it also had limited support because it was such an untried technology. By 1965, the nuclear energy industry was beginning to ramp up, and utilities had ordered up to 12,315 megawatts of nuclear-generated power, which was enough electricity to roughly supply about 12 million residential customers.[40]

Ingram was a mathematician who was to become Brower's deputy in the Grand Canyon campaign. He first met Brower at a conservation conference in Santa Fe, New Mexico, where he asked a question that impressed Brower so much that he invited the young man to lunch. It would be another year before Brower would offer the post of southwestern conservation representative to Ingram, a step that transformed Ingram's life. It was a hectic job, whether he was working on his own in the Southwest or more often by Brower's side in meetings in Washington, D.C., and elsewhere. Several key meetings in Washington in early 1966 stood out for Ingram. Although the sessions featured staff members from the major environmental organizations that had fought Echo Park, he left them unimpressed. He explained later: "These were all longtime Washington insiders, and they were not going to win the fight; we were. It was up to us young—I hate to say it—young middle-class revolutionaries to redefine the issue and defeat the dams."[41]

Brower assigned Ingram to work with Litton to draft legislation to create a new, greater Grand Canyon National Park that would include all of the gorge's geologic features and ban any dams. Aspinall was planning new hearings in the spring of 1966, which Udall hoped would lead to passage of a bill. Aspinall indicated he would welcome only new testimony, new information. Brower worked with the MIT Trio to fit Aspinall's requirements, saving for himself the emotional arguments about the canyon.[42]

Then Brower got his first break. Richard Bradley had been unable to interest the *Atlantic Monthly* in a Grand Canyon article, but the editors of *Audubon* stepped in to fill that gap. The article "Ruin for the Grand Canyon?" was blunt, opinionated, and one-sided. "Surely the best-educated nation on earth can devise a better way to finance the $500-million water-diversion network than by building $750 million worth of unnecessary dams," Bradley wrote. "And if a law must be changed to do it, then let the law be changed. Better to amend an outworn concept than to ruin a two-billion-year-old geological and scenic wonder."[43] Then the *Reader's Digest* offered to reprint this article in its April issue. No other magazine at the time exceeded the *Reader's Digest* in sales. It also had an international reach, circulating in more than forty editions, thirteen languages, and Braille.[44] Its publisher, DeWitt Wallace, took a personal interest in the Grand Canyon dam issue and proposed sponsoring a conference at the Grand Canyon to commemorate Bradley's article.

Udall, Dominy, and others tried to block the *Reader's Digest* article and then the conference. Dominy pointed out that the very title "Ruin for the Grand Canyon?" was biased. He also presented a long list of factual errors, but Bradley responded with his own set of facts that satisfied the editors. He told Brower that Reclamation Bureau officials got enmeshed in their own "prevarications" and were helpless in escaping their predicament.[45]

Brower knew the conference could be important. He told a friend that this kind of heavy-duty support never happened to his side; the big money always went to the guys on the other side of the dispute.[46] He was right. Opponents discovered that it would cost $250,000 to mount a slick public-relations campaign to counteract the *Reader's Digest* article and book.[47] The conference would invite reporters to hear the case against the dams, and Brower and his staff struggled to come up with an agenda. They proposed that it begin late Saturday afternoon at the elegant El Tovar Hotel on the canyon's South Rim. There would be time for views of the canyon at sunset. On Sunday, Litton would ferry reporters in his small Cessna 195. Litton had learned to fly in the army, piloting small glide planes. As a pilot, he was fearless. He would fly passengers to Toroweap, one of the most spectacular outlooks on the North Rim with views directly down to Lava Falls.[48] Then the plans went awry.

The *Reader's Digest* had hired the J. Walter Thompson agency to handle public relations, and their advance man, Larry Holder, got his instructions mixed up and invited prominent dam supporters to the conference. Morris Udall was soon on a plane from New York with eastern reporters who would present his side and the governor of Arizona. Stewart Udall publicly announced they were coming and that they wanted to be included among the speakers. Brower expanded the schedule on the understanding that none of these invited guests would be on the agenda.[49] Much of the Arizona press had supported the dam projects, and news stories in the days before the conference in Arizona began to raise alarms about the conservationists' plans. Ben Avery, a columnist for the *Arizona Republic*, said he was suspicious about what was planned, and another Arizona newspaper said local water officials were anticipating taking a "low blow" from conservationists.[50]

Brower did not understand how much he had lost control of the conference until he came down the first morning for breakfast. He found Dominy's huge scale model of the Grand Canyon sitting in the middle of the dining room where the reporters were going to sit. Brower got so upset that Ingram thought his boss was going to throw the entire model out the glass windows that overlook the canyon. A low-level Reclamation employee warned Brower that the model was government property and he would be liable for its loss. Brower eventually calmed down, and the model stayed.[51]

Hilda Burns, another J. Thompson employee, under pressure agreed to dramatically alter the agenda so that dam supporters could have an equal say in the debate. Brower was disgusted. He called it a "wild west debate."[52] According to some reports, Brower and Morris Udall faced off at one point, and Udall got the better of the exchange. Brower began crying at the podium, complained George Steck, a Sierra Club member. Udall said neither reservoir could be seen from anywhere on the rim, and a number of antidam advocates corrected him, crying, "Toroweap!" Udall conceded they were right. Another pro-dam speaker asked how many people would really prefer to see the Glen Canyon Dam dismantled. The generally antidam audience roared yes in response.[53]

Senator Barry Goldwater decided at the last moment to come Sunday morning, but his arrival was delayed, and then Brower was forced into another debate with him. The flights to Toroweap were late, and some reporters missed them, so the press coverage was compromised. From a journalistic standpoint, the conference had improved significantly from its original one-sided presentation to a lively debate between key figures, but Brower left convinced that Dominy and his forces had succeeded in "torpedoing" the conference.[54]

He was wrong.

Reporters had finally discovered the story, and major in-depth stories, with stunning photos of the Grand Canyon, were soon coming out of the *Denver Post*, *Rocky Mountain News*, *Desert News*, *Arizona Republic*, *Kansas City Star*, *Christian Science Monitor*, *National Observer*, *Daily News*, and *New York Times*. Even the shorter story featuring Goldwater had its positive moments for the conservationists when Goldwater said he could support a dam at Bridge Canyon but did not want to see Marble Canyon destroyed. Charles Callison of the National Audubon Society told Brower that no matter what had happened indoors in the El Tovar, the key was what happened later outside. It was the flights over the canyon and Martin Litton's lectures on the rim that made the difference, he said. They did more than all that was said at the El Tovar to put the issue into perspective for the visiting journalists and to emphasize the unique and irreplaceable nature of what would be destroyed.[55]

The Sierra Club bill to expand the Grand Canyon National Park was introduced in Congress on April 4, 1966, although its prospects were poor. Brower met with his MIT team in San Francisco in early May to prepare for the new round of House hearings scheduled to begin May 9. They flew first class to Washington and waited and waited to be called.

Aspinall preferred to have the conservationists go last. Brower had high hopes for the testimony that his experts had prepared, but if he was looking for the kind of knockout punch he had delivered in the Dinosaur fight, he was disappointed. The hearings turned nasty when a member of the Subcommittee on Irrigation and Reclamation, Craig Hosmer of California, tried to trip up Carlin, the RAND expert. Carlin had a slight speech impediment, and Hosmer, a dam supporter, began to interrupt and disrupt Carlin's testimony. Finally, John Saylor, one of Brower's few allies on the committee, stepped in and apologized to Carlin "for the abuse that you have taken."[56]

At the hearings, Brower did draw attention to complaints that he had been making for months that Udall and Dominy had "muzzled" the Park Service staff at Grand Canyon. As evidence, Brower had a memo from George B. Hartzog Jr., who had succeeded Conrad Wirth as Park Service director, prohibiting employees from distributing Sierra Club information that opposed the dams. He had a second memo from the acting regional director indicating that Hartzog's memo should be destroyed after it was read.[57] Park rangers had privately been telling Brower and others in the club that the intimidation was quite high. George Stricklin, Grand Canyon park superintendent, took Brower aside during the *Reader's Digest* conference and assured him that *Time and the River Flowing*, although not on display in the Grand Canyon book store, was available for sale from a back room to customers who asked for it. Udall told Brower that no employee was prohibited from expressing an opinion, but while they were on the job, they needed to abide by Interior Department policy.[58]

Meanwhile, both at the hearings and afterward, dam supporters attempted to portray Brower as a racist. Months before the hearings, Aspinall had written Udall urging him to "find an Indian" who would counter the conservationists. The Hualapai tribe demanded several concessions if it were to support the dams, including renaming Bridge Canyon Dam the "Hualapai Dam." Dam supporters agreed, and George Rocha, chief of the Hualapai tribe, testified at the hearings that the dams would help pull his people out of poverty. Henry Dobyns, a prominent anthropologist from Cornell University, supported that view at the hearings. He urged his "fellow liberals" to "see the wisdom of the conservation of Indians as well as rocks and ducks."[59] Several Arizona newspapers reported that environmentalists had tried to recruit Dobyns to testify on their behalf and that he not only refused but also accused the conservationists of racial discrimination for stifling the tribe's economic prospects.[60]

Brower and the Sierra Club had their own expert, Stephen Jett, a geography professor at the University of California at Davis, and an Indian tribe, the Navajos, who opposed the dams. Jett, speaking after Dobyns, noted that the Navajos numbered more than 100,000 compared with the much smaller Hualapai tribe. The Navajos were upset because they had never been consulted about the dam project. They were so irked that the tribal government eventually opposed the project, effectively stymieing the campaign to brand Brower a racist.[61]

In his appearance at the hearings, Brower stressed that the choice was not between putting dams in the Grand Canyon or denying Arizona the water it needed in the future. "We do not want people to go without water," Brower said. "We are concerned that the Grand Canyon, the most important, perhaps, of all of our national parks, should be unimpaired. We believe that it can be."[62]

Was the tide finally changing? At the close of the hearings in May 1966, *Newsweek* reported that authorization for the dams seemed to "face an uphill struggle." Or, as the magazine quoted a congressional insider, "The conservationists will fight this one hard, and you know that the conservationists don't lose many fights around here."[63]

Brower, who for years had tended more to editing books than to running conservation campaigns, seemed galvanized by the fight. In April 1966, he sent out a special directive to his Grand Canyon Task Force warning them that the *Reader's Digest* was being inundated with letters from dam supporters. *Digest* editors were upset, reported Brower, and it was time to hear from the good guys instead of the bad guys. In May, he published a special Grand Canyon edition of the *Sierra Club Bulletin*. In June, he sent Ingram and others to Washington to lobby individual congressional members. Ingram believed the results were mixed, although he did come away with one priceless anecdote. Morris Udall had also been lobbying his colleagues in the House, and one colleague passed on a comment he supposedly made to Udall: "You want me to build dams in the Grand Canyon, the Speaker wants me to vote to expand the Capitol. Sorry, Mo. I can only vote for one desecration at a time."[64]

Brower needed more ammunition, and he was working to get it. His next tactic would be so brilliant it would not only seriously weaken Udall's dam project but also go far in defining Brower and his role in conservation history.

13. Losing While Winning

It took a redwood to kill the dams in the Grand Canyon. Or, to put it more precisely, it took a forest of redwood trees to do it.

The campaign that had begun more than a decade earlier when the winter storms struck California's Bull Creek was still running at a high level in the mid-1960s. By December 1965, the conservationists who wanted a new Redwood National Park were fighting not only the loggers but each other as well. The Sierra Club proposed a 90,000-acre park in the Redwood Creek area of northern California. A second conservation organization, Save the Redwoods League, backed a park plan in the Mill Creek area, which was smaller and thus politically more palatable because of the timber industry opposition. Stewart Udall was torn between the two competing proposals and was leaning toward the league's more compact park. Another powerful backer of Mill Creek was Laurence Rockefeller, whose family had already spent millions to protect many of the small redwood groves.[1]

It was time to do something dramatic. David Brower had always been impressed with the advertisement published in the *Denver Post* on October 31, 1955, that had caused supporters of the Dinosaur dam projects to capitulate. The ad, directed at the dam supporters who were meeting in Denver on November 1, warned that conservationists were prepared for a campaign that could go on for years. So Brower arranged for running a full-page ad in support of the Redwood Creek Park on the day Udall was to meet with park backers.[2] It ran in five newspapers and was titled "An Open Letter to President Johnson on the Last Chance *Really* to Save the Redwoods." It featured photos of old-growth trees and a clear-cut stand and warned that the virgin forest of Redwood Creek would be felled within two years unless Udall acted.[3]

Udall held up the ad at the meeting with conservationists. He was so angry that he was red in the face. Later, after he calmed down, he pleaded with Brower to be flexible because the cost to acquire the land for the park would be enormous. In March 1966, the Johnson administration introduced a bill in Congress that sided with the smaller proposal advanced by the Save the Redwoods League.[4]

Despite that setback, Brower was intrigued by other reactions to the advertisement. In the three months after the redwood ad, the club received $7,481 in donations, 1,845 responses, and 772 applications for Sierra Club membership. He heard that President Johnson had received more than 3,000 responses. The ad was produced by a San Francisco advertising agency run by Howard Gossage and Jerry Mander. Brower would learn from both of them about how to write an effective advertisement for the environmental movement. Mander taught him that the ad must start with a compelling headline, followed by text that could never afford to have a dull sentence. Gossage told him about the "water cooler" principle: the most important aspect of any ad was that it had to be so dynamic it forced people to talk about it.[5]

Brower began planning a Grand Canyon ad. Despite advice from the advertising professionals, he argued with them on the right approach. Brower and Jeff Ingram eventually wrote one version, Mander wrote a second, and the *New York Times* did a split run so that both ads could be published. It was easy to track which one was more effective. Brower looked at the results and admitted defeat. "The upshot of it all was that Jerry Mander's ad outpolled mine two-to-one," he said later. "So I conceded that the professional knows what he is doing, and the amateur doesn't."[6]

The ad ran June 9, 1966, and pro-dam supporters were furious. Morris Udall on the House floor condemned the "inflammatory attacks." He said: "I must say that I have seldom, if ever, seen a more distorted or flagrant hatchet job than this." Udall was angry that the ads did not say both dams were outside the national park. He also felt that it was grossly unfair for the ads to say that the Grand Canyon would be flooded. The reservoir depth would be no more than 500 feet. One ad made that clear, the second did not give a measurement, but, as Brower pointed out, it clearly did not say that the canyon was going to be filled rim to rim.[7]

At 4:00 P.M. the day after the ad was published, a federal marshal appeared at the Sierra Club's San Francisco offices. He delivered a notice from the Internal Revenue Service (IRS) stating that because of the newspaper ads and other political activity, contributions to the Sierra Club would no longer be tax deductible pending an IRS investigation.[8] It took hours for the news to reach Brower, who was in New York, but he knew almost immediately what to do. He announced that the federal government was trying to censor the Sierra Club, and the story literally exploded.

As Brower remembered the situation, "We got editorials; we got headlines all over the country. Why were we losing our tax status? Because we were trying to save the Grand Canyon in the public behalf. The papers over the country were irate, and the public was irate on this thing."[9]

The maneuver also raised the specter that the government was trying to squelch its opposition by abusing its tax powers. The *Fort Lauderdale News* editorial was headlined "Free Speech Apparently Does Not Apply If Objecting to Action of Great Society," and the *Wichita Eagle's* was headlined "The Smell of Retribution." The *Albuquerque Tribune* called the IRS statement a veiled threat to stop the dam opposition, and *Sports Illustrated* complained that the Sierra Club "is being slowly throttled by the Internal Revenue Service." An editorial cartoon that ran in many papers showed a Sierra Club protester at the end of a Grand Canyon cliff, with an IRS agent taking a photo and saying, "Back, just a little further . . . back." The *New York Times* thundered that "under the guise of strict tax regulation [the IRS] is making an assault on the right of private citizens to protest effectively against wrongheaded public policy."[10]

Dam supporters and the IRS were immediately on the defensive. Sheldon Cohen, IRS commissioner, said he was personally active in conservative causes, but in this case the Sierra Club had gone too far. "We simply had to find out whether the organization is devoting a substantial part of its activities to carrying on propaganda or otherwise attempting to influence legislation," Cohen told the *New York Times*.[11]

Speculation about who ordered the investigation pointed to figures ranging from Wayne Aspinall, Stewart Udall, Morris Udall, and Treasury Secretary Henry Fowler to the president. According to a report that Brower had picked up, when Johnson was asked what to do, he reportedly replied, "Enforce the law, of course." Brower claimed that Morris Udall eventually admitted that he had called the IRS and that he had met an IRS assistant commissioner in the bar at the Congressional Hotel, where he demanded, "How the hell can the Sierra Club get away with this?"[12]

Morris Udall, Cohen, and Treasury Undersecretary Joseph Barr offered another story. Barr told the *Arizona Republic* that he received a phone call from Cohen on the day of the ad. Cohen told Barr that it appeared that the Sierra Club had violated the federal ban restricting nonprofit organizations from lobbying. Barr said he was surprised by the call because the Treasury Department normally did not discuss such cases. Cohen, in several accounts, always maintained that the IRS began the inquiry on its own. Later in the day on June 10, Barr said he encountered Morris Udall at the Democrat Club on Capitol Hill. "The first thing Mo asked me was about the Sierra Club tax deductible status on account of those advertisements," Barr said. "I told him, 'Boy, we're way ahead of you.'" Barr surmised that someone overheard a portion of the conversation and got the impression that Udall was ordering him to have Cohen begin the case. Udall denied that he ever began the investigation, and he was seconded by his brother, Stewart. Morris Udall did write to Cohen on June 10 asking whether contributions to the Sierra Club that were being solicited in the ad were tax deductible. The letter, however, did not ask about investigating the club's tax status.[13]

Brower could never prove that Morris Udall had confessed, although Edgar Wayburn also claimed to have heard the admission. According to Brower, "He [Morris Udall] did tell me that he thought it was the greatest mistake he had made in the Grand Canyon battle because the threat to the Sierra Club was headline news all over the country. People who didn't know whether or not they loved the Grand Canyon knew whether or not they loved the IRS."[14]

After the first wave of response from the press, the second arrived in the form of letters from the public attacking the government. The letters came to congressional offices, the White House, the Interior Department, and especially the IRS. Stewart Udall estimated that Interior received at least twenty thousand letters; at Congress, Morris Udall called it a "deluge," and one official contended that the mail came in dump trucks. California senator Thomas Kuchel called it "one of the largest letter writing campaigns I have ever seen." The letter writers were upset about the threat to the Grand Canyon, and they were worried about free speech and due process. Many also wrote the Sierra Club asking to become members, and the membership rolls grew. In August alone, an unprecedented eight hundred joined. Also by August, approximately $12,000 in donations had arrived, more than paying for the $10,500 cost of the ads.[15]

Brower worked to keep the controversy going, running additional advertisements about the Grand Canyon and the redwoods. The most famous ad during this campaign was titled "SHOULD WE ALSO FLOOD THE SISTINE CHAPEL SO TOURISTS CAN GET NEARER THE CEILING?" (figure 14). The question was composed in response to dam supporters who said it would be easier to appreciate the canyon walls on a power boat atop a Grand Canyon reservoir. Brower said the ads were a surprising political stimulus. They generated an emotional appeal that prompted private citizens to write letters, come to hearings, and do whatever they could.[16]

No one should have been surprised by the IRS action. The Sierra Club had been worried about its tax-exempt status for years. By 1954, the IRS had already examined the Sierra Club's books four times to ensure that the club was meeting its requirements as a tax-deductible organization. That year, a Supreme Court ruling further complicated the tax situation. The court redefined lobbying and in the process sharply limited a nonprofit's ability to be politically active.[17] The result for activist groups such as the Sierra Club was confusion. As one tax lawyer wrote Brower, the tobacco industry could lobby, but health organizations concerned about smoking could not; real-estate boards could oppose schools, but education

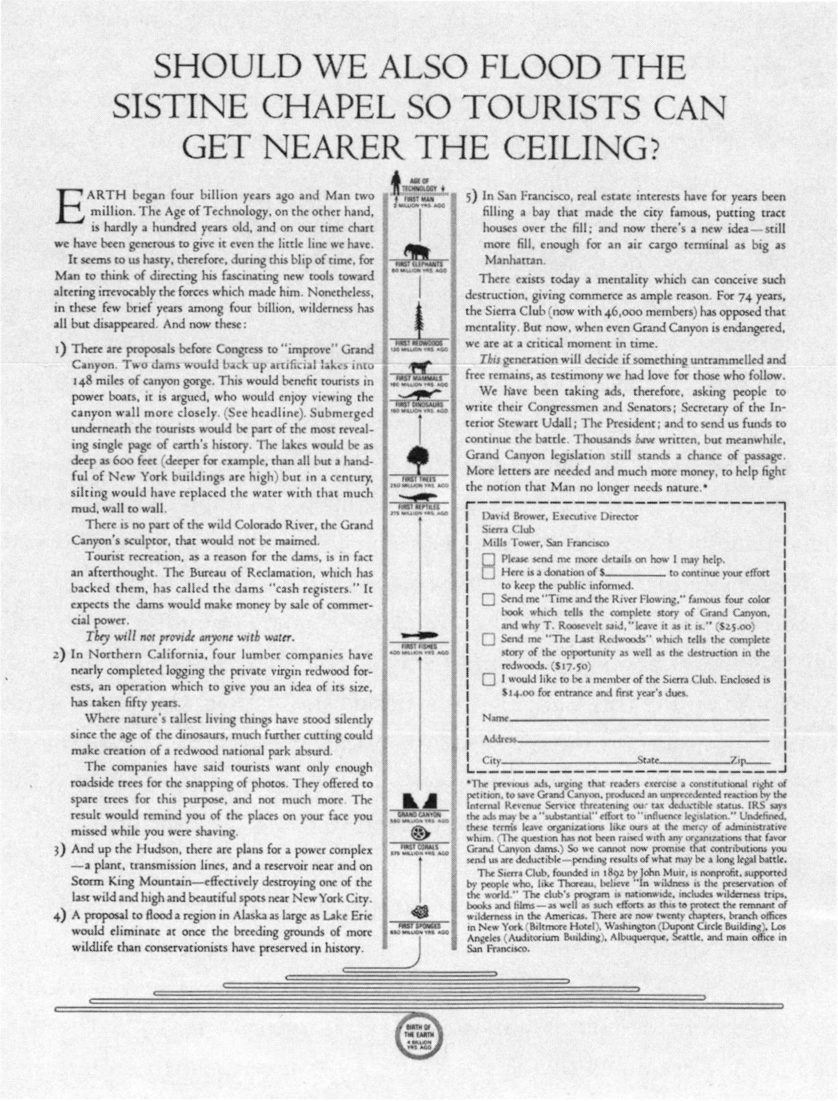

SHOULD WE ALSO FLOOD THE SISTINE CHAPEL SO TOURISTS CAN GET NEARER THE CEILING?

EARTH began four billion years ago and Man two million. The Age of Technology, on the other hand, is hardly a hundred years old, and on our time chart we have been generous to give it even the little line we have.

It seems to us hasty, therefore, during this blip of time, for Man to think of directing his fascinating new tools toward altering irrevocably the forces which made him. Nonetheless, in these few brief years among four billion, wilderness has all but disappeared. And now these:

1) There are proposals before Congress to "improve" Grand Canyon. Two dams would back up artificial lakes into 148 miles of canyon gorge. This would benefit tourists in power boats, it is argued, who would enjoy viewing the canyon wall more closely. (See headline). Submerged underneath the tourists would be part of the most revealing single page of earth's history. The lakes would be as deep as 600 feet (deeper for example, than all but a handful of New York buildings are high) but in a century, silting would have replaced the water with that much mud, wall to wall.

There is no part of the wild Colorado River, the Grand Canyon's sculptor, that would not be maimed.

Tourist recreation, as a reason for the dams, is in fact an afterthought. The Bureau of Reclamation, which has backed them, has called the dams "cash registers." It expects the dams would make money by sale of commercial power.

They will not provide anyone with water.

2) In Northern California, four lumber companies have nearly completed logging the private virgin redwood forests, an operation which to give you an idea of its size, has taken fifty years.

Where nature's tallest living things have stood silently since the age of the dinosaurs, much further cutting could make creation of a redwood national park absurd.

The companies have said tourists want only enough roadside trees for the snapping of photos. They offered to spare trees for this purpose, and not much more. The result would remind you of the places on your face you missed while you were shaving.

3) And up the Hudson, there are plans for a power complex —a plant, transmission lines, and a reservoir near and on Storm King Mountain—effectively destroying one of the last wild and high and beautiful spots near New York City.

4) A proposal to flood a region in Alaska as large as Lake Erie would eliminate at once the breeding grounds of more wildlife than conservationists have preserved in history.

5) In San Francisco, real estate interests have for years been filling a bay that made the city famous, putting tract houses over the fill; and now there's a new idea—still more fill, enough for an air cargo terminal as big as Manhattan.

There exists today a mentality which can conceive such destruction, giving commerce as ample reason. For 74 years, the Sierra Club (now with 46,000 members) has opposed that mentality. But now, when even Grand Canyon is endangered, we are at a critical moment in time.

This generation will decide if something untrammelled and free remains, as testimony we had love for those who follow.

We have been taking ads, therefore, asking people to write their Congressmen and Senators; Secretary of the Interior Stewart Udall; The President; and to send us funds to continue the battle. Thousands have written, but meanwhile, Grand Canyon legislation still stands a chance of passage. More letters are needed and much more money, to help fight the notion that Man no longer needs nature.*

David Brower, Executive Director
Sierra Club
Mills Tower, San Francisco

☐ Please send me more details on how I may help.
☐ Here is a donation of $_____ to continue your effort to keep the public informed.
☐ Send me "Time and the River Flowing," famous four color book which tells the complete story of Grand Canyon, and why T. Roosevelt said, "leave it as it is." ($25.00)
☐ Send me "The Last Redwoods" which tells the complete story of the opportunity as well as the destruction in the redwoods. ($17.50)
☐ I would like to be a member of the Sierra Club. Enclosed is $14.00 for entrance and first year's dues.

Name_____

Address_____

City_____ State_____ Zip_____

*The previous ads, urging that readers exercise a constitutional right of petition, to save Grand Canyon, produced an unprecedented reaction by the Internal Revenue Service threatening our tax deductible status. IRS says the ads may be a "substantial" effort to "influence legislation." Undefined, these terms leave organizations like ours at the mercy of administrative whim. (The question has not been raised with any organizations that favor Grand Canyon dams.) So we cannot now promise that contributions you send us are deductible—pending results of what may be a long legal battle.

The Sierra Club, founded in 1892 by John Muir, is nonprofit, supported by people who, like Thoreau, believe "In wildness is the preservation of the world." The club's program is nationwide, includes wilderness trips, books and films—as well as such efforts as this to protect the remnant of wilderness in the Americas. There are now twenty chapters, branch offices in New York (Biltmore Hotel), Washington (Dupont Circle Building), Los Angeles (Auditorium Building), Albuquerque, Seattle, and main office in San Francisco.

FIGURE 14 The Sistine Chapel Grand Canyon advertisement

Brower and the Sierra Club ran this advertisement in six publications (*Wall Street Journal, San Francisco Chronicle, Saturday Review, National Review, Harper's,* and *Ramparts*) between August 23 and October 1966 after dam supporters said a reservoir would make it easier for visitors to get a close-up view of the Grand Canyon by boat. The ad was one in a series that created political repercussions from the Internal Revenue Service. (Courtesy of the Bancroft Library, University of California, Berkeley) foundations could not support them; and lumber companies could fight in Congress to cut more trees, but conservation organizations were prohibited from making the counterargument.[18] If a group such as the Sierra Club did lobby—as it had been for years—it stood a good chance of losing its tax status.

After the ruling in 1954, the Sierra Club made two attempts to shield itself from the IRS. As the Dinosaur fight was escalating, it created an umbrella organization called the Trustees for Conservation. The organization was to be the Sierra Club's political lobbyist, actively engaging in what Brower technically was prohibited from doing.[19] But it was a cumbersome arrangement that never really worked for Brower. In 1960, he convinced the club to create the Sierra Club Foundation. It would not be a lobbying organization but would be created solely to raise contributions that could be tax deductible and that could be passed on to the more politically active Sierra Club. The problem with this arrangement was that in its first years the foundation had difficulty getting out its message, so fund-raising lagged. It also put the parent Sierra Club in greater jeopardy because Brower and the board were not giving up on their most important campaigns.[20]

Even among accountants, lawyers, and government officials, no one seemed to quite understand how the tax and lobbying laws could legally intersect. In 1965, Brower sent the club's Washington lobbyist, Bill Zimmerman, to a two-day conference on the issue. Zimmerman reported back that even Treasury and IRS officials disagreed on major points of the law. No one seemed sure of anything.[21]

Confusion continued after the IRS investigation began. Mike McCloskey met a local IRS agent during a hike in late June 1966. The agent told McCloskey that the local office was embarrassed by the way the case had been forced on them. In a meeting in June, Brower was given to understand that the review would take only a few weeks, but by September Cohen at the IRS told him to prepare for a far longer review period. Brower pointed out this would be difficult for contributors because if the review dragged, on no one would know if their contributions to the club were tax deductible. Some people in a situation like this might act cautiously at this point. Not Brower. He again publicly attacked tax officials, complaining that the club stood to lose up to $100,000 a year in donations (it was also still receiving large donations from its continuing advertisements). By the fall, the investigation had broadened, and IRS agents wanted to see everything—financial ledgers, payroll accounts, column inches of newspaper coverage, and film footage.[22] It would be December before the IRS would conclude its investigation.

Meanwhile, Stewart Udall was trying to recoup his losses. He announced in late June that he was willing to eliminate the downstream Hualapai Dam, formerly named Bridge Canyon Dam, in return for support of the upstream Marble Canyon Dam. Brower promptly rejected the offer. "Doing only half of the destruction to the Grand Canyon is hardly a solution," he responded.[23]

And the losses continued. Morris Udall learned that Walter Reuther, who headed the United Auto Workers, was coming out against the dams. If organized labor, which normally supported job-producing legislation, was against the project, it was clearly in danger. Udall wrote to Reuther and met with John Oakes of the *New York Times* and with the editors of the *Washington Post* to try to control the damage. In Arizona, the *Tucson Daily Citizen* distributed five hundred copies of its editorial "Snowed by the Sierra Club" to members of the American Society of Newspaper Editors. The *Arizona Republic* and *Arizona Daily Star* distributed coupons permitting readers to urge Congress to support the dams. They claimed that seven thousand of these coupons were received in Congress. The Southwest Progress Committee produced a film, *Grand Canyon, the Ever Changing Giant*, which it attempted to air on local television around the country.[24]

In Washington, D.C., and on the road, speakers attacked Brower and the Sierra Club in the harshest terms. "The Sierra Club has transgressed far beyond the limits of propriety in lobbying against the Marble and Bridge [Hualapai] Canyon dams, and they should be made to pay for such transgressions," said Representative Wayne Aspinall. "It is a nasty, indecent, and ignorant attack." Senator Frank Moss of Utah declared that "extremists" had "twisted" the truth by suggesting the reservoirs would fill the Grand Canyon. Senator Barry Goldwater told reporters at the National Press Club that the "Big Lie" was being repeated so often, it might be declared the truth. In Los Angeles, Floyd Dominy said, "These people, carried away by their single purpose, have failed to recognize the great gray area between total preservation and total development, neither of which is contemplated."[25]

None of this effort did any good. By the end of August 1966, the bill for all practical purposes was dead for the year. Moreover, political support for the two dams was becoming increasingly tenuous. Senator Clinton Anderson, a New Mexico Democrat whose support was critical for passage of the legislation, told Morris Udall in a meeting on August 31 that there would be no action on the bill that year. According to one report, Anderson was acting even though "he hates Brower and the Sierra Club."[26]

What had happened? As late as May 1966, Morris Udall and others believed that they had the votes to get the bill with both dams passed through Congress and onto the president's desk. There are two theories on why the legislation failed. Some, a minority, say it was simply that the politics had changed for a variety of reasons. Senator Henry Jackson and other Pacific Northwest politicians remained wary of Stewart Udall's regional water plan and continued to fear that it would ultimately lead to a raid on the Columbia River. Jeff Ingram, Brower's deputy in the Grand Canyon campaign, spent long hours in the offices of Pacific Northwest congressional representatives working against the project.[27] Tied to this theory are two unlikely sources that have been credited with decisive roles. One was Northcutt "Mike" Ely, a lawyer who for many years represented California water interests and was a power broker on any Colorado River deal. According to some accounts, it was Ely who blocked key votes from the California delegation in the House, effectively killing the legislation on the House side even before Anderson scuttled it in the Senate. Byron Pearson, a prominent historian on the Grand Canyon, said Ely's stopping of the legislation temporarily averted a court case from being implemented that would have allocated water to Arizona at the expense of California.[28] Another unlikely ally was Sharon Francis, an aide to Lady Bird Johnson. Brower always believed that he had a friend in both Francis and the First Lady, and dam supporters equally felt that they had an enemy in Francis. "Sharon Francis is about as 100 percent a preservationist as they come," Orren Beaty told his boss Stewart Udall in July 1966, "and I deeply regret her influence in Mrs. Johnson's office." Brower went further in describing Francis's role: "Sharon Francis convinced Lady Bird that these dams were no good," he said. "Lady Bird convinced Lyndon."[29]

A second theory with stronger credence credits the environmentalists' campaign and, more specifically, Brower and the newspaper advertisement campaign regarding the Grand Canyon dams. The response from the IRS and the reaction from the public reached the point that the two dams became politically untenable. The movement "came of age" at that moment. It was a sea change, states writer Marc Reisner, prompted by Brower's ability to get the public's attention on the issue of wild rivers as opposed to dams. Although dam supporters railed over what they called Brower's lies, the issue was no longer how much of the canyon was going to be flooded. The public, argues Reisner, "wanted no dams, period."[30]

Even Aspinall grudgingly credited Brower in a speech in the fall of 1966 about what had gone wrong. He said that the "misleading" and "massive campaign, aided and abetted by such influential publications as *Life*, the *New York Times*, and the *Reader's Digest*, has been effective."[31]

Brower was willing to take the credit, although he could be modest when asked about it. In June 1977, he spent two weeks rafting down the Grand Canyon with a tour company arranged by Ron Hayes, a river outfitter and one-time radio and television personality and actor. The company was in Marble Canyon when Hayes introduced Brower and said that without Brower they would not be going down this river. "You've got to spread the credit pretty wide," Brower responded. "I was there and I worked hard and I can tell you that as I fly over and look at Echo Park and I realize that I had a role there and as I come down here and realize that I had a role in no dams here, I feel pretty good about it. But I also feel a great debt to the people who made this possible. You can be a quarterback and call a couple of signals, but you had better have ten other people on the team, and we had thousands."[32]

As prospects for constructing dams in the Grand Canyon eroded, hard feelings grew. In November 1966, both Brower and Aspinall were invited to the annual meeting of the National Reclamation Association and crossed paths at an Albuquerque hotel between sessions. Newspaper photographers, seeing them near each other, asked if they would pose. Aspinall, clearly angry at the thought of standing beside Brower, refused. "You've been telling a bunch of damn lies to the newspapers, and now you want your picture taken with me," exclaimed Aspinall. He pointed a finger at Brower and accused him of being "an utterly unreasonable man who would never compromise on anything."[33]

Dominy struggled to concede that Brower had bested him. One former Bureau of Reclamation employee said, "If you even suggested to Dominy that Brower was winning, he would have fired you on the spot." During the fight over the dams, a frustrated Dominy ordered aides to begin tracking Brower by showing up at his speaking engagements, heckling him when they could, and reporting back his remarks. The reports were not always helpful. One, about a talk in Denver, indicated that Brower's speech had been highly emotional. "It was completely lacking in any kind of substantiating data, and he appeared a far less formidable opponent than anticipated," the report stated. More neutral news reports said that Brower received a standing ovation after saying he would not oppose dams being built in the Grand Canyon—as long as the bureau built another Grand Canyon somewhere else.[34]

Stewart Udall was stung by Brower's attacks but brushed them off and never publicly complained. His brother remained angry with the Sierra Club and with Brower in particular for a very long time. "I want to continue to work with the Sierra Club," Mo Udall said, "but I frankly think that Mr. Brower's effectiveness in the cause of conservation is nearing an end." Such hard feelings bothered Dick Leonard, who in the past had always defended Brower, no matter how fierce the accusations. Leonard believed that Brower had gone too far, that he had insulted not only Stewart Udall but the president. Brower's problem was that others in the Sierra Club agreed with this assessment.[35]

In the throes of defeat, Stewart Udall now had to grapple with how to save his grandiose regional water plan and especially how to get the water to Arizona that he believed his home state deserved. The solution, as hard as it was for Udall, was to eliminate the dams. In August, he directed a deputy to begin looking into alternatives. The answer he received was that to produce the money the Central Arizona Project needed, a coal-fired electric generating plant was preferable to a nuclear facility. The siting for a conventional plant would move swiftly, but the regulatory hurdles for nuclear would take at least ten years. In October 1966, Udall publicly announced he was considering alternatives to the dams; by December, he told Johnson of a new plan; and on February 1, 1967, he made his decision official.[36]

Even though a coal-fired plant could and eventually would create significant air pollution in the Grand Canyon, Brower at this time supported the plan. Both Dominy and Aspinall argued strongly against the coal plants, with Dominy pointing out that dams did not pollute the way a coal plant would.[37] During hearings in May 1967, Morris Udall forced Brower to look at photos of strip-mining operations. He asked Brower if that kind of damage was preferable to a dam. In that exchange, Brower attempted to avoid making a choice by contending that the power plants were not necessary. "We are not advocating the alternative steam plants," Brower said. "It is an attempt—and I don't think it is a bad one—to find some way to get the Bureau of Reclamation off the hydroelectric horse, which is becoming rather spavined these days. I think this is a good thing. We don't like strip mining any better than you do."[38]

The problem was that Brower and his MIT Trio had in the past advocated such alternatives. As Brower was to concede in 1977 when he visited the Grand Canyon, he had supported nuclear power for years until he began to waiver in the 1960s. When it came to the Grand Canyon, he said he "was arguing primarily for the coal alternative. We had a number then, there were 800 billion tons of coal in the Colorado basin states. And it was an idle resource and we should use it instead of the hydro." Even when Morris Udall showed him the strip-mining photos, his "thought was, 'Well that is not very important country compared to Grand Canyon.'"[39]

The Navajo Generating Station would be built 16 miles from the Grand Canyon by a consortium of local utilities and the Reclamation Bureau. Pollution from the station would billow into the sky, and on some days visitors standing on one rim had difficulty seeing across to the other rim. It got so bad that environmentalists sued and forced the federal government in 1991 to reduce the air pollution by a projected 70 percent. Visibility improved. However, with the plant consuming 24,000 tons of coal each day, it remained a significant contributor to the growing concerns about the global warming of the planet in the early twenty-first century.[40]

Politicians decided the fate of the Grand Canyon; the IRS would determine the fate of the Sierra Club's tax status. The review was exhaustive, although for a while Brower and others believed they might be able to overcome the IRS's objections. In the past, the IRS had expressed concerns that lobbying could only be a small percentage of a nonprofit organization's budget. The decision on the Sierra Club arrived December 16, 1966. District IRS director Joseph Cullen described in detail in a twenty-three-page letter the club's "substantial legislative activity" over the past thirty months. Cullen did not calculate how much of the Sierra Club's activities were political. He instead gave an exhaustive summary ranging from the decision to hire a Washington lobbyist to its work on behalf of legislation and its many public attempts to sway public opinion. The club, he concluded, could not keep its tax-exempt status.[41]

At a press conference, Brower declared that the Sierra Club would contest the IRS decision. "Substantial legislative activity is a meaningless charge unless the IRS defines these terms," he said. "It adamantly refuses to do this in its attack on the club for defending Grand Canyon. If, as seems apparent, the service is determined to class as legislation anything that leads eventually to legislation, then the service has contrived a big enough umbrella to cover almost any human activity that seeks law and order—and is now holding the umbrella over whomever it chooses."[42]

Despite the strong language, an internal disagreement was percolating inside the Sierra Club leadership on how to fashion the IRS appeal. Phil Berry, the young Brower acolyte, was now a lawyer and becoming increasingly active in club affairs. He favored a quick administrative appeal that would seek to get the IRS headquarters in Washington, D.C., to overrule the decision by the San Francisco office. It would be a major victory and quickly overcome what had been a hasty, poorly organized defense up to this point, he contended. Brower wanted a more definitive ruling that would force the IRS to describe in detail how much lobbying was permitted. He did not trust administrative rulings because they could be easily altered, and, ultimately, he wanted the courts or Congress to establish more permanent guidelines that would free the Sierra Club.[43]

Brower had been hunting for a large East Coast law firm to handle the appeal, but the Sierra Club board settled on Gary Torre, a former law clerk for William O. Douglas and a senior member of a California law firm whose specialty was tax law. Virtually from the beginning, Torre and Brower did not agree. Torre favored the administrative ruling touted by Berry, arguing that it was going to be difficult legally to get the type of ruling Brower sought and that it could be extremely expensive.[44] Torre prevailed and wrote a ninety-three-page brief arguing that the Sierra Club engaged in conservation activities but that doing so was not the same as lobbying. It had spent 2.6 percent of its budget on the Grand Canyon and Redwood Park campaigns and perhaps 10 percent overall on conservation. But Torre was disingenuous with his numbers, arguing that such tactics as the newspaper political advertisements could be considered educational as opposed to political.[45]

Brower despised the argument and the brief. "That was a complete cover," Brower said. "It didn't fool anybody. It certainly didn't fool me. And I had no intention of the Sierra Club retreating from what John Muir founded it to do." He was not sure if he wanted Torre to win or lose. Winning meant that Brower would lose his First Amendment case, his attempt to force the IRS to explain what constituted speech that could be taxed as opposed to speech that was exempt from the government's reach. But losing might force the Sierra Club to conclude that further efforts would be fruitless and needlessly expensive.[46]

By the spring of 1968, the club was still waiting for a final decision from the IRS, and Brower worried that he was running out of time. Then he learned that Torre was meeting with Treasury officials in Washington in late May to discuss the case. Torre had invited Edgar Wayburn, who was serving his second term as Sierra Club president, to accompany him. Wayburn had declined. Torre did not want Brower on the trip; they had clashed too often. Brower pressed Wayburn for permission to go; he even coaxed Pat Goldsworthy, a loyal friend and now board member, to advocate in his behalf. The answer repeatedly was no.

Finally, on May 27, the day before Torre's hearing in Washington, Brower, Wayburn, and former club president Will Siri had lunch at the Sir Francis Drake Hotel in San Francisco. There, Brower told the two men that he wanted to go to Washington and press for a more grandiose gathering. He wanted to invite all major conservation leaders as well as representatives of the Johnson administration and Congress to gather and work out a permanent solution on how environmental groups could both lobby and remain tax exempt. He worried that if Torre prevailed, it would not be just a temporary victory, but a disaster. The club needed a more permanent solution. Brower said he was willing to take an overnight flight to Washington to make the meeting.

Wayburn and Siri reminded Brower that the club had hired Torre and that it still had confidence in its lawyer's argument. They had read the brief, and they believed in it. "Dave, it's not necessary," said Wayburn. "We have an attorney we as the client have engaged, and this is a conference between our attorney and the IRS." Brower didn't go to Washington, but he was not willing to do nothing.[47]

Torre went to the Treasury Department offices the next morning, made his arguments, and returned to the Watergate Hotel to have lunch with his wife. They were heading to a wedding in New Haven, Connecticut, but while dining he was interrupted by a telephone call. The caller was the IRS commissioner who had heard Torre's testimony earlier in the day. He had received a telegram from Brower. Perplexed, Torre returned to the Treasury offices. He read the telegram and was amazed. Brower made many of the same arguments in the telegram that he had made at lunch the day before, and he had sent the telegram not only to the IRS but also to key officials in the administration, Congress, and various environmental organizations. There was also language in the telegram about Torre and his role before the IRS. "Basically it said that the present attorney of the Sierra Club is not qualified to appear," said Torre. "He doesn't know what he is talking about, and let us have a constitutional convention on the environment."[48]

Torre went from amazed to angry and then livid. "This was a very serious libelous comment about an attorney appearing before a federal agency, or a court, or anyone saying he's pretending to represent somebody he's not qualified to represent," he said. He called his office and demanded that Wayburn send a second telegram affirming that Torre did represent the Sierra Club and disavowing Brower's wire. Wayburn read Brower's telegram and agreed. As president of the Sierra Club, he believed he had no choice but to back his lawyer over his executive director.[49]

About a year later, speaking in his defense, Brower maintained that he had not been trying to interfere with Torre or his case. He only wanted to protect the case from going astray. He wanted to control the protest and any appeal to get the legislation that he believed was necessary. He also said that he was using the authority the full board had given him as executive director and that the telegram he sent was consistent with the Sierra Club's intent regarding a final solution to the IRS tax problem.[50]

He was wrong. The telegram was a defiant action meant to undermine both Torre and the board.

Congress settled the Grand Canyon water plan before the IRS ruled on the Sierra Club. Two weeks before Brower tried to sabotage Torre's court case, the House of Representatives, after five years of consideration, enacted Stewart Udall's water plan. The dams in the Grand Canyon were not part of the legislation, and the Grand Canyon National Park boundaries were enlarged. Despite the animus that Brower had built up during the fight, he was still an invited guest at the White House on September 30, 1968, when Johnson formally signed the bill.

The IRS denied all the Sierra Club appeals. Brower was also correct in his assessment that the club's board would conclude that it had spent enough money on the issue. It instead turned to an organization that Brower had set up earlier, the Sierra Club Foundation. The separate organization legally funneled money to the club, although for years there was friction between the two groups. The foundation was run by a board of directors made up of former Sierra Club presidents who were often critical and wary of the younger, more aggressive staff. "We used to sing for our supper at each meeting," recalled McCloskey, who ran the Sierra Club in the 1970s. "The staff would stand up, and we would have our begging cup out, and we would do a ritual song and dance to justify the program, even though it was the same program as before." Trust eventually replaced distrust, and it was agreed that about 95 percent of the funds the foundation was collecting should go to the club's conservation efforts without such "song and dance" presentations.[51]

By sending that telegram to the Treasury offices and key officials, Brower likely undermined any chance the Sierra Club had of winning its case before the IRS. Wayburn was once asked why Brower sent it. What was going on between Brower and Torre? Was it more than a philosophical disagreement?

Wayburn paused before answering and then said, "Dave wanted to fight the IRS to the end; in fact, we all did. Torre, after reviewing the entire problem thoroughly, said that he had doubts if we could succeed short of filing a lawsuit against the federal government, and that would cost us, in his estimation, at least $100,000. It might not be successful, and didn't we want to spend that money on conservation?" But the bottom line was that the telegram "was an act of direct insubordination in which he undercut the board's selected representative, and it could have caused us a tremendous amount of trouble."[52]

In blocking the dams in the Grand Canyon, Brower had won the greatest victory of his career, one even sweeter than the one at Dinosaur ten years earlier. But it was difficult to savor success when he was angering so many of his friends.

14. Diablo and Galápagos

California's coastal Diablo Canyon and the
Galápagos Islands west of South America
are separated by more than 3,000 miles of
Pacific Ocean, and they appear to have little
in common. And yet for David Brower in the
mid-1960s they created similar opportunities to further the conservation cause at the
expense of increased disruptions within the
Sierra Club.

Diablo was the site of a proposed nuclear
electricity-generating station, and some
Sierra Club members were instrumental—
surprisingly and ironically—in helping the
developer, Pacific Gas & Electric (PG&E),
find it. Brower was not involved in those
efforts, and he became increasingly outraged
because the project was going to be built on
one of the last scenic and undeveloped canyons on the California coast.

The issue with the Galápagos was more
subtle but in some ways nearly as threatening to Brower. His decision to work secretly
toward producing a book about the islands made famous by Charles
Darwin for their unusual species angered many Sierra Club leaders.

Without the Diablo controversy, Brower probably could have calmed tempers about his handling of the Galápagos book. But the Diablo issue instead exacerbated the friction and the growing factionalization, pitting Brower and his supporters against a coalition of what had been his best friends, including Dick Leonard and Ansel Adams. It precipitated the greatest rupture in the history of the club, greater even than the split caused by John Muir's fight for Hetch Hetchy at the dawn of the twentieth century. Further, Diablo was only the preliminary round, a preview of what was ahead for Brower and his one-time allies.

Later, when the questionable safety of nuclear energy became a rallying cry for environmentalists, Brower was one of the movement's leaders. In 1978, he helped rally more than three thousand protesters at Avila Beach near the Diablo site in an action organized by the Abalone Alliance, an antinuclear organization. Later that day, 487 of the protesters were arrested. A year later, before a crowd estimated at up to forty thousand, he praised the Abalone Alliance and protesters for their efforts to stop Diablo.[1] But in 1966 when Brower raised his first objections about Diablo, they had nothing to do with safety. "We were split over not whether there should be reactors," said Brower, "but simply where. I wanted the reactors in a place that was already developed, instead of taking a relatively unspoiled piece of coast." Fred Eissler, a Sierra Club board member and critic of nuclear energy, unsuccessfully warned Brower of the plant's safety dangers. "I thought that he was just being too excitable," said Brower, "that Pacific Gas and Electric Company had the experts, and they would take care of the safety matters. They shouldn't be allowed, however, to bother unspoiled country."[2]

PG&E's efforts to build a nuclear plant in California began in the mid-1950s at Bodega Bay, 50 miles north of San Francisco. It was to be erected on a grass-covered granite headland that curled into the Pacific, a bluff that served to protect a small, isolated fishing village and former Russian sealing community. Much of that peninsula had previously been considered for a state park, although the true opposition to the plant grew after it was learned that the site was one-quarter mile off the San Andreas fault. The utility would spend $4 million on the Bodega project between 1958 and 1964.[3]

Brower wanted to fight the Bodega plan. The Sierra Club had already gone on record as opposing new power plants on undeveloped areas of the coast. Brower had assigned a young staff member, David Pesonen, to Bodega. Pesonen told members of the Sierra Club Executive Committee that the Bodega plant could be defeated by showing it would be a hazard during an earthquake. Leonard erupted. "Don't you dare mention public safety," he said, pointing his finger at Pesonen. "The Sierra Club can talk about scenic beauty, and maybe the loss of scenic beauty, but not about public safety. That's not our job."[4] To reinforce that, the board agreed to oppose the plant, but not to fight it.[5] Pesonen quit the Sierra Club, created a grassroots campaign, and drove PG&E out of Bodega Bay. His campaign impressed and yet worried Brower. "I was a little uncomfortable about David's sort of showmanship—releasing balloons and showing where radioactivity would drift if it drifted," Brower commented later. "I didn't think that was going to be important because I thought that if they built something there, at least they would build it well."[6]

Stifled at Bodega, the company purchased 1,100 acres at Nipomo Dunes just south of San Luis Obispo, California, where it proposed building up to five nuclear reactors.[7] After escaping Pesonen, the utility now had to contend with Kathy Jackson. In her late fifties, five feet two inches tall, and matronly, this mother of six adopted children had been a Sierra Club member for years and had developed a devotion to the Nipomo sand dunes near her Paso Robles home.

The thousands of acres of rolling sand dunes at Nipomo are a geological anomaly, a unique, ten-thousand-year product of the most recent ice age. Although the geology there may have been as fragile as it was at Bodega Bay, the politics were far different. Unlike in the northern California fishing village, the politicians of San Luis Obispo County embraced the plans for the generating station as an economic windfall. The dunes were already zoned for heavy industry, and portions had already been exploited by squatters, dune buggies, and some industrial uses. And yet the dunes had survived, and especially where breakers pounded the shore they were quite spectacular. Jackson tried to counter the image of the dunes as a wasteland by conducting hikes throughout Nipomo. She developed a flair for courting the local press. Those efforts climaxed in January 1965 when she, Sierra Club president Will Siri, and a throng of reporters hiked the Nipomo shoreline. Siri was strongly dedicated to both science and the environment. He was a biophysicist and active proponent of nuclear energy as well as an experienced international mountain climber; he had been the deputy director of a recent Mount Everest expedition. Standing at a scenic outlook of the beach, he exulted in the beauty of the breakers crashing nearby. "I didn't know it looked like this," he said loudly. "It is magnificent."[8]

By winning over Siri, Jackson acquired the clout of the nation's greatest environmental organization, and she finally began to achieve some parity with the behemoth utility company. Siri said the dunes should host a park, not an electricity-generating plant. Quiet negotiations were soon under way between Siri, representing the Sierra Club, and PG&E officials. The talks were coordinated by Doris Leonard, who had begun a consulting business and was the wife of Sierra Club board member Dick Leonard. Out of those conversations, engineers found a new potential site, Diablo Canyon, which is about 15 miles northwest of Nipomo on a privately owned promontory where public roads did not penetrate. The talks expanded to include some of the Diablo owners.[9]

Jackson was mostly out of the negotiations, but she supported any alternative that kept the plant out of her dunes. One day, she led a hike along the shoreline to the mouth of Diablo Canyon. "We had looked briefly over our shoulders and had seen nothing but closely cropped, closely grazed hillside," she recalled later. "We had not seen oak trees nor woodland, so we had taken it for granted that the canyon winding back, inland from where we stood, was probably as barren as the coastal marine terrace where we stood."[10]

PG&E accepted the Diablo site, but it insisted on a reassurance that the Sierra Club would not interfere and oppose this site too. Siri took that request to the board of directors meeting on May 7, 1966. In retrospect, many would regret what happened next. Siri was adamant. He virtually demanded that the board endorse Diablo. In return for getting Diablo, PG&E would give up Nipomo and help make it a state park. Siri, as Edgar Wayburn noted, could be "a dominating, emotional man who feels strongly, and when he feels strongly about things he can be most convincing."[11] Nothing of any value, Siri maintained, would be lost at Diablo, while everything at Nipomo could be preserved.

No one on the board or the staff had been up Diablo Canyon to see what it looked like. The nearest view had been Jackson's, and when she gave her report, she described a sluggish, muddy creek coming out of a deep gully. Diablo, she said, was nothing but a "treeless slot."

Brower was surprised by Siri's demand. He wondered if there was time to visit the site. Siri had mentioned that PG&E did not need an answer until the end of the month. By then, Brower said, the board's Executive Committee could meet and make a final decision. Eissler, the nuclear critic on the board whom Brower often ignored, said that recommending the Diablo site was the equivalent of approving the nuclear plant. He also wondered whether the local chapters had weighed in on the issue and questioned if the utility would ever give up its dunes property.

On the board, Wayburn was bothered that Siri had become so involved in the issue and then was demanding a rapid vote. He also knew that there were few untouched places along the California coast. But he also had confidence in Siri, so he backed the recommendation. Adams had photographed Nipomo in the years before dune buggies had arrived, and he wanted to save it. Leonard had been well briefed by his wife, who would soon be appointed to the PG&E board of directors. He voted in favor. The resolution passed by a vote of nine to one, with only Eissler voting against it.[12]

Martin Litton had not made the meeting; he had been overseas. No one on the board had a greater appreciation or understanding for the geography of California than Litton, and he was shocked by the decision. He had flown his small plane along the California coast long enough to appreciate the few remaining untouched settings. The promontory that included Diablo was the last remaining native California coastline not crossed by a highway or a train track. Back home in late May, Litton felt sick when he learned of the decision. Diablo, Litton told Brower, was equal to Nipomo Dunes or even Point Reyes.[13]

Litton decided to hike the canyon. Brower was enormously busy in the summer of 1966, so he could not accompany him. When Litton reached Diablo, he found a pristine coastal gorge with groves of live oaks. He measured the spreads of two of the oaks at 123 and 129 feet, equal to the largest known measurement of any live oak. A few days later, Litton wrote an angry letter to Sherman Sibley, president of PG&E. He told Sibley that the proposed plant would ruin a priceless portion of the California coast and that the recent vote by the Sierra Club board had been "fraudulently obtained." Copies of Litton's letter were soon circulating throughout the Sierra Club. Adams demanded that Litton apologize, or he would sue Litton for libel. Litton clarified that the fraud had been committed by PG&E, not by anyone on the board. The club was a victim, not a perpetrator.[14]

During the summer of 1966, other Sierra Club leaders toured Diablo and came away with differing reactions. Conservation director Michael McCloskey was troubled by what would be lost, but Wayburn felt that it was a good site for a nuclear plant. Jackson, making her first trip into the canyon, was shocked and disturbed to see such a beautiful canyon. "I really had misgivings that maybe I had been the cause of PG&E looking into an area that maybe should be left alone," she said. Yet she was not willing to trade the dunes for Diablo, even if it were a beautiful canyon. "I realized that this was one of them, but that this was not the only one, and it was not the last of its kind," she said.[15]

By September, when the Sierra Club board next met, Brower was upset that he had not fought the resolution back in May. "I saw that it was a gross mistake, and I began to argue strenuously against it," he recalled. But a resolution calling for a one-year moratorium on Diablo and other planned nuclear plant projects drew only three votes. Many board members expressed private misgivings, and the board had a history of reversing earlier stands. But this board did not want to admit publicly that it had made a mistake.[16]

Brower now faced a choice. If he continued to press the board to reverse its position, he risked antagonizing its members and splitting the organization. But backing down would allow a beautiful place to be needlessly sacrificed. McCloskey believed that Brower's only option was to back the original decision. Although McCloskey, too, harbored misgivings about Siri's recommendation, he concluded that leadership should not be second-guessed. "You need a leader, and if your mode of operation is to shoot down your leaders and question their judgment and never be able to make up your mind when you face a deadline, you might as well withdraw from the field of public affairs," he said.[17]

This was not Brower's style, however, and he increasingly sided with Litton. He began reminding colleagues how he had failed before. "We cannot allow another Glen Canyon, no matter how small we think it is," he said.[18] Recalled Wayburn: "The opponents felt they had to make it the club's overwhelming issue. As they got the very full support of Brower, Dave then used everything he had at his command to condemn the directors of the Sierra Club for going along with PG&E." Although this was how Brower usually fought, in this case he was not opposing loggers, dam builders, or the federal government—he was fighting the Sierra Club and much of its leadership. He risked splitting the staff, the board of directors, the entire organization. "That put Brower in the position of being the great conservationist opposed to those weak-minded companions of the trail," added Wayburn. "It caused a great deal of emotional disturbance."[19]

Richard Sill tried to stop Brower. Sill had been one of Brower's early climbing partners. An astrophysicist at the University of Nevada at Reno, he had joined the Sierra Club to fight to preserve Nevada's Great Basin desert. He rose in the ranks of the Sierra Club Council, an advisory organization created in 1956 that would be composed of representatives selected by each of the club's chapters. The council was dominated by conservative old-timers who often disagreed with Brower's tactics. Sill was smart and articulate, but some found him odd and erratic. Phil Berry said that on one occasion when Sill was at an Executive Committee meeting and on the losing side of an issue, he adjourned, irritated, to the next room while the meeting continued. He found a dessert that was ready for the full group and ate it all. "Dick Sill," concluded Berry, "was a nut."[20]

Sill wrote to Litton and Eissler asking that they halt their public campaign opposing Diablo. He called it "disruptive" and warned that it was causing "factionalism" and "political ineffectiveness." Besides undermining the Sierra Club and its board, their opposition could result in destroying Brower's career and silencing an effective conservation leader.[21]

Brower responded directly to the letter, and his reply explained why he was so determined. He said that if his career were lost, the blame should not go to anyone other than himself. People like Litton and Eissler needed to speak up, as did Brower, to stop the types of ploys being employed by the enemy. That ruse, said Brower, involved opponents splitting the organization and muting Brower and other dissidents. In other words, the problem was that PG&E, not Brower or his allies, was dividing the organization. And Sill, by asking Brower and his friends to back down, was making the situation worse.[22]

The split, however, was not entirely of Brower's doing. As Sierra Club leaders argued publicly and privately through the fall of 1966 and into the winter of early 1967, they seemed less intent in listening than in expressing their point. They were talking past each other. "There were also enormous conflicts of personalities involved," said Berry. "You take a purist such as Martin Litton, who had little patience with 'procedures,' and you take someone like Dick Leonard. They could argue furiously with one another without seeming to understand each other." Brower could not bring them together, and he did not see that as his role. "Diablo was never considered calmly and coolly at a point of maturation where it could be decided on the merits," added Berry. "It was always affected by these strong personal antagonisms."[23] Berry would soon be thrust in the middle of this dispute, and he struggled to understand the two sides. "I think both sides ultimately were found to be right, one on the substance and one on the procedure. And if you want to describe how to get a real donnybrook going in the Sierra Club, look for those two elements."[24]

The dispute came down to a comparison of which site was more ecologically valuable, Nipomo or Diablo, as opposed to saving both. Diablo was either one of California's last unspoiled coasts or a common, overgrazed oak woodland and chaparral canyon. Nipomo was either a unique sand dune sanctuary or a beach so fouled already by nearby industry as to be marginal. More time was spent debating Diablo than Nipomo. Botanists dueled over the value of the Diablo site. The Sierra Club commissioned its own scientific study, and in February 1967 the noted ecologist the club had retained called Diablo Canyon "the finest remaining natural canyon that is still reminiscent of early California." The board ignored that finding. At a hearing in San Luis Obispo, PG&E showed photos of a beautiful shoreline canyon, and both Litton and Eissler testified that it could only be Diablo Canyon. PG&E then revealed that the photo was a nearby shoreline canyon that Kathy Jackson had directed them to.[25] The longer the dispute lingered, the more entrenched each side became.

Brower's mistrust, which could sometimes border on the paranoia, reappeared. He and Pesonen, who was still active in antinuclear causes, became convinced that their phones were being tapped by PG&E. They tested that theory. Pesonen called Brower and asked when a fictitious white paper on PG&E would be available. Brower told him it would be done by 3:00 P.M. At that hour, a visitor arrived and asked the Sierra Club receptionist for a copy of the report. Others reported similar experiences, including one staffer who left a press release on his desk that was never circulated, yet PG&E called to respond to that particular press statement. Some began to wonder if Sierra Club members were colluding with PG&E. Berry doubted whether anyone actively worked with the utility company, but he believed the tapping theory because he said it later happened to him when he became president of the Sierra Club. "I would begin and end each telephone conversation by telling them to go to hell," he said.[26]

Diablo was now being raised at every board meeting and rebuffed at each one. Frustrated, opponents wrote a petition signed by more than fifty club members seeking a membership vote on the Diablo issue. The wording was basic. Members would be asked to choose between supporting a moratorium on a power plant at Diablo Canyon until further shoreline planning was completed or favoring construction of the plant along the lines of the board's May 1966 endorsement. The board did not like the wording, and over the strong objections by some of the petitioners it adopted new wording for the referendum: club members could either support or oppose the board's decision to protect Nipomo and to consider

Diablo a reasonable alternative for a power plant.[27] Diablo opponents were opposed to even mentioning Nipomo in this debate, and they argued that it was not relevant.

By now, the fight had spilled into the open, with numerous news stories about this growing schism in America's most well-known environmental organization. George Marshall, who had succeeded Siri as president, wanted to control the story as much as possible. He was startled after the January meeting to discover that the staff under Brower had sent out a press release about the upcoming referendum. Marshall was opposed to overturning the board's decision on Diablo and upset by the publicity, which he believed only promoted the opposition viewpoint. He ordered the staff to stop publicizing the issue.[28]

Hugh Nash, the editor of the *Sierra Club Bulletin*, proposed that each side in the Diablo issue submit its arguments by January 25, 1967, so that they could go in the February issue, well before the vote scheduled in April. Siri and Adams insisted that each side be limited to only two pages. Brower would come to believe that his opponents preferred that the membership have as little information as possible before the election. Nash, who was strongly opposed to Diablo, came up with an alternative strategy. He told Litton he thought he could outmaneuver the "bad guys." Nash would write an introduction and give it very prominent play so that it dominated the February *Bulletin*.[29] Litton submitted his version on time, but Siri and Adams asked for an extension to January 31 and then missed that deadline. "About two weeks after the extended deadline, I was fit to be tied," recalled Nash. Nash was convinced that Siri was sabotaging any attempt to publicize the referendum. "Brower agreed to let me print a limited edition *Bulletin*," said Nash, "for distribution to several hundred club leaders, containing everything but the Siri material."[30]

It was called the "half-*Bulletin*," and it seriously backfired on Nash and Brower. The full ramification of what Brower had done surfaced at a board meeting. "It created quite a sensation," recalled Richard Searle, a Sierra Club member. Searle remembered Brower telling board members earlier that the staff would never engage in one-sided reporting on Diablo. "But the next thing, somebody held up an issue of the *Bulletin* and said, 'I have it here.'"[31] Many club members were angry, and that response overshadowed Nash and Brower's explanation.

The half-*Bulletin* triggered more animosity. After Siri produced his submission, Marshall decided to make changes in Nash's "neutral introduction." There was nothing neutral about it, in Marshall's opinion, but Nash countered that Marshall's revision was nakedly biased in support of his own views.[32]

It is unclear how many members read the *Bulletin* before they voted and how many might have changed their minds after reading it. But it is clear that much of the Sierra Club establishment opposed the referendum. Ultimately, the vote came down to an issue of loyalty to the board leadership as opposed to deciding what was best for the Diablo landscape. The Los Padres chapter was the closest to Diablo, and chapter officials at first were outraged by the board's decision and adopted a resolution opposing it. Marshall replied to the chapter that although such disagreements were healthy, they could jeopardize the club's effectiveness. The chapter then rescinded its original motion and voted to back the board. Fifteen of the nineteen other chapters passed similar support statements. The Sierra Club Council voted twenty-seven to one to back the board.[33]

Both sides campaigned. Board supporters got much of their chapter support by explaining at local meetings why it was important not to back down. Litton and Brower organized a front group called the Committee to Clarify the Diablo Issue and published an eight-page newsletter titled *This Is the Issue* that strived to explain the opposition to Diablo and the board. However, Brower's group was hamstrung by a lack of a membership mailing lists.[34]

Never in its seventy-five years had the Sierra Club taken a membership vote on a controversial issue, and the result was a sharp rebuke to Brower. Fewer than one-third of the 50,000 members voted. Yet among those who did, the response was overwhelming in upholding the board, with 11,341 voting in favor and 5,225 opposed. Brower did only slightly better on a second resolution designed to allow him as the executive director to vote as a board member. There were 9,059 for the change and 6,994 opposed, but this resolution needed a two-thirds majority, so it too failed.[35]

Diablo had helped Brower push for the second resolution that allowed him to vote with the board. Eliot Porter, now a board member, argued that the board was being too restrictive on what Brower could do, not only with Diablo but also with other crucial issues. The restrictions were beginning to take on the appearance of a personal vendetta, said Porter, and if they continued, it could be disastrous for the Sierra Club.[36]

Porter was especially worried about his project to produce a book about the Galápagos Islands. In October 1963, the Publications Committee had told Brower to stop all work on the Galápagos book project. He had quietly continued to search for a way to produce it, however, and at the Publications Committee meeting on January 21, 1966, he announced that an expedition team headed by Porter would be leaving the following month for the Galápagos and that they planned to produce an Exhibit Format book to be published the following year. Several members of the Publications Committee were dumbstruck by this sudden and unexpected announcement. Pressed, Brower provided details. Porter had offered to provide a $10,000 loan against his royalty earnings to help finance the expedition to the South American islands. In addition, the Conservation Foundation was supplying staffer John Milton to provide technical assistance on the biological issues. Brower had also been spending some money from his discretionary account over the years to keep the project alive. None of these details reassured committee members, who pointed out that no one but Brower had authorized a loan that should have been approved by the board or the Executive Committee. In addition, spending such a large amount of money would now pressure the Publications Committee to authorize a book it had already rejected.[37]

Marshall, who sat on the Publications Committee, asked what would happen to the loan if the book were not published by the Sierra Club. Brower replied that any unspent money would be returned, and he was confident that Porter could sign a book contract with another publisher to get his money returned. More importantly, several committee members wanted to know why Brower had ignored its orders and if similar decisions were also being disregarded. Brower told them he felt that the project was important and that although the committee controlled the books program, he was not responsible to them, but to the club president. Brower said that if the committee wanted to have tighter control on the books, it should meet more frequently. That was how the subject was left, with the committee giving no indication of taking any action on the Galápagos book or against Brower.[38]

Brower and one of his assistants, Bob Golden, scrambled in the next three weeks to complete all the necessary details to get the Sierra Club party to the islands by the planned February 15, 1966, departure. The Galápagos are 600 miles west of Ecuador near the equator. Over time, they have drawn, as Brower would write in the foreword to the projected book, "Spanish sailors, English pirates, Yankee whalers and scientists from all over."[39] When Brower, his son Ken, Porter, and others arrived in Quito, Ecuador, which was the nearest South American port from the islands, they encountered a political revolt. Traces of tear gas still remained in the streets when Golden was making some of the final trip arrangements. Brower wanted to return to the Galápagos in April, although the Grand Canyon campaign and the *Reader's Digest* conference made that impossible.[40]

From Porter's perspective as a photographer, the trip was a spectacular success. He made thousands of images of birds, tortoises, iguanas, and other species. Many of the birds were so unacquainted with humans that Porter was able to get quite close. In total, 140 images would eventually be selected, and Brower realized that the book could be two volumes.[41]

But the trip had been expensive, costing $19,356, including $1,062 for Brower's journey to Ecuador. Brower had told Publications Committee members that Porter had agreed to cover any excess in expenses from future royalty payments. When Marshall learned that the expedition had cost an extra $9,000, he ordered that Porter's royalty checks for other Sierra Club books be stopped until the organization could collect its $9,000. What made this move so extraordinary was that Porter at the time was an elected member of the Sierra Club board. Marshall admitted he was embarrassed because it appeared that Porter had been caught in the middle of "certain unauthorized acts." Porter was furious. He told Marshall that he had never authorized that additional royalties would go for any shortfall. Even if he had, he said that the Galápagos costs had nothing to do with what had already been earned from other projects. Part of what was happening here was the Sierra Club leadership's increasing unease with Brower's blasé financial management style. Marshall pointed out that the expedition had continued even though there was no guarantee that the Sierra Club would ever publish a book about the Galápagos. It was Brower's expedition, not the Sierra Club's. The Sierra Club was responsible for paying back Porter's loan, but it had no other publications or financial obligations, Marshall said.[42]

Underlying all this squabbling was a secondary issue. Brower's career had been spent in pushing the Sierra Club beyond its traditional boundaries. Not only had he transformed the Sierra Club from a provincial California club into a national organization, but he had remade it from a social and hiking club into a vibrant, modern environmental and political force. Now he was pushing the boundaries even further. It was one thing to fight for the Grand Canyon or the redwoods; it was another to produce a book about an archipelago off the coast of South America thousands of miles away. Brower had by now been talking to Russell Train, the president of the Conservation Foundation, about doing a series of books about wild places around the world. Although there was no binding agreement, by February 1966 Train privately outlined a specific plan to Brower calling for his foundation to provide the scientific competence and the Sierra Club to offer administrative, photographic, and publishing experience.[43] This was not something many Sierra Club leaders were interested in, and it would take the Sierra Club years to recognize the necessity of moving beyond the boundaries of the United States.

Meanwhile, Brower struggled for nearly a year to solve the financial problems of the Galápagos book, which included not only the cost of the expedition but also the money to publish the book. The Sierra Club was cash strapped. But Publications Committee members were mollified in November 1966 when they saw the first photos, which were spectacular. The committee told Brower he could now officially work on the book, but he would also have to find the money to publish it.[44]

Brower's problems were mounting: intemperate remarks, lax management, Diablo, and Galápagos were only a few of them. It was one thing to fight with Siri or Marshall, but now he was increasingly at odds also with Leonard and Adams (figure 15). In January 1967, Porter wrote to Marshall questioning whether Brower's enemies were engaged in a vendetta. It was not a private communication. Within days, it was being circulated to other board members and to Brower.[45]

The situation was reaching a point where Adams could be easily provoked on the issue of David Brower. After reading Porter's letter, he reacted with a volley of pent-up disgust and anger. He said Brower had brainwashed Porter. Brower had created major financial problems for the club. He had spent money without authorization. He had committed the club to costly and at times embarrassing commitments. Adams wrote four, single-spaced, typewritten pages detailing Brower's faults. For years, Adams said, he would talk and reason with Brower after an episode caused Brower to get upset. Brower had always complained that the others were wrong, that he was always right. He would repeatedly complain that other factions were out to get him. The fears were absurd, stated Adams, but over the years they had grown into an obsession. Now, Adams claimed, Brower was blinded by all these delusions. Two weeks later, Adams publicly addressed the Sierra Club board and escalated the conflict. He said that the Sierra Club was faced with a cancerous growth of dissention and irrationality that had to be lanced. He called for Brower to "be relieved of his duties" and urged a moratorium to be placed on publishing any further books.[46]

Adams's call shocked Brower. "I was stunned when it came up first, because I didn't think that the friendship we'd had for so long had deteriorated to that point," he commented later.[47] But it had.

Adams did not get a second on his motion, but his complaints were soon followed by a public letter from seven former presidents, primarily a legion of the old guard that had always had reservations about Brower. According to Dick Leonard, they accused Brower of making "biased, emotional, and irresponsible statements," and they urged that the organization "not continue to compromise with integrity."[48]

In the wake of so much expressed opposition to Brower, a secret meeting was held in late April at Leonard's house. The participants came up

FIGURE 15 Ansel Adams: friend, enemy, friend

Brower met Ansel Adams while on a hike in the Sierra Nevada in 1932. Adams, who was ten years older, became a mentor to the young Brower in the 1930s. He was a strong defender of Brower while serving on the Sierra Club board in the 1950s, but by the 1960s he was one of Brower's biggest critics. They became friends again in the 1970s. (Courtesy of the Bancroft Library, University of California, Berkeley) with a plan not to fire Brower directly but to shuttle him out of power. The club would be split into two organizations, one based in New York that would be responsible for publications and the second remaining in San Francisco that would be responsible for the conservation program. Brower would have to move to New York. The conspirators, who included Leonard and Adams, counted votes on the fifteen-member board. They were confident that they had enough to pass the proposal when the board met on May 6, 1967.[49]

Adams arranged to meet Brower in Carmel on April 28. He outlined the plan to Brower and told him that his only choice would be to take the New York position or be fired. Dazed, Brower drove on to Santa Barbara, where he had to give a speech. He was stunned. He had to go to Washington, D.C., and while there, after consulting with many friends and staff members, he decided to fight. On May 1, the popular San Francisco gossip columnist Herb Caen published this report in the vernacular style that was characteristic of his columns: "The rift inside the Sierra Club is deeper than the Grand Canyon, sweet. The talk is that at the executive board meeting Friday at the St. Francis, an attempt will be made to oust Dave 'Mr. Conservation' Brower, the club's executive secretary for the past fifteen years. Apparently Mr. Conservation isn't conservative enough for a certain faction." By May 2, the story had been picked up nationally. Pesonen, the former staff member who had fought PG&E at Bodega Bay, had leaked the news to Caen.[50]

Brower wrote a conciliatory letter to the board, and by the time he returned home from Washington, phone calls were coming in and telegrams and letters arriving, the vast majority expressing outrage over the move to send Brower to New York. Asked Paul A. Wilson of San Francisco, "Is the fight for conservation so nearly won that you can already begin to bicker with your executive director over the division of the spoils?" "Fire Dave Brower?" asked Dudley and Penelope Yasuda, also of San Francisco. "You must be joking!"[51]

The board meeting on May 6 was anticlimatic. Members unanimously passed a resolution in support of Brower's continued leadership. The *San Francisco Chronicle* reported that some members tried to pretend the reorganization plan had never existed and that at times the meeting sounded like "a Dave Brower testimonial dinner." Brower was gracious in victory, telling reporters that the meeting had been "a successful search for accord, for getting differences of opinion on a constructive basis, as they usually have been in the club."[52]

Brower had survived, and he emerged from the conflict even more confident. He was not through fighting for Diablo or working to get the Galápagos book published or getting involved in whatever other environmental issues needed his aid. He was more hell bent than ever that he was right.

15. Conflict

Compromise is essential to operating a pluralistic society but it is no value whatever in an environmental organization. In conservation organizations, it's our business to stand for something we believe in, not to trade off, not to seek the benefits of access, or of being insiders, but to say what we think and to fight for it.

DAVID BROWER, IN STEPHANIE MILLS, "AN ARCHDRUID—OR SOME DAMN THING"

The Sierra Club's Diamond Jubilee was celebrated in glittering style at the Fairmont Hotel in downtown San Francisco on December 9, 1967. Planners had hoped for 600 to attend; 1,302 arrived. David Brower sat at the head table and was among the speakers extolling the changes to the nation's most powerful environmental organization over its seventy-five years. Membership had passed 50,000 earlier in the year, the budget was greater than $2 million, and its voice could be heard from coast to coast. Book sales would reach 1 million in 1967, and the club was on the cusp of major victories in the Grand Canyon, the redwoods, and the North Cascades campaigns. Much of that could be credited to Brower, for his aggressive conservation campaigns, his books, and, as Ed Wayburn put it, his "brilliance and total commitment." He was the hero of the American conservation movement.[1]

And he was increasingly also a villain, not just to the enemies who had now fought him for years over dams, forests, wilderness, and parks but also to many onetime supporters. Brower was still spending recklessly, ignoring direct orders from the board, and planning grandiose Sierra Club ventures to Europe and beyond. No one knew it, but that trend would intensify in 1968. If some were entranced by Brower's daring and audacity, others shuddered and worried that the club was going bankrupt. No one was indifferent.

Ed Wayburn had reluctantly agreed to return as president at the May 1967 meeting in which Brower had narrowly overcome efforts to fire him. That decision reduced the tension between Brower and the club president, be it Will Siri or George Marshall. However, Brower had not entirely escaped his enemies. The board at the same May meeting agreed to continue to explore the idea of creating an independent corporation to publish the books, the Publications Reorganization Committee. They put Charles Huestis, an old climbing buddy of Siri's and an enemy of Brower, in charge of this new committee.[2] Brower worried that it would mount a new effort to get him dismissed, but Wayburn thought it might be a solution to keeping Brower within the Sierra Club. Many board members recognized Brower's brilliance, said Wayburn, and he hoped that the reorganization would insulate the organization financially. "In spite of the fact that Dave repeatedly said the books were making money," said Wayburn, "the board was having to sell securities in order to get enough capital to support the publications program, and the net worth of the Sierra Club kept dropping."[3]

The club's success in convincing Stewart Udall by January 1967 to drop the Grand Canyon dams had cleared staff and resources to pursue the Redwood and Cascades Park campaigns that had remained blocked for years. Regarding the redwoods, the Sierra Club had in October finally settled its differences with Laurence Rockefeller and the Save the Redwoods League, and the united conservationists had endorsed a compromise plan for a new park. This new proposal called for creating a national park in the most important areas out of what had been the two competing plans at Mill Creek and Redwood Creek. Even more promising, by the end of 1967 the Senate passed a bill along the lines of the compromise plan. But the House was balking at the legislation, and area timber companies remained staunchly opposed. Most alarming was the reaction of Georgia-Pacific, which was felling redwoods in the Emerald Mile, one of the last, great remaining groves of virgin redwoods, a 300-foot wall of old-growth forest bordering Redwood Creek.[4] One of the biggest challenges was that much of the land for the new park was privately owned by logging companies. The North Cascades had the opposite situation: virtually all the property was owned by the federal government. But that was also one of the biggest impediments to conservation there. The Department of Agriculture and the Forest Service were unwilling to give up the land to the Department of the Interior and the Park Service. The issue was also becoming a clash between Henry Jackson in the Senate and Wayne Aspinall in the House. Jackson wanted a new national park for his home state, Washington; Aspinall wanted to ensure that the land was harvested by loggers and miners. The Senate would pass a bill, and Aspinall would block it in the House.[5] One conservation issue that Brower did not have to deal with personally, however, was Diablo Canyon now that the membership had endorsed the decision the Sierra Club board had made in 1966. Although Martin Litton, Fred Eissler, and others remained strongly opposed, they simply did not have the votes to move the issue.[6]

Brower believed that the Sierra Club's best chances in these and future fights was to build a greater, grander Sierra Club. In November 1967, he unveiled a five-year plan that envisioned the club going from 60,000 members and a $2 million budget to 225,000 members and $20 million. Chapters would increase from twenty-one to fifty, and the six Sierra Club offices would rise to twelve, including Los Angeles, the Midwest, and London. He also wanted the Sierra Club to become a land manager. It owned 1,400 acres. In five years, Brower wanted it to control and preserve 100 million acres.[7]

The expansion was going to be led by the newly created John Muir Institute. Many of its aims coincided with the ideas expressed in the proposal that Brower had written a few years earlier for a Ford Foundation grant. It was to be a research center concentrating on environmental education and the preservation of threatened natural resources. Board leaders were consulted, and Wayburn agreed to be listed on the incorporation papers. The $5,000 seed money to get the institute started came from Brower's discretionary fund.[8]

In addition, an expanded book program would also draw new members and revenue from around the world. Brower called his new series Earth's Wild Places, and he forecast that it would publish one hundred titles on places from the Sahara to the Amazon to the Arctic. The books would be published in English and the native language of the area profiled. He had already completed a detailed plan for the inaugural book, *Islands of Wilderness*, which would describe in prose and photos ten of the earth's geologic regions, from mountains to oceans, taiga, and steppe.[9]

There was one major impediment to all these plans—money—but Brower never worried about where the money would come from. He didn't have to because until now book sales and new member's dues had at least somewhat kept pace with the growing staff, conservation, and books programs. The financial picture at the beginning of 1968 appeared healthier than in the past, but just to sustain the staff and the ongoing campaigns the club needed at least $2.8 million. The only way to produce that much revenue was to publish more books. That was too great a risk, argued Ansel Adams and Dick Leonard during budget talks in late 1967. They preferred cutting back expenses and book production. But over their objections, the club balanced the budget in 1968 by agreeing to produce four new Exhibit Format books. The burden of producing so many books would fall on Brower. Even some of Brower's supporters wondered if he had the energy for it.[10] Brower, however, was excited because one of the books was to be the long-delayed Galápagos book. Not only had he found a way to pay for it, but he had also come up with the money in such a way that the Sierra Club was going to be forced to expand overseas.

The breakthrough had occurred earlier in 1967 when Sally Walker, who had emigrated from England to Portal, Arizona, gave the club £22,634, which came to $67,000 when exchanged. After conferring with Brower, she directed that the money be used to publish the Galápagos book. There was one catch—the money had to be spent in England. Brower realized that the donation would justify opening an office in London. His seemingly inexhaustible energy quickly kicked in. In early September 1967, he told the Publications Committee about Walker's gift and said he wanted to explore printing operations in the United Kingdom. In October, he found a London printer, Garrod & Lofthouse. By January, he introduced to Sierra Club leaders the young, handsome Allan Horlin as the Sierra Club's first employee in its new London office.[11] Brower had moved so rapidly that the news of all the steps he had taken came as a surprise to some. Ansel Adams remembered getting a call from Dick Leonard.

"Do you know we're in London now?" asked Leonard.

"What do you mean?" replied Adams.

"Well, I've just been informed we have a London office with a staff."

"What do you mean?" said Adams, still not understanding.

"Well, Dave decided that was necessary."[12]

Around ten years later, when Brower was asked if he felt that he had the authority to open the London office, he replied, "We didn't have a motion specifically authorizing the thing, just as we didn't have motions authorizing a great many steps that I took. There were things that I was able to do under what I felt was my administrative discretion."[13]

Because most of the money to set up the London office had come from Brower's discretionary account, the board in February stipulated that Brower could no longer spend any of that money without first seeking the club president's approval.[14]

That stipulation did not slow Brower down. Now he was flying not just cross-country but also over the Atlantic on a regular basis. He needed to set up the London operation and help young Horlin, who had hired his brother, Robert, to assist in the office. Neither brother was experienced in publishing or administration. The printer Garrod & Lofthouse needed help; it could not find the right paper stock for the Galápagos book. By April, Brower was going beyond London, flying to Copenhagen, Amsterdam, Antwerp, and Vienna to talk with publishers about the new international Sierra Club book series. He was rarely home, although he did make it back to Berkeley by May 1, 1968, for his and Anne's twenty-fifth wedding anniversary. The family stayed up late that night, with champagne, cake, and talk.[15]

It was a good time to celebrate, to reflect, and to commiserate. Besides the European travel in April, he had spoken to adoring audiences in

Portland, Maine; Jacksonville, Florida; Ithaca, New York; and elsewhere. He had checked in with the printer in New York and the lobbying efforts in Washington, D.C. Books were selling, and television programs seemed the next best bet. It all seemed so sweet. And yet it could turn sour so swiftly. Newspaper columnist Herb Caen in March had heard rumors that Brower might get fired.[16]

Brower decided he would regain control of his discretionary account. He chose to get the money from the burgeoning book program. For years, the club had not signed contracts with authors or photographers. Financial arrangements could be haphazard. No one complained until 1966, when one of the photographers of the Mount Everest book protested that he was being grossly underpaid for his work.[17] The club narrowly avoided a nasty lawsuit, and the board ordered Brower to come up with a formalized, written contract system. In the late winter of 1967, the Sierra Club finalized a contract that was common in the publishing industry. Under the terms, authors and photographers would be paid royalties earned from the sale of each book. More than one hundred contracts were sent out by early spring of 1968.[18]

In May 1968, Phil Berry discovered an unusual rider attached to two of the contracts, for the Galápagos book and a book about New York's Central Park, also planned for 1968.[19] Both indicated that Brower, as the editor, would receive a 10 percent royalty payment for each book. Such an arrangement was not totally unheard of, but it was extremely unusual, and Brower had not clearly communicated it to anyone.

Berry was dumbfounded. "I frankly did not know what the hell to do," he recalled. "I remember losing sleep over it over a number of nights. I asked my wife about it; she did not have any advice that was very cogent. I decided I had better tell somebody." He conferred with Brower, Wayburn, and August Fruge, who was still chair of the Publications Committee. "I told Brower, I said, 'Dave, you can't do this.' And I am still bewildered on why Dave did it. Partly it was that he was beginning to be a great man, and great men should be paid more, and sometimes great men could be blinded."[20]

Brower later contended that he had informed Berry by sending him a copy of the contract to review back in February that year. "I sent the contract over to him to see if there was anything wrong with them [*sic*]," said Brower, "and he said nothing about what was wrong with them and then sprang this on me."[21] But the royalty payment was a rider attached to the standardized contract, and Berry would contend that Brower never sent him that particular document to review.[22]

The issue came to a climax the evening of June 8, 1968, when the Sierra Club Executive Committee discussed it for about forty-five minutes. Leonard took notes and tried at times to record the conversation as closely as possible (figure 16). Brower, who did not fare well in Leonard's portrayal, would dispute the accuracy of Leonard's account. It began with Wayburn asking Brower for an explanation.

FIGURE 16 Dick Leonard and other Sierra Club leaders

Brower had been recruited in 1939 to help lead Sierra Club high trips, expeditions into the Sierra Nevada and beyond. That summer he posed with club leaders, including (*left to right*) Oliver Kehrein; William Colby, who had been recruited by John Muir and led the high trips for years; and Richard Leonard, sitting next to Brower, his longtime climbing friend. (Courtesy of the Bancroft Library, University of California, Berkeley)

"The royalty is for creativity," replied Brower. "The photographs and the text of the authors do not sell themselves. It takes a great deal of skill to make these books as successful as they have been."

Wayburn pointed out that Brower was paid a salary for that.

"I am not paid for it! I am badly underpaid!"

"But Dave," responded Leonard, "you cannot raise your own salary."

Wayburn said that he might be willing to adjust Brower's salary, but that was separate from this issue.

"It is not extra pay," said Brower. "I know that it would not be right for me to take this money myself, and I am not doing so. Every one of those royalty checks will be paid back into the Sierra Club."

Leonard said it made no sense for the Sierra Club to pay Brower and for Brower to then pay the Sierra Club. Brower said that he received honorariums for his speeches and he always turned them in to the club. Wayburn persisted: What did Brower need with the royalties?

"The purpose is to provide a discretionary fund," said Brower, "for the executive director."

It was pointed out that Brower already had such a fund.

"I don't have a discretionary fund," said Brower in what Leonard maintained was a hysterical tone. "It was taken away from me in February and turned over to the president."[23]

Brower had a reputation for being scrupulously honest when it came to money. With the exception of board member Richard Sill, no one thought he had been trying to embezzle. "I don't think anyone believed in their heart that Dave was trying to steal," said Berry. "We all believed it was foolish."[24]

Brower for years had operated the discretionary account as a slush fund with no controls or oversight. In 1967, he spent $30,587 from a discretionary account budgeted for $25,000. It was how he paid Gary Soucie's $9,300 salary when the board had not wanted to hire the Sierra Club's Eastern field representative. And how he paid for $3,005 in rent, supplies, and equipment for a New York office that the board said it could not afford. And how he paid $5,000 to the new John Muir Institute.[25]

Brower's spending was coming to light at an unfortunate time. Book sales had not fared as expected in 1967, and by the spring of 1968 the

Sierra Club was facing a major cash-flow dilemma. Creditors were calling, asking when they could expect payment. Brower took a call in March from Joshua Barnes of Barnes Press. Barnes was the club's biggest obligation, and it owed the printer $165,000. Brower told Barnes that the club was about to sell $100,000 in stocks and that half of that money would go to the printing company. Brower was optimistic that the financial situation would improve, and he felt that the telephone conversation had gone well.[26] But if Barnes had attended the May 1968 board meeting, he would have heard a direr forecast from the club treasurer, Siri. The club's deficit had unexpectedly grown to $140,000. However, a greater concern was the club's cash-flow situation. So much capital was now tied up in the books that the club was having difficulty paying its bills. New dues from more members and more sales from existing books would not solve the problem, said Siri. The club needed capital, a great deal of it, from outside funding sources. Adams blamed the financial problems on mismanagement in the publications program. Leonard worried that the situation might be even worse, and he began to do his own review of the finances.[27]

Meanwhile, the rumors and stories about Brower's management problems were growing. Dick Sill was telling friends a story about how three years earlier several Sierra Club leaders had confronted Brower over unpaid bills. Siri said Brower had pulled out of his desk $189,000 in unpaid bills from printers, book binders, and others. The club had had to struggle to pay them off. Wayburn discovered that Brower was sending multiple staff members to conferences when board members were refraining from travel because of money concerns.[28] It was also in May 1968 that Brower defied Wayburn and sent the telegram to the IRS that conflicted with lawyer Gary Torre's case.

The situation only worsened that summer. Brower repeatedly flew to London to manage the increasing problems that were developing both with the Horlin brothers' management of the Sierra Club office there and the publishing of the two-volume Galápagos book. Brower's critics were saying that he had hired Allan Horlin because the young man had impressed him when he sold Brower a Volvo. Brower said Horlin did not sell cars; he worked in Volvo's European Delivery Plan office. Robert

Horlin had been hired because he worked for a firm that did color printing on metal. There were even more problems at Garrod & Lofthouse. Print galleys were coming back with numerous problems. By July, Brower thought he had the problems under control. To make doubly sure, he brought his son Ken to London to manage the technical problems and his secretary, Anne Chamberlain, to manage the office there. Before he left, he stressed the urgency of not falling behind on the two-volume book.[29]

Brower returned to London on July 30 to discover that his son's corrections had been ignored, the galleys had even more errors, and some of the color photos were fuzzy, so he recruited David Hales, who had extensive printing experience. Hales would recall his time in London as five difficult weeks, perhaps the worst he ever experienced. Brower stayed in London for a week, and then, as planned, he and Anne departed on August 7 for a two-week trip through Europe. No trip could have been as poorly timed. Brower would get periodic reports from London as he traveled. When he returned to London on August 20, one volume had been published, and Hales was struggling with the second. Brower stayed up until 4:00 A.M., making what changes he could. He left the next morning for California, still fearing an "imminent printing disaster."[30]

The Galápagos books did get published; the two-volume set was on bookstalls in the United States by the fall, which was behind schedule but not too late for the Christmas sales. However, sales of both the Galápagos and the Central Park books were disappointing, and the other planned Exhibit Format books that were supposed to generate the club's needed revenue had to be postponed.[31] Their failure was the beginning of a disappointing reversal for Brower. Never again would his glossy, big, expensive books be as popular as they had in the past. Even the public-relations campaign in the United Kingdom built around the Galápagos volumes fizzled when Brower was late for a press reception, and virtually none of the reporters who went to the reception wrote a story.[32] A bigger problem was that both the books and the London office had cost far more then Brower had anticipated. The printing company that set the type estimated that 145 galleys had been set and that Brower and his staff had made changes to every single galley, often numerous times. The Galápagos books' final production cost was $235,000. Even if one made the dubious calculation that the two-volume set should cost the equivalent of producing two books, the expenses were still twice what they had been in the past for similar books.[33] In October, Brower reluctantly fired the Horlins, lamenting in his letter the many lost opportunities to make the London office work. However, he did provide Allan Horlin with about $2,000 so that he could shoot a film in the Seychelles Islands.[34]

The toll that summer from Brower's handling of the royalties contract, the London operation, the financial situation, and the dispute with the IRS was great. Brower needed to avoid further controversy. He instead dived into yet another disagreement—this one involving Diablo Canyon.

In May 1968, a new slate of pro-Brower supporters (and anti-Diablo partisans) were elected to the board. Now there were enough votes on the board to reverse the May 1966 Diablo position, and on June 11, 1968, Brower asked Sherman Sibley, the president of PG&E, to drop the Diablo site and investigate alternative locations. Two weeks later, eight of the fifteen board members wrote to Sibley making essentially the same plea.[35] Sibley, backed by Wayburn and even by some newspaper editorial writers, expressed amazement that the club would raise an issue that the membership vote had settled. Brower responded that he was compelled to do everything he could to save a site that was irreplaceable.[36]

It took until September for the Sierra Club board to meet and muddle through the controversy. Berry, one of the new board members in favor of changing the board's position, introduced a motion that the board regretted bargaining away Diablo and that on principle it opposed building electric generating plants on coastlines. After the motion was approved, no one could figure out what it meant. Berry said that even though he favored changing the position favored in 1966, his motion did not alter the board's earlier position. "Oh Phil, of course it does," replied Wayburn, who had voted against the resolution. Berry began to realize that somehow he had made a colossal blunder. "I managed to get myself on both sides of the same issue," he said. "My mistake was in failing to realize that those two positions could not be drawn together."[37] The bewilderment over Diablo would continue for three more months until Berry proposed and the board accepted that the September resolution be rescinded and that the Diablo issue be taken to a second membership vote in the spring of 1969.[38]

Brower had been waiting for the Publications Reorganization Committee to report, and that report finally came at that same meeting in September 1968. But rather than indicating that Brower should be fired, the committee proposed demoting him. It recommended creating a new position, president, that would trump the role of the executive director. The Sierra Club had always had a president, who chaired the board, and under this new arrangement that part-time position would be renamed "chairman." Brower could become an executive vice president, responsible for conservation and publication issues. A second executive vice president would handle administrative and finance issues, in which Brower had shown very little competence. If Brower did not like the recommendation, Adams was even less satisfied with it. He told the board at that same meeting that he had come to the reluctant decision that Brower needed to be fired. But Adams's motions to dismiss Brower failed twice, with only Leonard and Sill supporting them.[39]

The summer had been difficult for everyone, but for those who already had feared the worst, it was especially intolerable. Now they became emboldened. It was as if a dam had been breached. Brower's critics began their drive against him in the weeks after the September meeting. Adams and three other board members, joined later by several chapter leaders and members of the Sierra Club Council, began calling for a special meeting. They wanted a formal inquiry into Brower's management of the club. Adams wanted Brower and other staff members fired. He said the Sierra Club should hire a business manager to run the organization. Also, every board member should resign, and there should be a special election to elect a new board. The campaign was very public, and newspapers were soon running stories with headlines such as "Adams Asks Ouster of Chief" and "Turmoil over Sierra Club Leadership."[40]

Under the bright glare of television lights, the Sierra Club board of directors convened on October 19, 1968, in a ballroom of San Francisco's Sir Francis Drake Hotel. The event was fraught with high drama, high emotion, high tension. Many in the crowd considered the inquiry a quasi-judicial tribunal. Others believed that it was a kangaroo court. Wayburn gaveled the meeting into session. Leonard, Sill, and Adams produced three reasons why Brower should be dismissed: he attempted to divert book royalties, failed to follow orders involving the Galápagos book, and was financially irresponsible. Even though each of these issues had already been hashed out repeatedly over the past months, participants spent hours discussing them again. Adams made the case for the Galápagos book, Leonard for the royalties issue, and Sill for the far broader charge of financial irresponsibility. Sill would get carried away, as noted in chapter 3, making public accusations that Brower was a homosexual and as a result had lost his job at the University of California Press. Brower had asked David Sive, a board member and lawyer, to defend him at the meeting. Sive, however, felt that he couldn't because Brower's accusers had spent weeks honing their arguments, whereas he had been given the details only that morning. "This whole proceeding of trying to act as a board of inquiry at a point like this just completely baffles and shocks me," he said. Sive wanted time to review the charges, to hear more of the case, to prepare a proper defense. Yet Brower was far more eager to talk, to make a public defense, and to warn his accusers. The club had not been in financial danger, he said, but it soon might be. If his accusers kept stressing such a false point, creditors might lose faith, and the club indeed could be imperiled. Board members took their turns. Opinions rambled, tempers rose. The meeting ended inconclusively. Brower was allowed to prepare a defense. The question of whether the board of inquiry would meet again was left unanswered.[41]

The meeting had been a true exercise in futility. Brower would spend weeks preparing a detailed defense of the charges against him. They were accepted in December, but the board of inquiry would not reconvene.[42] Neither the inquiry nor the reorganization had much of a future. Other events would soon brush them aside. Yet some things had changed.

Emotionalism was trumping civility, and the trend was escalating. New arrivals coming to board meetings were shocked by what they were witnessing. James Moorman, the representative from the new Southeast chapter, arrived for the first time in 1968 to witness "a vicious meeting. I can still remember the intensity of it," he said. "Brower was quiet through the whole thing. He was the eye of the storm. A lot of people said things which were very emotional and to me, upsetting, because this was not what I was interested in at all." It got so bad that a staff employee was asked to investigate whether a social psychologist should be hired to be on call at meetings. James Elliott Bryant was the new chairman of the Rocky Mountain chapter. The conflict was transforming him from a young, idealistic admirer of the Sierra Club to a disillusioned and disappointed detractor. He told friends that a sickness had overtaken the combatants and that the board was tearing apart the Sierra Club.[43]

And yet those in the center of the conflict simply could not curb their behavior. The failure to reason may have in part been a reflection of the late 1960s, a time of heightened political conflict where consideration of altering viewpoints was often lacking. Or it may have resulted from simply too much familiarity among the participants. Polly Dyer, Brower's long-time friend and supporter, felt that too many board members were jealous of him and what he had achieved. That label was applied specifically to Adams, but Adams was already a famous photographer. Rather, the case seemed to be that he felt betrayed by his brilliant former apprentice. "I have not forgotten the attitude of intimidation in which the executive director threatened that if any action is taken against him," said Adams in September 1967, "the membership would rise in protest and take action against the present board." Others who could divorce their passions from the continuing confrontation blamed the debacle on emotion. "Dave was not out to be dictator," said Berry, himself a former Brower protégé. "Some people have used that word, but I think it is overdrawn. No, I think the problem was his inability to show respect to people who deserved it because of their intelligence, their good motives, and their history of service to the club."[44]

In the midst of all of these trials, good news blossomed. After years of deadlock, Congress in the fall of 1968 passed a flurry of conservation bills. They included the creation of the Redwood and North Cascades National Parks as well as a Southwest water plan that did not authorize any dams in the Grand Canyon. The key seemed to be the Grand Canyon

legislation. Once it was approved, opportunities opened for other measures. In the North Cascades, 674,000 acres of mostly Forest Service property were transferred to the Park Service, creating the new park. Then
Congress fashioned two new recreation areas that bordered the park and
designated another 520,000 acres as a wilderness area directly east of the
park. That meant that 1.2 million acres would be protected from logging,
and most of that land would be protected as wilderness. The final plan
was close to the once highly ambitious vision that a young David Simons
had proposed a decade earlier. The North Cascades Conservation Council
celebrated with a victory banquet in Seattle on October 10, 1968, and
presented a special invitation to Simons's parents.[45]

If only the victory in northern California could have been as sweet.
The bill authorized a record $92 million to acquire 58,000 acres, which
included some of the finest remaining virgin redwood groves. But no buffer zone was established in the Redwood Creek basin. Brower went to
the dedication of the new park that fall and had difficulty hearing the
speakers because of the roar of the nearby chainsaws.[46] Over the next
years, logging companies would fell 95 percent of the timber in the watershed, creating massive erosion that threatened the park. The Sierra Club
and other conservation groups fought and failed to get federal officials
to protect the buffer zone and the tall trees that were in danger. Finally,
after ten years, Congress in 1978 passed legislation to enlarge the park,
agreeing to spend up to $359 million. By then, the definition of what
constitutes a national park had to be redefined. Decades later the forest is
returning, but it will take centuries to bring back the mighty behemoths
that had once stood there.[47]

The later tributes to Brower's career and legacy would point to his
major role in creating both the Redwood and North Cascades National
Parks. This claim is both true and not true. Others did much of the work,
but to many people Brower was the Sierra Club. But that meant that he
embodied both the best and the worst of the club. By late 1968, Adams,
Leonard, and Sill had now failed repeatedly to oust Brower from his position of authority. The next move would be up to Brower. He was not the
type to end a conflict in indecision.

16. Campaign

David Brower made his move in January 1969. It was the most difficult decision of his life. He announced that he would run for the Sierra Club's board of directors in the April election and resign as executive director if he were elected. The nation's most prestigious environmental organization was in crisis, Brower declared, beset by "allegations, bitter attacks, infighting, and damage to members' faith in their organization." It was time to "get the club moving forward again." He wanted to launch the international book series, expand overseas, oppose Diablo Canyon, and do whatever else was needed to protect the environment. His offer to resign was a gamble. The bylaws prohibited the executive director from sitting on the board. But if Brower and his supporters were elected, they would constitute a majority of the board, and they could begin to change the rules.[1]

It was a gutsy decision. "There were a lot of seconders, but no movers," he commented ten years later. "I thought that one of the ways to get these things done would be to be there and to make a presentation to the board and to move what I thought would accomplish it, and then just hope for a second." Too many board members sat "there like lumps," waiting for critics to kill yet another of Brower's ideas. "I wanted an end to this business of having the motion put by the person who wanted to eviscerate what I wanted to do," he said.[2]

It had been more than three months since the board of inquiry had convened its gothic-like hearing and more than a month since Brower had delivered a detailed written response. Yet the conflicts between Brower and his enemies continued. More bills were coming in from London, expenditures that Brower had authorized with no one's knowledge. One problem was the money Brower had given to Allan and Robert Horlin so that they could go to the Seychelles, India, and Africa. He wanted a European-African-Asian version of Terry and Renny Russell's highly popular book *On the Loose*.[3] The club's finances were tight, but Brower was still spending on publications no one knew anything about.

The latest, mailed to every member, was called the *Sierra Club Explorer*. Brower had wanted to promote the new two-volume Galápagos books in the October issue of the *Sierra Club Bulletin*, but board president Ed Wayburn blocked the full plan. Brower warned Wayburn that without proper promotion the club was going to lose book sales. Wayburn told Brower to find some other way to market the book. Brower took the material already set in type and without consulting anyone produced the special edition of what he called the *Sierra Club Explorer*. The cost was $7,166, about $2,000 more than printing one edition of the *Bulletin*.[4]

Wayburn was running out of patience. "I had allowed part of the ads for the book to go into the [October] *Bulletin*, but not the extensive copy [Brower] wanted," said Wayburn. "These were acts of direct insubordination." In January 1969, the board told Brower never to produce another *Explorer*.[5]

The *Explorer* also reported on another of Brower's projects, the new international series Earth's Wild Places. Despite the setbacks in 1968, Brower still planned to publish one hundred titles. In the next five years, he declared, "I estimate that about $80 million will have been spent in spreading around the world the message the series will contain about the essential need for wild places and diversity." He was also considering high-quality television programming, beginning with a documentary tentatively titled "Farewell to a First Lady," a tribute to Lady Bird Johnson. The program would honor her commitment to conservation issues, and she considered but declined the proposal.[6]

August Fruge, who was still chairman of the Publications Committee, remained opposed to these expansions of the publications program. "We have to make publishing the servant of conservation," said Fruge, "and I'm afraid that a huge program of this kind means that publishing is undertaken for its own sake and not for others."[7]

The plan to expand overseas was also in trouble. Opponents cited a technical reason—the club's original incorporation papers described it as only a domestic organization. Lawyers said bylaws and incorporation papers should be amended before starting ventures outside the United States.[8] Some of Brower's supporters found these concerns ludicrous, but the opposition was firm, and it superseded Brower. Well into the 1970s, two factions on the board consistently opposed expanding internationally: one was a politically conservative bloc against any change, the second a more politically liberal bloc that believed such expansion was a form of environmental imperialism.[9]

In December 1968, the board considered the international book program and set some ground rules. Brower was permitted to search for funding. In the meantime, the Publications Committee would review the international book proposal and make a recommendation to the full board. Brower was not to take any further steps that committed the Sierra Club to the international project.[10] The Publications Committee met on January 10, 1969, and the meeting did not go well. Fruge and his allies had arranged the meeting knowing that none of Brower's supporters would be there. Brower worried that the committee was going to kill his plans. He became increasingly upset and angry. "This matter is important, I think it is of vital importance," Brower stressed at the meeting. "I think it is one of the most important things we have to do." Exchanges became testy, speakers interrupted each other, and arguments strayed and became personal. Fruge, exasperated, addressed Brower.

"I'll tell you very bluntly what I think, I think you'd get a lot farther if you would ever give an inch," he told Brower. "If you would just give a little once in a while, you'd get a lot farther."

"The story of my life," responded Brower.

"Dave, I know the story of your life."

"No, you don't, you don't."[11]

Fruge had known Brower for twenty-five years; he had supervised him at the University of California Press and had helped him get the job at the Sierra Club. For nearly a decade, they had been battling at these sessions.

Brower did tell the committee that he was paying $600 a month out of his discretionary fund to assist authors working on several of the books. He planned to publish three in 1969.[12] Meanwhile, he made no mention at this or other meetings about his work for the past months on a secret project to advertise the book series. Rumors had begun that Brower was up to something secret and big. One report was that an advertisement would be published in the *New York Times*. When Wayburn heard these rumors, he sought out Conservation Director Michael McCloskey, who had been briefed by Brower. McCloskey demurred. "I felt that I could not betray my boss," said McCloskey.[13]

The next morning, January 14, Wayburn was home sick in bed with the flu when he got a call from a newspaper reporter. Had Wayburn seen that morning's *New York Times*? Wayburn had not.

"Well, there's a full page-and-a-half ad on Earth National Park signed by Dave Brower, and I understood that this was not allowed without your permission," said the reporter.

"I knew nothing about it," said Wayburn.[14]

It was an unusual ad. Below the banner headline "Earth National Park" was a small photo of the earth taken from space and a lengthy address from Brower about the need to preserve a fragile earth. He said that the Sierra Club was doing its part by embarking on a new international publishing program "to export the view that it is now the entire planet that must be viewed as a kind of conservation district within the universe." Attached were coupons, including one seeking donations to support the Earth's Wild Places publication program.[15]

Wayburn was so angry that he wanted to fire Brower right then and there. He told Phillip Berry, a lawyer and board member, "This is the last straw. This is in violation of two different board policies and is very deliberate."[16]

Berry urged a suspension and not a firing. Brower was back East,[17] and by the time he returned to San Francisco, Wayburn had calmed down. Wayburn asked Brower for an explanation. Brower contended that he had not overstepped his authority. And yet, as Brower himself would admit later, "I was skating on fairly thin ice." He had directed an employee to notify Wayburn the night before the ad was to run, but that never happened. Brower admitted he had not consulted with Wayburn because the board president would have cancelled the ad. He nevertheless argued that it was within his discretion as part of his promotion responsibilities. "The thoughts enunciated in it were essentially what the Sierra Club was working for in all of its fields," he said. "It wasn't anything new there."[18]

On January 28, Wayburn suspended Brower from making any financial commitments for the Sierra Club. Brower said Wayburn did not have the authority to do that, only the board. Ansel Adams told Wayburn not to pay the bill for the ad. Richard Leonard saw the ad as "the breaking point" in Brower's diminishing support.[19] But what would the board do?

It met on February 8. Wayburn asked the board to back him up. Brower again argued that the ad was within his authority. Others joined the debate. At one point, Brower declared that he had consulted Wayburn about the ad. Wayburn was adamant that he had not.

"Well, I don't think you remember this at all," said Brower.

"You had every opportunity to tell me," replied Wayburn.

The board tentatively agreed with Wayburn by a seven to six vote.[20]

Brower thus lost the first major skirmish in his quest to take total control of the Sierra Club. In his journal for February 8 and 9, he wrote in the margin that he had wasted nearly three days.[21]

Or had he? McCloskey argued that Brower had purposely provoked the Earth Island ad controversy. "It was clear to me at the end from discussions with him that he wanted to provoke a showdown; he wanted to have it out once and for all as to who was running the organization," said McCloskey. "Either I run it, or you run it. There is no in-between ground."[22]

At that February board meeting, Brower requested a leave of absence until the election was completed. It was unanimously approved.[23] Two opposing slates were forming. The Committee for an Active Bold

Constructive Sierra Club featured Brower, and two current board members, Fred Eissler the antinuke advocate, and Dave Sive, the lawyer from New York whom Brower had asked to represent him at the board of inquiry hearing. They added Brower's old friend from Seattle, Polly Dyer, and a young man from Washington, D.C., George Alderson, who had been doing volunteer work for Brower but whom very few people knew. The opposing anti-Brower slate, Concerned Members for Conservation (CMC), featured Ansel Adams and August Fruge. They added Maynard Munger, a young volunteer from San Francisco, and Raymond Sherwin, a California Superior Court judge. Wayburn was also up for reelection, and both sides considered adding him to their slates until the Earth Island ad controversy. Then there was only one choice, and the CMC added him.[24]

Campaigns messages were stark, and they all centered on Brower. The CMC slate said Brower was financially irresponsible, unresponsive to directions, and dictatorial. The CMC candidates were endorsed by all twelve of the club's living former presidents and key members in nineteen chapters around the nation. Endorsements were especially strong in California. Brower and his allies charged that their opponents were lying and censoring the truth to end Brower's successful and aggressive conservation message. If the CMC won, the Sierra Club would revert to its days "as a society of companions on the trail" when conservation was unimportant.[25]

Brower's biggest advantage was that he was the most famous conservationist in the nation. For years, club leaders had been afraid to go too far in challenging him, fearful that without the charismatic leader the Sierra Club would fail. Brower's verbal abilities worried opponents such as Adams. "Dave is a most compelling person," he said a few years later. "He can tell you that two and two is five. If you don't watch out, you may believe it."[26]

In the beginning, however, Brower found himself cut off from communicating directly to most of the 78,000 Sierra Club members. Brower ally Hugh Nash edited the *Sierra Club Bulletin*, but Wayburn controlled what could be published. When Nash wrote about Brower's decision to take a leave of absence and run for the board, Wayburn severely edited the notice, but no one cut Wayburn's lengthy article about why he had suspended Brower. That discrepancy outraged Eliot Porter, who told Wayburn his editorial censorship was biased and inequitable.[27] Brower had even greater difficulty accessing the chapter newsletters because over the years he had antagonized so many chapter leaders. Many of those newsletters were reporting the controversies in ways that rarely reflected well on Brower.

Stifled, Brower responded with a threat. He hired a local lawyer, Henry Siegel, who sent letters to at least twelve editors of chapter newsletters warning them to stop these libelous reports or face a lawsuit. This notice only inflamed emotions, produced more stories, and prompted a legal defense fund to combat any Brower lawsuits. Leonard told the threatened chapter leaders that Siegel's letter was only a bluff designed to suppress unpleasant facts that Brower did not like.[28]

Brower was more successful in wooing the press, eventually garnering a fair amount of television and newspaper coverage both in California and throughout the nation. United Press International portrayed the conflict as a struggle "between traditionalists who favor the 77-year-old group's image as a hiking club, and the progressives who believe the club should vigorously attack all environmental pollution and decay." The *Seattle Times* reported that Brower's candidacy was "being strongly opposed by a slate of conservative old guardsmen who, he said, all reside in the San Francisco area and lack his slate's wide geographic distribution." The story made no attempt to get the other side. Sometimes Brower's paranoia would kick in, such as when he "hinted" to a reporter for the *Wall Street Journal* that outside forces such as lumber companies, public utilities, and government agencies were somehow responsible for the dispute. He offered no proof.[29]

He also traveled, carrying now a far more personal message in meetings, speaking engagements, and debates with Sierra Club opponents. Between February 21 and April 2, Brower went up and down California, back and forth across the country, stopping in New York for appearances on the nationally broadcast Martha Dean and Arthur Godfrey radio programs.[30] Rather than attacking his opponents, he toiled over long position papers defending his handling of the London office, the IRS appeal case, his management and financial record. Sometimes he would go on the offensive. At a March press conference, he said that a major oil spill in January off of the Santa Barbara coast might have been prevented, but a year earlier Sierra Club conservatives had not allowed him to legally challenge offshore oil leases.[31]

The CMC slate chose its most famous member, Ansel Adams, as its spokesmen. But Adams could be strident, threatening to ask the California attorney general to investigate Brower, the dictator. Other times, he was too gentle. Bruce Kennedy, a CMC partisan, complained that in one recent press interview Adams sounded like a "reactionary fuddy-duddy," the same type of fool whom Brower had been depicting. After another press skirmish, CMC ally Tom Jukes said that opponents were using a peashooter against an attacking rhinoceros. Meanwhile, Brower was become adept, said Jukes, in brainwashing every reporter.[32]

The key for the CMC was not the public campaign but the quiet, behind-the-scenes discussions of Brower's failings by people of influence within the club. Such discussions were especially effective in California. Richard Searle believed that many members were most likely confused by all of the accusations. But the CMC's advantage was that so many former leaders were on its side. "It probably tipped the scales very predominately," he said. "The membership probably thought, 'I don't know what all the details are, but it's obvious to me that these are responsible people.' So I think the endorsement served to a great extent in influencing the membership."[33]

Some CMC campaign statements were flat wrong, however. One accused Brower of spending $1,800 for a Christmas party. The actual bill was $161, and the party was tied to the promotion of the new Central Park book. The CMC also contended that one or more New York publishers were bankrolling Brower's campaign, a rumor that it never documented. The most misleading charge came from Dick Leonard, who claimed that the books program had lost hundreds of thousands of dollars but kept amending the figure, however, with losses ranging between $130,000 and $327,000 over a period that ranged from one year to five years.[34]

The clashes between the two sides were not always mean and nasty. A two-page play script called "Prometheus Unboundaried" began to make the rounds. It starred the Great Hero. No matter what the problem, the Great Hero would arrive to save the day. The author was not listed, although it was obvious that Brower was the Great Hero.

"I sent that first play to seven people," said Phil Berry, the lawyer who as a child had worshipped Brower and who wrote the script. "In a few weeks it had gone to a couple of thousand."

At the next board meeting, Great Hero Book Club buttons were being exchanged. Few yet knew the author's identity. Brower thought that it was funny, and he approached Berry. "Phil, I want a signed copy from the author," he said.

Berry gave him one.[35]

But light-hearted moments were the exception. The fight was hard on the Sierra Club staff, in particular those loyal to Brower. Brock Evans, the Northwest field representative, said he would be "transfixed" at board meetings by how his hero "was getting beat a little bit . . . inch by inch." He added, "I never believed for one instant that Dave was doing any of the things that they said he was doing. I thought it was all a pack of lies and all bad stuff."[36]

So did Hugh Nash, the editor of the *Sierra Club Bulletin*, who clashed often with Wayburn. Finally, Wayburn suspended Nash for a month without pay. That action prompted yet another press conference in which Brower attacked Wayburn and Nash contended that the suspension was politically motivated. Employees were warned to stay out of the fight. Some did, but others openly campaigned for Brower.[37]

Then Wallace Stegner attacked Brower, and suddenly the fight became far more personal. Stegner originally had no wish to become embroiled in the dispute. Originally recruited into the club by Brower, he had become good friends with Adams. As Adams became more critical of Brower, so did Stegner. By 1969, Stegner had been off the board for several years when Brower called one day seeking help. He reached Mary Stegner. She told him that her husband was busy at the moment writing a new novel. According to the Stegners, Brower had never been that impressed with fiction writing, a point he repeated again while on the phone with Mary. In an increasingly harsh voice, he suggested that Stegner should instead be helping to save the planet. Stegner was furious when his upset wife relayed the exchange. The result was a letter to the editor of Stegner's local newspaper, the *Palo Alto Times*, which the CMC reproduced and distributed widely. Stegner emphasized that the dispute was not between progressives and conservatives and that the charges against Brower were serious. Brower had once been a great crusader, but he "has ceased to be what he was. He has been bitten by some worm of power."[38]

Stegner's column cut not only Brower but also his family. Both Anne Brower and Ken Brower wrote replies. Anne's letter was published in the *Palo Alto Times*. She said the attack was not written by Stegner the historian, who would have checked his facts, but by Stegner the creative writer, who was uninformed. "And thus Wallace Stegner, the man, without first-hand knowledge, joined in a public attack on the integrity and the career of an old friend," she wrote. She would leave it to others to determine whether Stegner had "been bitten by the worm of envy."[39]

Young Ken Brower, now in his mid-twenties, was horribly confused by the attacks. In a three-page typewritten letter to Stegner, he asked why Stegner had not sought out his father to get his side of the dispute. Ken had grown up knowing and admiring this great author and others who were now publicly vilifying his father. Now he was wondering why his father's style so infuriated those old friends. He could only conclude that they were envious and resentful of his father and the freedom he had to attack special interests. The resentment was apparent, he wrote, and he hoped that it was only envy and not venality.[40]

It was a difficult time for the entire Brower household. The children were old enough now, in high school or college or in Ken's case working for the Sierra Club, to be caught in the conflict's rhetoric. It was because of the children that Anne Brower insisted that she and her husband seek out a lawyer. She wanted the letters threatening action for libel to stop the smears, the anger, the pain.[41]

Joining in this fight could not have been easy for Anne Brower, her husband's quiet confidante. Now that the children were older, she was traveling more with him, attending banquets, standing by his side, but always slightly in the background. If she could help, Anne Brower the editor stayed up nights helping Brower on the latest book, an unnamed collaborator. She had never publicly interfered in club business—until now. She had attended the September 1968 board meeting when Leonard, Adams, and Dick Sill had verbally attacked Brower. The Leonards, the Adamses, and even Sill had been friends with the Browers. Some of those relationships went back more than twenty-five years. After that meeting, she wrote to Leonard, demanding an explanation for his action. Leonard demurred, explaining that he was not free to supply answers or supporting documents, as she requested. That answer was difficult to accept, she told Leonard, because he was uninhibited in discussing the charges with mutual friends. She told him that she was disgusted.[42]

She wrote other letters, sometimes to strangers, sometimes to one-time friends. Each of the recipients had criticized Brower, and she wanted to know why. One went to Nancy Newhall, the editor who worked with Brower on the early Exhibit Format books. Newhall had remained close to Adams, and she had harshly and publicly complained about Brower. Newhall wrote back to Anne, commending her loyalty but urging her to seek out the truth. The facts were well documented, said Newhall; they were not merely fantasies, smears, or vituperations of a personal nature.[43]

Never would Anne Brower see the assassins making these assertions, accept them, or absolve them. She told another accuser of her growing bitterness and anger about the letters and articles about her husband that included distortions and innuendos. Not only did they vilify, but they also made no attempt to correct the false accusations or get the other side. How could they all be so blind to what her husband had created over the past seventeen years? What more could they ask than what Brower had already given?[44]

The Browers had not had a conventional relationship. Brower had strayed, not just sexually but physically. He was in New York, Washington, or London or simply on the road often more than half the time, year after year. He never sought riches; the couple's house was unkempt but comfortable. He was opinionated and exasperating. When he was home, he worked nights and weekends, sometimes toiling the entire night at home or the office. But Anne Brower remained loyal, faithful, and devoted.

It would take years for Anne to forgive these critics, far longer than it would take her husband. Daughter Barbara could remember the couple driving to Carmel to see Adams, and Anne's refusal to get out of the car. "She is totally loyal to him," Leonard said a few years after the clash in 1969. "So loyal that she wouldn't talk to Ansel or me for at least two years after our first motion to dismiss in 1968. Dave would, but Anne Brower wouldn't."[45]

The election results were supposed to be in by April 12, 1969. They were not. Nor did they arrive April 13, 14, or 15. Brower had come home from his campaigning. Anne needed him; she was ill. As the waiting lingered, the speculation and rumors built. Both sides had vowed to quit if they lost. Each planned to begin a rival organization. On April 15, two election judges told Brower that the turnout had been high. One guessed that it was close, the second that it looked bad for Brower and his slate. The same day, Brower issued a statement announcing that the election was so unfair that he was changing his mind about resigning. He had thought that the campaign would be fairly fought over the issues, but membership never had the opportunity to hear those issues. Therefore, he would not resign.[46]

Finally, on April 16 at 10:27 A.M. he got the call. It was bad. The entire CDC slate of five candidates had won. Wayburn had received the most votes, 28,120—double Brower's sixth-place finish. Neither Brower nor anyone else on his slate had collected enough votes to serve on the board. The Diablo question was soundly thrashed by a three-to-one margin. Brower spent the afternoon and evening talking to reporters and friends. Even though there had been warnings, he was shocked by the results. He wondered if it was too late to make a deal, to trade, to somehow stay on at the Sierra Club. But in his journal he questioned if he could work in a "whorehouse."[47]

On the tenth floor of Mills Tower at Sierra Club headquarters, the phones did not quiet until a little after 4:00 P.M. John Flannery, chief aide to Wayburn, felt a sense of gloom in the air. It was, he realized, as if a family member had just died. Flannery, like many others, wondered how anyone could fill the gap that must inevitably occur with Brower's departure. Was the nation running out of individuals who would oppose destruction of the environment without compromise? Flannery asked. Probably not, if one talked to the victors. Conservation was bigger than just one man, he had heard them say. He wondered if they were right.[48]

In the following days, Brower received hundreds of phone calls and letters. Some supporters were quitting the Sierra Club in protest. It is impossible to understand how people can succumb to mudslinging, wrote Paul and Liz Wilson. Dale Jones of Seattle said his heart had gone out of the entire conservation movement with this defeat. Newspaper columnists and editorial writers debated whether the club would unite or splinter. Would the Sierra Club survive this war?[49]

Brower debated his options. On April 25, he decided that he could compromise, that he could live in the whorehouse. He wrote to the board and the Publications Reorganization Committee that he would do whatever he could to unite the club in the eyes of the board and the public. He still had the same goals as the board; he had only differed on the "procedures." He offered seven specific concessions of conciliation on Diablo, the books, advertisements, and conferring more often with volunteers.[50]

His compromise offer came too late. The anti-Brower forces held their first meeting at Siri's house on April 19. They agreed that Brower would be fired at the next board meeting on May 3. Dissident employees would be fired; loyal or neutral staff would be retained or rewarded. On April 25, after a second caucus four nights later at Leonard's house, Wayburn called Brower. It was the same day that Brower sent his concession plea to the board. Brower was in New York. Wayburn reported that when he had proposed Brower be kept on to run the publications program, the new board laughed it off. Wayburn suggested that Brower be given six months' severance pay. Or five minutes' pay, someone responded.[51]

It was after 11:00 P.M. that night when Brower wrote to Ken and described the phone conversation. Brower was in New York because he was making contingency plans with friends to begin a new environmental organization. He was flying to Washington, D.C., in the morning for more discussions. He needed a name for the new group. He told Ken that he was thinking of writing a book on the debacle. And he was thinking of filing a lawsuit to force disclosure of what he called "the truth."[52]

On the morning of May 3, more than three hundred people gathered in the Empire Room of the Sir Francis Drake Hotel in San Francisco. Brower had written his letter of resignation. His new environmental organization still did not yet have a name. Anne Brower had suggested "Friends of the Earth," a takeoff from "local friends of the library." Some thought the name awkward, but it would stick and would be announced in a few days.

That morning, the *San Francisco Chronicle* reported that a massive redwood had fallen in Yosemite. There, on the front page of the newspaper, was a photograph that Adams had taken years earlier. Adams held up a copy of the newspaper and pointed it out to onlookers. The giant tree had been hollowed at its base and inside the tunneled portion was an old Pierce Arrow automobile. Adams pointed to a figure standing beside the tree, little more than a blur. It was a young David Brower, Adams said. It had been taken in the 1930s when Brower often assisted Adams, as he had that day. The headline above the photo: "A Fallen Giant."[53]

The anger, tension, and emotion of so many past board meetings was missing on this day—at least until Martin Litton grabbed a microphone and in resentment declared, "This election has been rife with perjury, calumny and fraud."[54] Litton was the exception; otherwise, civility prevailed. It came time for Brower to give the traditional report from the executive director. He stood. He was still an imposing figure at six feet plus, the hair now silver, the face still handsome. The room was silent; the crowd, respectful. He read from his typewritten notes. His tone was flat, but his words were still sharp. He told them he would have liked to have remained; he had even suggested a way to make that happen. It was not to be. He would be moving on, forming a new environmental organization. "We cannot be dilettante and lily-white in our work," he declared. "Nice Nelly will never make it."[55]

The "nice Nelly" remark grated on Leonard's ears, but when Brower finished and the applause began, Leonard and others on the board joined in. As the audience rose, so did the board members, connecting respectfully, politely, but not enthusiastically. Brower listened a moment. He shook hands awkwardly with Mike McCloskey, who was slated to replace him. Then Brower left the room. Many in the room had tears in their eyes.[56]

Brower's letter of resignation was offered and accepted by a vote of ten to five. Phil Berry, who was elected president, promised that the Sierra Club would not back down in its conservation efforts. "I do not foresee any change whatsoever in the militant conservation stand of the Sierra Club," Berry told a reporter. That night the Sierra Club held its annual banquet amid the elegant Victorian splendor of the Claremont Hotel in the Oakland hills. Although Brower was leaving as the boss, he had vowed to remain a Sierra Club member. Handed the microphone at the banquet, he proclaimed, "There's lots of work to do—let's."[57]

Before the board directors went their separate ways the next day, Brower's young staff of conservation activists—Jeff Ingram in the Southwest, Brock Evans in the Northwest, Gary Soucie in the East—had a private meeting with them. "We just said, 'Look, we're with you if you can just commit to us that this is going to be the same strong Sierra Club without Dave Brower,'" Soucie said.

Added Evans, "We don't want to work for anyone who won't fight."

"Of course we are going to fight," Evans remembered board members saying. "You'll see us, you watch what we do. We had to do this. We regret it too, but we had to do it, and we'll fight too. We'll show you."[58]

For some, this assurance did not matter. Soucie left the Sierra Club to help lead Friends of the Earth; Ingram also quit. The apartment in New York that Brower had rented on behalf of the Sierra Club was closed, as was the London office. Brower took the concept of the international book series with him, and the Sierra Club struggled financially for several more years.[59] Evans stayed on, eventually heading the Sierra Club's Washington office. His naiveté about the Brower conflict gradually began to dissolve.

"Over the years to follow, just every now and then, I would hear another reference about still trying to pull ourselves out of the financial hole," Evans said. "So it finally dawned on me that, gee, that was true. We were in a hole."[60]

Yet for years afterward, the contention continued that the election in 1969 had been a battle between the conservatives and the progressives, between fighting and compromising—a clear-cut, black-and-white conflict. Both Brower partisans and supposedly more neutral reporters, writers, and filmmakers proclaimed this storyline so often that it took on a life of its own.[61]

Anne Brower once told a friend that Brower truly believed that the issue was between a forceful club for the environment and a lesser, more diminished organization. He did not have the political savvy, she said, to pretend anything differently.[62]

The 1970s and the arrival of Earth Day marked a new surge of environmental activism. It was virtually impossible for the Sierra Club to retreat from protecting wilderness in Alaska and elsewhere, fighting air and water pollution, and eventually developing an energy policy that opposed nuclear power.[63]

It was never easy for Brower to discuss his departure from the Sierra Club. For years, at virtually every speech he made, he would be asked, How did it happen? How had he left the Sierra Club?

"Well, I was fired," he would reply. "I had to walk the plank."[64]

Brower wrote three personal books about his life and his values, including two that were labeled autobiographies. At the close of the third book, he mentioned that he suspected he was fired primarily because of his opposition to Diablo. "Until that moment, my other faults, which were beyond number, were tolerated, even forgiven," he wrote.[65] The number of pages in the three books total more than a thousand. That solitary reference, however, was the only time he wrote about why he had to leave the post, the calling, that he loved more than any other.

17. Echoes

David Brower began again.

He called his new organization Friends of the Earth, and it was clearly designed to be David Brower's organization. He removed all of the bureaucratic encumbrances and rules that he loathed. It was originally based in New York, far from those vexing California-based critics of the Sierra Club. It had no insufferable chapters or chapter officers to answer to. The board of directors was composed of loyal Brower boosters, there only to provide advice, not to direct Brower. "What I wanted to do was what our policy was," said Brower.[1] Friends of the Earth was small, but it was agile, quick, and single-minded. Finally, Brower believed, he no longer had to answer to anyone.

It would take years before he realized how wrong he was.[2]

Brower unveiled the new organization at a press conference on September 16, 1969, four months after he left the Sierra Club. It received some attention, but reporters were also intrigued by a different and secondary Brower organization, the League of Conservation Voters. The group, which Brower created and then spun

We need a new conservationism. It's got to mean more than using up resources at a slower rate. We've got to do something about restoration, hanging on to the things we cannot replace and we have to restore as best we can.

It takes quite a bit of confidence to try to restore nature. Nature knows what to do. At least we can get a start instead of getting in the way. Conservation, preservation, restoration, restoration of our human system.

DAVID BROWER, IN DAVID KUPFER, "FINAL INTERVIEW: DAVID R. BROWER" off independently, would successfully work over the years to elect environmentally friendly politicians. Brower also talked at the press conference about how he would be a director on the John Muir Institute, which would work closely with Friends of the Earth. "The earth needs a number of organizations to fight the disease that now threatens the planet: Cirrhosis of the environment," declared Brower.[3]

It was an exciting time. Brower had snatched the John Muir Institute from the Sierra Club. A young New Mexican, Max Linn, headed it. Tax-deductible donations would go to Muir, and then Linn and the Muir Institute would help fund Friends and its aggressive environmental campaigns. The two organizations would be so symbiotic that salaries and budgets would be intermingled. Brower convinced Robert O. Anderson of the Atlantic Richfield Company to give the Muir Institute $80,000. McCall Publishing Company was so impressed by Brower's literary alchemy that it provided $30,000 in advances for the new international publication series that Brower would edit. He turned those funds over to Friends.[4] Although the Friends headquarters was in New York, Brower also created a San Francisco office and staffed it with his loyalist disciples fired from the Sierra Club. He recruited Joe Browder, a young television reporter turned environmental campaigner, to head the Washington, D.C., office. In its first eighteen months, Friends of the Earth made a name for itself by taking on and winning the fight to prevent the federal government from building the controversial Super Sonic Transport aircraft, a fight that other environmental organizations hesitated to join.[5]

Earth Day in April 1970 signaled a watershed moment. Conservation was evolving into environmentalism. Brower was there for the first Earth Day, and his agenda at Friends of the Earth mirrored the movement's transition. The organization embraced the spirit of the past, fighting to protect endangered species including the California condor, wolves, snails, and others; it opposed whaling, dams, the fur trade, and the Alaska pipeline; it fought for the Alaskan wilderness—no, it fought for all wilderness and all roadless areas. At the same time, its agenda also expanded into areas once *verboten*. It opposed toxic chemicals, air pollution, water pollution, overpopulation, nuclear energy, the B-1 bomber, pesticides, and herbicides; it was for the bottle bills, waste recycling, energy conservation, and hypercars.

By the twenty-first century, energy, the state of fossil fuels, and their impact on the atmosphere would be of paramount concern, so in the 1970s Brower was still clearly ahead of many of his contemporaries. He embraced others who were also ahead of the pack: physicist Amory Lovins and the energy soft path, the concept that energy conservation trumps all alternatives; Paul Ehrlich and his worries over population growth; Donella Meadows, who preached on the limits to growth; and Lester Brown in his advocacy for world peace.

Brower thought big and fought bigger. He worried that the earth was on the cusp of nuclear holocaust, that the world stock of resources was being irretrievably vandalized, that world peace was stretching beyond our reach. His answer? Hold a conference and title it "Fate of the Earth," or make signatories embrace a document and call it the "Magna Carta II," or celebrate America's bicentennial by embracing a better earth.

Many of his ideas fell like seeds on granite, but his failures only prompted him to strive harder, to reach higher. He was traveling more, throughout Europe as well as to Russia, Japan, and New Zealand. Nothing seemed beyond the scope of the environmental cause anymore, certainly not nuclear disarmament, which finally became his paramount concern. The way to stop nuclear proliferation, he claimed, begins by opposing nuclear energy, not just at home but also abroad.[6] Brower was convinced that as nuclear technology was introduced into nations around the globe, it would lead to the development of weapons. Thirty-five or forty countries would soon be armed, he predicted, and by the 1990s it would inevitably lead to global atomic war. "The thing to do is to stop it," he said in an interview in the late 1970s. "There's no more important issue in my life now than to do everything I can, and see that Friends of the Earth does everything it can, here and abroad to stop the nuclear experiment before it's too late. There'll be no environment if we don't."[7]

Perhaps Brower's greatest accomplishment during this period was to recognize that environmental issues transcended national boundaries (figure 17). As the United Nations began exploring environmental

FIGURE 17 "Endangered Earth"
After Brower left the Sierra Club in 1969, and the conservation movement flowered into the environmental movement, he became increasingly concerned about the future of the earth. He created two environmental organizations, Friends of the Earth and Earth Island Institute, which carried out many of his concerns. (Courtesy of the Bancroft Library, University of California, Berkeley)

issues and hosting conferences devoted to the environment that were held in Stockholm, Geneva, and then Nairobi in the 1970s, Brower and Friends of the Earth staff were there both as advocates and as reporters informing others.[8] It quickly became clear that a U.S.-based organization was not sufficient to solve the world's ills. The solution—international chapters of Friends of the Earth. Brower, aided by international lawyer Edwin Matthews, began this international effort by helping a handful of Parisian allies form Les Amis de la Terre only one year after Friends of the Earth was established in the United States. Then they assisted in the formation of a Friends of the Earth chapter in the United Kingdom. The idea of a domestically based environmental activist organization in individual nations spread to the Netherlands, Honduras, South Korea, Uganda, and elsewhere. Surprisingly, Brower wanted no control of these groups. He only insisted that they use the name "Friends of the Earth" in either English or their native tongue. Each national organization would be autonomous, politically and financially. They did recognize that some issues crossed national boundaries, and so they met regularly, but none had a greater voice or vote than any of its neighbors.[9] These chapters filled a need for environmental activism in nations that had never before considered the concept. Greenpeace would follow and take yet an even greater hold on environmental consciousness internationally. It was only years after Brower's death, as global climate issues began to become increasingly paramount, that it became obvious how far Brower had been ahead of his peers. Friends of the Earth became so rooted in many nations that even today it remains a thriving and vibrant political, social, and environmental force. Few remember the name "Brower" or the role this man played in it.[10] They should, said Brent Blackwelder, a former director of the U.S. Friends of the Earth. "I would give Brower credit for the great global vision," he commented. "Brower was the one with the clarion voice, and it was a voice that carried internationally."[11]

What helped Brower the most was the growing recognition of him, aided in part by the publication of *Encounters with the Archdruid* by John McPhee in 1971, portions of which had already been published in the *New Yorker*. The story of Brower's battles over the Grand Canyon would be consumed not just by a first generation of readers but by young readers for years afterward in environmental studies classes on college campuses. Brower was adored by the young, who identified with his no-compromise stands, and his popularity would rise even further when he came to campuses and delivered his favorite stump speech, "The Sermon." Relying on the gospels of John Muir, Henry David Thoreau, Buckminster Fuller, and others, Brower in speeches year after year and eventually decade after decade wove his tale of the creation. He would say that if the earth is 4 billion years old and that time period is condensed to six days, we can say that humans arrived on the scene at three minutes before midnight on the last day, and Christ arrived one-quarter of a second before midnight. But humans are consuming resources at a rate comparable to an automobile going 128 miles an hour, and he said the car is accelerating.[12]

Brower was the evangelist, the apostle, the messiah drawing the young, who would become pilgrims to the cause. David Phillips was one of those acolytes, first meeting Brower while the great man was giving lectures at the University of Montana in Missoula. Phillips was about to graduate with a biology degree with an interest in endangered species. Over coffee, he described to Brower how he was interested in megafauna and mini-fauna, and as Phillips talked, Brower grew more excited. Finally Brower exclaimed, "We have to have that! We totally need that. You must come and work for Friends of the Earth." Months later when Phillips arrived in San Francisco, Brower was out traveling, and no one was expecting him. That was normal at an organization that specialized in what was called "build your own job." Phillips got a desk, but he worked for six months before he finally obtained a grant so that he could get on the payroll.[13] That was also how Friends, with only thirty thousand dues-paying members, could support a staff of more than sixty not only in San Francisco but in Portland, Maine; Lansing, Michigan; Ames, Iowa; Eau Claire, Wisconsin; Minneapolis; Essex, Illinois; Crested Butte, Colorado; Salt Lake City; Tucson, Arizona; San Diego, Santa Barbara, and Arcata, California; and Spokane, Washington. Ed Dobson said he worked as a "green slave" in Montana, dependent on others to pay his rent, light bill, and groceries. Boyd Norton was the Denver representative, compensated with medi-cal benefits and an occasional plane ticket to Washington, D.C., to give needed congressional testimony. The nucleus remained in San Francisco, where, according to Stephanie Mills, Brower recruited some great minds, "creative dissenters for whom friction is elixir." Mills said "the result was a surprisingly functional amalgam of the quixotic and [the] purposive."[14]

To those willing to take orders, especially those young, idealistic, and eager to please this patriarchal messiah, it all seemed so wonderful. Yet friction was growing among those who needed to find the money to satisfy Brower's appe-tite. This irritation would grow and eventually lead, fifteen years after he had created Friends of the Earth, to his firing. The scenario was eerily similar to his downfall at the Sierra Club in 1969. Once again, former friends and sup-porters were responsible. They accused Brower of financial irresponsibility, mismanagement, and failure to follow the board's orders.

"We still admired him, but he was impossible to satisfy," said Avis Ogilvy Moore, a board member and the widow of Stewart Ogilvy, one of Brower's staunch supporters in his final days with the Sierra Club. "We could never raise enough money, and we never could do all of the things he wanted done, and he would not take any limitations."[15]

Perhaps everything would have worked out if the book publications had succeeded. An early paperback, *The Environmental Handbook*, sold more than 1 million copies. But Brower's series of eleven international books, although splendid in their depiction of the Alps or Micronesia or other locations, failed to connect with readers. By now, the market may have been oversaturated; plus, the books were too expensive for the international market, and overseas destinations did not sell to the American public. A book on the Hawaiian island of Maui (1971) sold well, and *Wake of the Whale* (1979), describing the beauty and vulnerability of the world's great mammals, showed great promise.[16] But Brower's judgment sometimes frayed. After the Book-of-the-Month Club ordered 17,500 copies of *Wake of the Whale*, Brower increased the internal order to 63,000 copies, including 20,000 that would carry no text because he planned to fill it in later, depending on demand from particular countries. That didn't work. "We had never been able to sell more than 2,000 copies of a hardback on our own, but that didn't matter to Dave," said Bruce Colman, who worked in Friends of the Earth's publishing division.[17]

Brower had believed that at least 60 percent of the Sierra Club's growth came from book sales and that was how he wanted to build Friends of the Earth. But the deficits from the books kept growing, and by 1982 the organization's finance people were estimating the books had cost Friends $350,000 over the previous five years, which was about the same amount as Friends of the Earth's cumulative deficit. Over Brower's fierce objections, the book program was shut down.[18]

A few years earlier, in 1979, Brower had relinquished his position as full-time president for what was supposed to be a part-time role as chairman of the board. The problem was that he never did anything partway. After that change, Friends of the Earth went through a half-dozen administrators over the next few years. Most resigned because Brower was constantly countermanding their orders. Avis Ogilvy Moore, who remarried after Stewart Ogilvy died, was a charter board member of Friends of the Earth. She said the new administrators complained that after they instructed a staff member to do something, the employee would then "wear a pathway in the carpet over to Brower's office to find out what to really do."[19]

Brower still did not worry about finances, budgets, or deficits. Dan Gabel sat on the board of the Friends of the Earth Foundation, which was created to raise money for the parent organization. He remembered many arguments in which Brower wanted to launch a project when the money was not there. "His favorite expression," said Gabel, "was, 'I'm not in charge of the money; you're in charge of the money. You go get me the money.'"[20]

When Brower did seek financial help, he sometimes employed a highly unorthodox route—he borrowed from his friends. At the Sierra Club, Brower had sought loans from supporters to produce some of the early Exhibit Format books. Those loans were at least secured to a product, although the value of the books was admittedly a matter of opinion. But shortly after creating Friends of the Earth, he began soliciting loans from supporters to keep paying the rent and the payroll. Not only were the loans unsecured, but most of them were also not interest bearing. The amounts from individual lenders ranged from $100 up to $30,000. Hundreds or perhaps even thousands of loans were made over time.[21] Some were repaid; many were not.

In 1989, several years after Brower had left the organization, Friends of the Earth merged with the Environmental Policy Center. Joe Browder and former Washington Friends of the Earth staff had created this organization in 1972 after splitting with Brower. Auditors initially thought Friends of the Earth had a deficit of $300,000. After the merger, however, more records were found, primarily showing the small loans from supporters, and the deficit bloomed to $800,000. Blackwelder, who was involved in that merger, said that many of these lenders were now elderly or had died, some could not be tracked down, and even when they could be found, there was not enough money to pay them back. The newly reorganized

Friends of the Earth survived, barely, by cleaning up the books and wiping away or defaulting on the loans.[22] Experts in nonprofit management said that the solicitation of such unsecured loans is virtually unheard of in their business, and at least one questioned its legality. However, another expert added, "I do marvel at the ingenuity of this idea."[23]

Brower estimated that he contributed hundreds of thousands of dollars in speaking fees and other income to Friends of the Earth, in addition to loans that he forgave.[24] But none of it was enough to keep the organization solvent. By July 1984, when his differences with the board finally erupted, too many people had had enough. "He had been difficult to deal with all of the time," recalled Moore. "We got to realize that this was not going to change, and we could not do enough to help him, and it was not going to work with us."[25] The deficit then was $700,000, and plans were developed to drastically cut the staff. Brower responded by placing a full-page ad in the organization's monthly newsletter, *Not Man Apart*, protesting the budget cuts. When the ad was discovered before the newsletter went out, the press run was halted, the board met and fired Brower as chairman of the board, and the story made the front page of the *New York Times*.[26] The bitter internecine struggle continued two more years. It featured lawsuits and an election in which Brower ran so that he could take back control of the board. He finally gave up when he lost. Harold Gilliam, a San Francisco newspaper writer who had eulogized Brower's departure from the Sierra Club in 1969, sought to explain why it had happened again. The reason, he said, was Brower's "style of leadership[, which] may be traceable to his early years as a pioneer rock climber in Yosemite and on Sierra peaks. His high-exposure routes have taken him places most people would be unwilling to follow and have enabled him to reach heights no one else had previously attained."[27]

Brower was now seventy-two years old. A second firing was not supposed to happen at this time in his life. It was 1984, and the environmental movement had flowered in an unimaginable way from more than thirty years earlier when Brower had become executive director of the Sierra Club. He was now the movement's elder statesman, revered by the young, a veteran of the banquet and lecture circuit, sought throughout the world. His visions were grander now, his aspirations higher. But ever since he had been fired by the Sierra Club, his accomplishments could never meet his ambitions. Friends of the Earth, which he had now lost, had struggled, whereas the Sierra Club, under Mike McCloskey's solid and stolid leadership, was soaring, with membership rising to 300,000 and beyond. Everyone knew the Sierra Club; few had heard of the strangely named Friends of the Earth.

He began again.

Just as Brower had worked on creating Friends of the Earth before he was fired from the Sierra Club, he created another landing place before he was forced out of Friends of the Earth. Earth Island Institute had been established four years earlier, and it absorbed a key number of staff from the San Francisco office of Friends of the Earth. The institute's role was to be a laboratory for young environmental activists and scientists getting started. It was similar to the make-your-own-job concept at Friends of the Earth. Brower served as coach and cheerleader on projects that ranged from saving the rainforests to protecting dolphins from tuna fishermen. At the same time, the institute gave him the resources to travel the world, including three trips to Siberia's Lake Baikal as part of a campaign to protect the great Russian freshwater lake from pollution.[28]

If Brower retired in his early seventies, certainly no one could tell. In 1983, he was once again elected to the Sierra Club board. Because of term limits, he would serve on and off through 2000. "He's very engaged when you are talking about big picture and the issues," Sierra Club executive director Carl Pope said in an interview in 2000, when Brower was still on the board. "But he can grow impatient with the bureaucracy."[29] At meetings, Brower would often come unprepared, having failed to do his homework, and he annoyed others by pecking away at his laptop during meetings. He had never had any use for bureaucracy, and now he was on the board of an organization with nearly 500,000 members and a staff of more than 200. "Dave was largely disengaged as a board member," recalled Bruce Hamilton, a Sierra Club conservation director. In an earlier era, Brower had acted unilaterally without consulting local chapter leaders or board members. He could go to Washington and walk in to see the interior secretary or the forestry chief. Those days were gone. Board members held Brower in a mixture of respect and disdain. Brock Evans, who had once served as Brower's Northwest representative, had also been elected to the board. He remembered a time when the latest executive director had resigned and the board was discussing the need for an interim director. "I'll do it," Brower volunteered. There was a palpable silence. No one volunteered even an opinion before the board moved on to the next item on the agenda.[30]

But when Brower had a cause, no matter how hopeless, he still energized others. If he felt deeply about an issue, followers might show up with placards and march around the meeting room. The subject could be the need to oppose nuclear war or to prevent logging on public lands. Such stands placed him on the fringe of the organization. He complained that Sierra Club management was being dominated by a clique of perennial insiders who made too many compromises. He eventually joined a slate called the John Muir Sierrans that tried and often failed to elect more outspoken board members.[31]

Brower remained haunted by his failure at Glen Canyon, and he eventually was part of a movement that hoped to remove the dam and restore the river valley. In November 1996, he introduced a resolution before the Sierra Club board supporting the Glen Canyon Dam removal. Normal procedure called for such a proposal to be vetted through the bureaucracy, including the chapter in Utah. But Brower, eighty-four years old and as impatient as ever, did not want to wait, and the board suddenly and unexpectedly agreed that the old man should not have to wait. The resolution was adopted, delighting Brower and upsetting the Utah chapter.[32]

He was drawn into controversy again in 1998 when the membership asked if the Sierra Club should support a ballot initiative to limit immigration to reduce the population and protect the environment. The leadership had been trying to dodge the issue, which was seen as both explosive and racist. Brower did not avoid the issue; he had opposed overpopulation for more than forty years. Although he did not like the wording of the ballot initiative, he did support its aim. "If California continues to grow at the rate it has since I was born here," said Brower, "the population will be

500 million in the next eighty-five years. There will be no place to park. We'll all have to park in Oregon." The ballot proposal was defeated, but Brower vowed to continue pushing the population issue in the future.[33]

Bower was a writer and an editor, so it was only natural that book publishers would approach him about doing an autobiography. In 1982, he signed a contract with Morrow and directed that the initial $12,500 advance be deposited in Friends of the Earth's bank accounts. He then failed to write the book. When Morrow pressed him to repay the advance, he pointed out that they should be asking Friends of the Earth, not him, an argument the book publisher quickly rejected. Brower had no legal recourse, so he eventually paid back the money, but for years the very temerity of such a demand angered him. A small Utah book publisher, Peregrine Smith Books, finally rescued him by offering a new contract for his life story. Brower continued to struggle with the writing, although in the end he would produce not one but two books about his life, *For Earth's Sake* (1990) and *Work in Progress* (1991).[34] Both are less memoirs and more anthologies of old articles Brower had written and transcripts of testimony or speeches he had given. The strongest portion of either book comes at the beginning, where Brower writes about his early life and family. Reviewers noted these gaps but generally were kind. Roger B. Swain, writing for the *New York Times*, deplored Brower's failure "to examine old wounds, to recount mistakes and to relive defeats."[35] Brower found it easier to go on the offensive than the defensive, be the subject Glen Canyon or why his old friends at the Sierra Club had fired him.

As a public figure, Brower was often sought out by newspaper reporters, magazine writers, and even filmmakers. Many of the portraits were flattering, including two short documentaries on his life. He had very little control over many of these ventures. But when he had an opportunity to review a lengthy history of the Sierra Club, he fought very hard to change how the book portrayed him. In 1983, the club and Sierra Club Books hired Michael Cohen, a Utah-based historian, to write an authorized history of the club. Cohen had difficulty connecting with Brower. In 1988, Brower began reviewing the manuscript prior to publication, and he found dozens of what he called mistakes. Brower was especially upset that

Cohen was "trashing" his son Ken and that Cohen had described Brower's diversion of book royalty payments and other financial irregularities. He said that he had not diverted Sierra Club funds, that the charge of fiscal irresponsibility was an allegation, not a fact—one that he denied. By then, however, the draft had been reviewed and approved by historians and other Sierra Club readers. Cohen agreed to make some changes but not all of the ones that Brower wanted. Brower complained to Michel Fischer, the club's executive director, who concluded that the organization had two choices. It could either publish the book the way Cohen had written it or reject it outright. Fischer asked the board to make a final decision. As members pointed out, it was a no-win situation, and they reluctantly agreed that the book should be published.[36]

One of Brower's greatest attributes was his ability to move on and never to hold a grudge when it came to personal friends. With some, such as Dick Leonard, that was not difficult. Even when Leonard was firing Brower in the spring of 1969, he was saying it was not personal; he just wanted to save the Sierra Club. Both Leonard and his wife were early supporters of Friends of the Earth, and amends were made rapidly. For years afterward, Brower attended the Leonards' New Year's Eve party. When Brower ran for the Sierra Club board in 1983, he needed to file a petition signed by Sierra Club members. Leonard was the second to sign; Ansel Adams was the first.[37]

"Will you behave?" Adams asked Brower before signing.

"No," replied Brower.[38]

By then, Adams and Brower were also friends again, but the reconciliation had not been easy. In 1975, Adams told Brower that their conflict had been very difficult for him, a time when principles conflicted with emotions. By the Reagan administration, however, the two were comrades again, especially because Adams loathed Interior Secretary James Watt, who was also a favorite target of Brower and Friends of the Earth. Adams died suddenly on Earth Day, April 22, 1984. Leonard died in 1993, lingering painfully for months, blind and bound to a wheelchair. Brower would eulogize both men in public and print, but he somehow could never find the strength to visit Leonard during his long good-bye.[39]

Friends came and went, but not family. Ken Brower's career as a writer and an editor of books and articles was molded in a pattern based on father and mother. Robert Brower had a love for words, but he also had his troubles, living for a while in the Browers' basement after his marriage collapsed. The two younger children, Barbara and John, had often been left behind by their father on treks with the two older boys. As they got older, that changed. In 1976, the younger siblings had the opportunity to accompany Brower to Nepal. Sponsored by Friends of the Earth, the trip centered around a month-long, 200-mile hike at elevations up to 18,000 feet in the Himalayas. Barbara and John had been recruited because of family fears about the sixty-four-year-old Brower's fortitude. They needn't have worried. Brower the old mountain climber delighted in the journey. Barbara had had her own apprehensions. "I had to drag myself to get to that trip," she said, "and, of course, it was a life-changing experience." Barbara, who would make a career in academia as a geographer, would return to Nepal on a Fulbright grant. She would also discover her father's love for world traveling. She eventually produced several granddaughters for him, and Ken sired a son he would name David.[40]

No one remained more constant than Anne. Over the fifty-seven years of marriage, Brower's ceaseless quests, energy, and obsessions would grow and eventually dominate their lives. His strong presence could be seen in the old redwood-sided house near the rim of the Berkeley hills, which a writer in 1995 was allowed to tour. One of the most prominent features was the more than one hundred file boxes stacked in the living room under the grand piano and behind the couch, blocking the basement windows and any other possible repository. There were also the rocks, hundreds of them—granite, limestone, quartzite, agate, sandstone, and more. Brower had picked up some as keepsakes, but friends had sent far more from sites around the world. On the patio were the four garbage cans filled with golf balls that Brower had picked up on walks near a golf course. The bedroom door was plastered with name tags from conferences, and they proclaimed, "MY NAME IS DAVID BROWER." Anne told the same writer in 1995, "I'm very uninteresting compared to David. Everything has always been connected to his work. I've lost my identity."[41]

In illness and health, they remained a couple. In 1986, Anne needed a hip replacement. In 1995, Brower collapsed and was hospitalized so that he could be given a pacemaker that would assist his now irregular heartbeat. Anne developed Parkinson's disease; she needed a cane.

For Brower, the honors kept arriving. An international award, the Blue Planet Prize, was bestowed on him in 1998, and he turned over the $422,000 prize to Earth Island Institute. A California congressman repeatedly nominated Brower for the Nobel Peace Prize.[42]

In early 2000, he was diagnosed with bladder cancer. The prognosis was not good. Anne Brower, who would die one year after her husband, once said late in her life that Brower did not think about death. "I think at my age there's nothing more exciting that's going to happen to me except dying," she said. "But he believes that if he were to think about death, it would keep him from accomplishing things. He wants to work up to the bitter end."[43]

That he did.

Cancer was only another foe that he would fight to vanquish. Between stays in the hospital and rests at home, he continued to travel and to be his outspoken self. In March 2000, equipped with a tank of oxygen that he sucked in the unpressurized cabin of an aircraft, he flew to Utah as part of a program to restore Glen Canyon. At an April Earth Day program at San Francisco State University, he incited students to join the cause. "There's something you can do to keep this Earth from going straight to hell—each of you can make a difference, and you'd damn well better!" In May, he announced he was resigning from his seat on the Sierra Club board. "The world is burning, and all I hear from them is the music of violins," said Brower. "The planet is being trashed, but the board has no sense of urgency" (figure 18). On July 4, he was supposed to be at home undergoing cancer treatment, but he was instead at a new railroad station in Bakersfield, California, promoting rail traffic to Yosemite National Park. Brower wanted all automobiles kept out of the valley; visitors should come by rail. In late July, he was back in the hospital for surgery.[44]

In the fall, he endorsed Ralph Nader for president, spurning the appeals of Al Gore, who would go on to lose in one of the closest presidential races in history. He was in and out of the hospital, weak but insistent

FIGURE 18 Brower in the wilderness he relished

Brower led three environmental organizations and was a talented book editor, but he preferred to be in the wilderness. He began leading Sierra Club expeditions into the mountains while in his twenties and was still making hikes into the wild well into his old age. (Courtesy of the Bancroft Library, University of California, Berkeley) that phone calls be put through. In a frail voice, he would explain that he needed to vote for Nader; they had been friends for thirty years. Brower was home at the end, fragile but determined to complete his absentee ballot days before the election. He put an X near Nader's name, but he was too weak to sign his name, so he marked the signature line with an X.

He died on November 5. He was eighty-eight years old.[45]

Nearly a month later, about 1,200 people gathered in the Berkeley Community Theatre to remember Brower. There was saxophonist Paul Winter and writers Stephanie Mills and Bill McKibben. From the environmental community also came old-timers such as Martin Litton and younger converts such as Julia "Butterfly" Hill, who had lived in a tree for two years to try to save it, a perfectly Brower quixotic act.

One of Brower's oldest friends and colleagues, Tom Turner, prepared a sixteen-page booklet that included tributes to Brower, photos of him with presidents and the Dalai Lama, as a young man and an old man with Anne Brower. There were also examples of his writings, including something titled "Third Planet Operating Instructions," and quotations that had inspired him over the years. Lester Brown, an old friend and fellow environmentalist, wrote, "Every once in a while, someone walks across the global stage and changes it in the process. David Brower was a pioneer in every sense of the word, defining some of the key environmental challenges long before the rest of us were even vaguely aware of them."[46]

A few days before the memorial service, Brower's last published article appeared posthumously in a column in the *San Francisco Chronicle*. A longtime nemesis, the National Park Service, had proposed a new plan for managing Yosemite village, which had been ravaged by a winter storm. Naturally, Brower did not like the plan. It catered far more to humans and commerce than to nature. "It seems intent on converting this temple into a profit center, with pricey hotels, scant camping, few modest accommodations, wider roads to field bigger diesel buses, ecological roadside mayhem [and] atmospheric damage."

He closed in what was a fitting epithet for so much more than just this shrine of nature: "It is time," he said in farewell, "to remember what Yosemite is all about."[47]

Epilogue

I never wanted a level playing field. I'm a mountaineer.

DAVID BROWER,
*LET THE MOUNTAINS TALK,
LET THE RIVERS RUN*

In February 1995, writer Mark Dowie had lunch with David Brower. The setting was Sinbad's, Brower's favorite San Francisco waterfront restaurant, which featured deeply upholstered furnishings and a full bar. Dowie had finished a book on environmental history, and a copy had been sent to Brower so that he could write a book-jacket blurb. Brower wrote: "This is a very important book." That declaration was followed by a fourteen-page, single-spaced attack that ranged from correcting middle initials to challenging the author's central premise. The two men then met at Sinbad's. What followed, said Dowie, was a two-hour harangue. "I witnessed the legendary Brower hubris, the arrogance, the prickly temper, all arrayed against the wit, the blunt honesty and the preoccupation with beauty that made any meeting with David Brower memorable," he wrote. As Dowie listened to Brower, he suddenly realized that Brower believed he was being mistreated, not just in Dowie's book but in other histories. He felt that he was not receiving the credit he had earned.[1]

In some respects, Brower had reason to worry about his legacy. Some historians examining the environment movement have treated Brower as a minor character and have failed to give him the credit that he does deserve. They praise founders such as John Muir of the Sierra Club and Bob Marshall of the Wilderness Society as well as writers such as Aldo Leopold and Rachel Carson. This neglect is unfair. Brower was the preeminent conservation leader in the 1950s and 1960s while heading the Sierra Club, and he was a respected voice for another thirty years beyond that. But in an examination of Brower's place in history, the dilemma is less about what he accomplished and more about what he could have done. To a certain extent, Brower's career was a case of opportunities lost.

The year 1969 was a terrible time to be fired as the leader of the Sierra Club, the nation's rising conservation leader. The public's interest and concern about the environment was accelerating. and Americans wanted the government to do more, especially about the nation's foul air and water. According to polls, the percentage of people who wanted the environment listed as one of the nation's three top priorities had risen from 17 percent in 1965 to 53 percent in 1970. The first Earth Day celebration, April 22, 1970, drew some 20 million participants. Within two years, politicians had responded by establishing the Environmental Protection Agency and new laws to clean the nation's waterways and air. Membership in existing environmental organizations increased drastically, and as many as two hundred new national and regional groups and three thousand local organizations proliferated.[2]

Today the nation and the world have an established environmental ethos, an understanding that the natural world is not just to be exploited but to be both savored and protected. Progress has been made in guarding wilderness and wildlife, in reducing pollution, in monitoring toxic chemicals, pesticides, and other poisons, and in understanding how closely tied race and poverty are to environmental degradation. Brower popularized the concept of the conservation and environmental watchdog, and today the number of environmental advocacy groups that fulfill that role in the United States has been estimated at ten thousand or more. The vast majority are very small grassroots organizations dedicated to a local problem or issue. Yet the muscle to address broader themes and issues remains with national organizations. Groups such as the Sierra Club, The Nature Conservancy, and the National Wildlife Federation have memberships as high as 4 million, annual budgets of greater than $100 million, and staffs of hundreds.[3]

Despite these efforts, the environment is in great peril, and much of the public is frustrated by how little is being done compared with the seriousness of the threats facing the earth. One might point to any number of problems, but three are key—the atmosphere, the oceans, and biodiversity. Human-induced climate change is making profound changes globally that might last centuries. More alarming is the failure so far by the world's nations to end the increasing rates of greenhouse-inducing emissions that are poisoning the upper atmosphere. Related and potentially just as dire is the increasing acidification of the oceans, which is being caused by a buildup from greenhouse-gas emissions. Once a seemingly inexhaustible resource, the seas of the world are threatened by pollution, overfishing, and other abuse. Finally, species extinction is increasing at a rate that is at least one hundred times greater than what we have seen in the past. Experts estimate that our grandchildren will live in a world with half the number of species that we have now. Over a recent ten-year period, when pollsters asked the public what they thought of the state of the nation's environment, 55.4 percent rated it "only fair" or "poor." Although the public over the years in these polls has been most wary of large corporations' role in affecting the environment, it does not entirely trust the government or even environmental advocacy organizations, either.[4]

There is a leadership vacuum, and it begins with the heirs of David Brower. The nation's largest and most influential environmental organizations include the National Wildlife Federation, the Sierra Club, the Natural Resources Defense Council, the Environmental Defense Fund, Greenpeace USA, The Nature Conservancy, and the National Audubon Society. The leaders of these organizations are typically not hired for their charisma but as professional chief executive officers who will spend most of their time on administrative functions. According to a study of such groups done a few years ago, the typical CEO spent 70 percent of his or her time doing fund-raising, planning, board development, personnel management, membership development, public speaking, and media relations. The balance was devoted to researching and planning environmental campaigning.[5]

To many, the type of individual one would ask to run a vast, multi-million-dollar organization would have to be a professional, experienced CEO. Historian Stephen Fox suggested that environmental advocacy organizations are born by amateurs and eventually taken over by professionals who attempt to transform them into business organizations.[6] But increasingly there are questions about that approach. Nathaniel Reed, once a Florida State environmental official and later an environmental advocate, once said while Brower was still alive that Brower was "the last of the great amateur environmentalists." Never again, Reed said, would a major organization such as the Sierra Club take a chance on someone who had dropped out of college.[7] Anne Brower used to say that her husband got out of college before the system shaped him to conform. Leadership does not necessarily come from a Harvard MBA or an apprenticeship within the Fortune 500.

We have lost sight of the Brower style of environmental advocacy, probably because of the great proliferation of organizations that represent the environment. At the national level alone, dozens of organizations compete against one another both for the attention of donors and to address specific issues. Sometimes bridges are built, but those efforts are challenging and time consuming. Plus, some advocacy organizations are so distant from one another in philosophy, goals, and leadership that a unified front becomes virtually impossible. The Nature Conservancy, which partners with the world's largest corporations, including some polluters, is clearly not going to have much to do with the radical anticorporate platform of Earth First!. Jon Roush, an environmental consultant, studied environmental leadership and wondered if collaboration is even possible among the disparate groups that have common cultures. "With a pyramid of such diversity and fragmentation, it is fair to ask whether there is a single conservation movement and whether the idea of a national consensus is a quixotic pipe dream," he said.[8]

Multiplicity is not necessarily a negative; one of the great attributes of small grassroots organizations is their ability to concentrate on a specific issue. Further, history shows that environmentalists often fight among themselves and rarely unite. The early history of the conservation movement is characterized by division, beginning with the epic split between John Muir and Gifford Pinchot. Over time, that break splintered into further divisions: Audubon sought to protect birds; the Wilderness Society shielded wilderness; the National Parks Association defended parks. That fragmentation actually aided the ascent of Brower and the Sierra Club in the 1950s. It was Brower's genius as both a political campaigner and a master of media that catapulted the Sierra Club into a commanding position of leadership.

Yet by 1969, Brower was gone from the Sierra Club, just when the fledgling environmental movement could have used a strong leader. Many have earmarked April 1970 and the arrival of Earth Day as the movement's time of arrival. It was conceived by a U.S. senator, Gaylord Nelson, and organized by a twenty-five-year-old Harvard Law School student, Denis Hayes. National environmental organizations, including the Sierra Club, were wary or indifferent and played very little role in it, though. Michael McCloskey had replaced Brower as executive director at the Sierra Club, but he was hobbled in his actions by a board composed of volunteer amateurs who deeply distrusted the professional staff after the Brower donnybrook. That division would hold back the Sierra Club for years. Brower immediately understood the value of Earth Day; he participated in it and encouraged others, but his impact was muted. He no longer had the grand pulpit of the Sierra Club with its vast resources.[9]

It could have been so different. Earth Day could have not only coalesced the power of Brower and the Sierra Club but with a dominant leader also presented a more unified national agenda. Not everyone would have signed on. Such incidents as the toxic chemical contamination of the Love Canal neighborhood near Niagara Falls, New York, precipitated one of the many local grassroots organizations that arose in the 1970s. Such efforts were inevitable, healthy, critical to the life of the movement, breathing in new, fresh air. But singular and strong leadership could have been an advantage, with one dominant national environmental advocacy organization such as the Sierra Club led by Brower in the 1970s. Brower could have represented the environment the way Ralph Nader represented consumerism, Martin Luther King Jr. represented race, and Cesar

Chavez represented labor. Consider two examples of how Brower could have stitched together a stronger Sierra Club. By the mid-1960s, he began urging the club to move into land preservation and management, a field that would eventually become the hallmark of The Nature Conservancy. In 1969, The Nature Conservancy had eighteen thousand members and a staff of thirty. Today it has 1 million members and manages or owns 110 million acres in the United States and thirty-five nations.[10] Another key Brower talent was his recognition of the power of the mass media. In 1969, he envisioned a large media campaign, led by more books but also relying increasingly on television. The lead in this campaign was instead assumed by others ranging from the wealthy and established National Geographic Society to independents such as the oceanographer Jacques Cousteau. Such were the opportunities lost.

To don the mantle of leadership, to manage land holdings, or to run media empires, Brower would have had to make some adjustments in his leadership style. Clearly, he was too abrasive at times. He could not have succeeded by winning the Sierra Club election in 1969 and assuming clear authority over the organization. This step would have caused a mass exodus of many of the members he had already alienated and likely would have crippled or destroyed the organization's finances. Although some members did not like Brower's political tactics, it was his lack of financial oversight that especially upset friends such as Richard Leonard and Ansel Adams. No, the solution was to come to a compromise, an agreement, something Brower was later willing to do, but only after he had lost the election. He had misjudged his standing, and by the time he corrected this misjudgment, it was too late. The ideal solution would have been for Brower to remain at the Sierra Club as the political and creative genius. Someone, perhaps Michael McCloskey or Edgar Wayburn, could have been responsible for the club's administration. That person would have had to firmly veto Brower projects at times.

That famous Brower hubris and arrogance might still have unraveled such an arrangement. At times, Brower's greatest strengths were also his greatest weaknesses. But what was needed then—and is still needed today—in leadership is much of what Brower characterized. We have CEOs when what we need are firebrands. Brower was eccentric but charismatic; he could be maddeningly independent, yet he could unite divergent forces around epic campaigns. He was an amateur among the professionals, difficult for even his allies to understand, impossible for his opponents to read. He was a visionary, a wild man, passionate and utterly unselfish in behalf of his cause. He became a catalyst for the environmental movement as we know it today; he failed in carrying it much farther.

Although many historians have lost sight of Brower, one who did not was Mark Dowie. Despite Dowie's adventuresome luncheon with Brower that day in 1995, his environmental history *Losing Ground: American Environmentalism at the Close of the Twentieth Century* raises Brower's name a remarkable number of times—eleven. Only a handful of other individuals have more listings in the index, and three of them are named Reagan, Bush, and Clinton.[11]

In 1995, Dowie was amazed at the old man's quest. "Before me was a hero," he said about his lunch with Brower, "struggling for a legacy that he had already won."[12] A legacy of hundreds of millions of acres of wilderness protected for future generations, of a greater understanding of the concept of environmental stewardship, and of an organizational hierarchy that, no matter how fractured, is intent on carrying out Brower's ultimate goals.

Perhaps it was Russell Train, a one-time ally and then a foe when he later ran the Environmental Protection Agency, who best summed up the life of this quixotic environmental warrior. "Thank God for David Brower," said Train. "He makes the rest of us look reasonable."[13]

Notes

Introduction

1. Hal Wingo, "Close-up: California's David Brower, No. 1 Conservationist, Knight Errant to Nature's Rescue," *Life*, May 27, 1966, 37–42.

2. Richard D. North, "Obituary: David Brower," *Independent* (London), November 9, 2000; Adam Bernstein, "David Brower Dies; Transformed Sierra Club into Powerful Force," *Washington Post*, November 7, 2000; Richard Severo, "David Brower, an Aggressive Champion of U.S. Environmentalism, Is Dead at 88," *New York Times*, November 7, 2000; Paul Rogers, "David Brower, Nature's Crusader Dead at 88, Mountaineer, Fiery Sierra Club Director Leaves a Legacy of Conservation from Coast to Coast," *San Jose Mercury News*, November 7, 2000; Mike Taugher, "David Brower, 1912–2000, Nature Loses a Best Pal," *Contra Costa Times*, November 7, 2000; "The Earth's Defender," *San Jose Mercury News*, November 8, 2000; "Farewell to an Arch Druid," *Denver Post*, November, 8, 2000.

3. Stephanie Mills, *Whatever Happened to Ecology?* (San Francisco: Sierra Club Books, 1989), 109.

4. Newspaper advertisements memorandum, June 9, 1966–January 14, 1969, Carton 115, Folder 15, Sierra Club Members Papers, BANC MSS 71/295c, Bancroft Library (BL), University of California, Berkeley; see also David R. Brower, "Environmental Activist, Publicist, and Prophet," an oral history conducted in 1974–1978 by Susan Schrepfer, Regional Oral History Office, BL, 1979, 148.

5. Quoted in Mills, *Whatever Happened to Ecology?* 110.

6. David Brower to the Sierra Club board of directors, memorandum, October 19, 1956, Carton 13, Folder 11, Sierra Club Office of the Executive Director Records, BANC MSS 2002/230c, BL.

7. Quoted in Bob Beier, "Aspinall Raps Opposition to Project," *Albuquerque Tribune*, November 18, 1966.

8. David Brower, interviewed by Tom Turner, March 30, 1983, Carton 18, Folder 49, David Ross Brower Papers, BANC MSS 79/9c, BL, 1979.

9. Edgar Wayburn and Allison Alsup, *Your Land and Mine: Evolution of a Conservationist* (San Francisco: Sierra Club Books, 2004), 106.

1. First Fight

1. Roy Webb, *Riverman: The Story of Bus Hatch* (Rock Springs, Wyo.: Labyrinth Press, 1989), 96; Sierra Club, "Sierra Club Dinosaur Trips," press release, June 9, 1954, Carton 8, Folder 20, Sierra Club Records, BANC MSS 71/103c, Bancroft Library (BL), University of California, Berkeley.

2. Craig W. Allin, *The Politics of Wilderness Preservation* (Westport, Conn.: Greenwood Press, 1982), 91.

3. Webb, *Riverman*, 97.

4. Quoted in ibid., 97–100.

5. Sharon Toney, "Conservationists' Role in the Echo Park Dispute," n.d., unpublished essay, Superintendent's Office, Dinosaur National Monument, Dinosaur, Colo., 7.

6. David R. Brower, "Environmental Activist, Publicist, and Prophet," an oral history conducted in 1974–1978 by Susan Schrepfer, Regional Oral History Office (ROHO), BL, 1979, 115, 134–135; David Brower to Senator B. H. Stringham, June 24, 1953, Carton 5, Folder 18, Sierra Club Members Papers, BANC MSS 71/295c, BL; Jon M. Cosco, *Echo Park: Struggle for Preservation* (Boulder: Johnson Books, 1995).

7. Bernard DeVoto, "Shall We Let Them Ruin Our National Parks?" *Saturday Evening Post*, July 22, 1950, 42.

8. Harold C. Bradley, "Danger to Dinosaur," *Pacific Discovery*, January–February 1954, 5.

9. Edgar Wayburn, "Sierra Club Statesman, Leader of the Parks and Wilderness Movement: Gaining Protection for Alaska, the Redwoods, and Golden Gate Parklands," an oral history conducted in 1976–1981 by Ann Lage and Susan Schrepfer, ROHO, BL, 1985, 50.

10. Elmo Richardson, "The Interior Secretary as Conservation Villain: The Notorious Case of Douglas 'Giveaway' McKay," *Pacific Historical Review*, August 1972, 336–339, including quote from Davis.

11. Sierra Club, *Wilderness River Trail* (Dawson Productions, 1953); Brower, "Environmental Activist, Publicist, and Prophet," 115.

12. The most complete accounts of the Dinosaur dam controversy in the 1950s are Cosco, *Echo Park*; and Mark W. T. Harvey, *A Symbol of Wilderness: Echo Park and the American Conservation Movement* (Albuquerque: University of New Mexico Press, 1994).

13. Wallace Stegner, "The Marks of Human Passage" in *This Is Dinosaur: Echo Park Country and Its Magic Rivers*, ed. Wallace Stegner (New York: Knopf, 1955), 13.

14. Mary Risser (superintendent, Dinosaur National Monument), interview by the author, Dinosaur National Monument, Colo., June 21, 2010.

15. Franklin D. Roosevelt, "Enhancing the Dinosaur National Monument—Colorado and Utah," proclamation, July 14, 1938, Carton 64, Folder 14, Sierra Club Records, BL; David Madsen (former employee, National Park Service), affidavit, March 27, 1950, Carton 64, Folder 14, Sierra Club Records, BL. Madsen certified that the Park Service told citizens that dams could be built in Dinosaur.

16. Marc Reisner, *Cadillac Desert: The American West and Its Disappearing Water* (New York: Viking Penguin, 1986), 158. Shasta Dam in California at the time was the largest concrete dam ever built.

17. Ibid., 164–168.

18. Allin, *Politics of Wilderness Preservation*, 92.

19. Harvey, *Symbol of Wilderness*, xiv–xv; Brian McKeever and Sarah L. Pettijohn, "The Nonprofit Sector in Brief 2014," Urban Institute, October 2014, http://www.urban.org/UploadedPDF/413277-Nonprofit-Sector-in-Brief-2014.pdf (accessed February 6, 2015); audit statements, 1963–1969, Carton 25, Folders 1–9, Sierra Club Records, BL; David Halberstam, *The Fifties* (New York: Villard Books, 1993), x–xi.

20. Harvey, *Symbol of Wilderness*, 40–43, 151.

21. U.S. Department of the Interior, "Upper Colorado River Development Plan Recommended by Secretary McKay," press release, December 23, 1953, Carton 65, Folder 14, Sierra Club Records, BL.

22. Harvey, *Symbol of Wilderness*, 287; Roderick Nash, *Wilderness and the American Mind* (New Haven, Conn.: Yale University Press, 1967), 219; John McPhee, *Encounters with the Archdruid* (New York: Farrar, Straus and Giroux, 1971), 164; Cosco, *Echo Park*, xv.

23. Sierra Club, press release, February 10, 1954, Carton 8, Folder 21, Sierra Club Records, BL; Byron E. Pearson, *Still the Wild River Runs: Congress, the Sierra Club, and the Fight to Save Grand Canyon* (Tucson: University of Arizona Press, 2002), 138; "What You Can Do for Dinosaur and the National Park System," *Sierra Club Bulletin*, January 1954, 3.

24. Lawrence E. Davies, "Fight Is on to Save Big Dinosaur Area," *New York Times*, December 21, 1953; Collinson quoted in "32 Groups to Fight New Dam in West," *New York Times*, January 5, 1954; Toney, "Conservationists' Role in the Echo Park Dispute," 20.

25. Brower to Don Hatch (river guide), March 16, 1954, Carton 5, Folder 21, Sierra Club Members Papers; Pughe and Aspinall quoted in Cosco, *Echo Park*, 69.

26. Harvey, *Symbol of Wilderness*, 193.

27. Brower, "Environmental Activist, Publicist, and Prophet," 111; Brower to Charles Woodbury (Wilderness Society), January 13, 1954, Carton 64, Folder 23, Sierra Club Members Papers; Brower to Dinosaur Cooperators, October 6, 1953, Carton 5, Folder 19, Sierra Club Members Papers.

28. "Wild by Law," *The American Experience*, PBS, February 10, 1992, transcript, Carton 18, Folder 19, David Ross Brower Papers, BANC MSS 79/9c, BL.

29. Ralph Tudor, testimony, U.S. House of Representatives, Subcommittee on Irrigation and Reclamation, *Colorado River Storage Project: Hearing on H.R. 4449, H.R. 4443, and H.R. 4463*, 83rd Cong., 2nd sess., January 18–23, 1954, 14–46.

30. Brower, "Environmental Activist, Publicist, and Prophet," 116.

31. Testimony, U.S. House, Subcommittee on Irrigation and Reclamation, *Colorado River Storage Project*: the speakers for the conservation community included David Bradley (847); Stephen Bradley (835); George B. Fell (876); Ulysses S. Grant III (708); C. R. Gutersmith (865); Kenneth D. Morrison (857); Fred Packard (799); Mrs. Bryan Warren (865); and Howard Zahniser (881).

32. Quoted in Cosco, *Echo Park*, 72.

33. David Brower, testimony, U.S. House, Subcommittee on Irrigation and Reclamation, *Colorado River Storage Project*, 789–796.

34. Brower, testimony, in ibid., 824–835.

35. Howard Zahniser (Wilderness Society) to Richard Leonard (board member, Sierra Club), telegram, January 29, 1954, Carton 64, Folder 24, Sierra Club Records, BL.

36. Richard Bradley (physics professor, Cornell University) to Brower, April 1, 1954, Carton 65, Folder 4, Sierra Club Records, BL.

37. Harvey, *Symbol of Wilderness*, 200; Richard Bradley to Brower and Zahniser, March 26, 1954, Carton 70, Folder 7, Sierra Club Records, BL.

38. Brower to Judge Robert Sawyer, April 29, 1954, Carton 65, Folder 5, Sierra Club Records, BL; Harvey, *Symbol of Wilderness*, 201.

39. Luna Leopold (U.S. Geological Survey) to Brower, May 14, 1954, Carton 65, Folder 6, Sierra Club Records, BL; U.S. Department of the Interior, "Changes to Evaporation of Upper Colorado Dams," press release, May 14, 1954, Carton 8, Folder 19, Sierra Club Records, BL; Ralph Tudor to Congressman William Harrison, May 13, 1954, Carton 8, Folder 19, Sierra Club Records, BL.

40. Brower, "Environmental Activist, Publicist, and Prophet," 135.

2. Mountains

1. David R. Brower, *For Earth's Sake: The Life and Times of David R. Brower* (Salt Lake City: Peregrine Smith Books, 1990), 8–10.

2. Ibid., 134.

3. Ibid., 135.

4. David Brower, interview by Tom Turner, March 30, 1983, transcript, Carton 18, Folder 49, David Ross Brower Papers, BANC MSS 79/9c, Bancroft Library (BL), University of California, Berkeley; City of Berkeley Landmark Application, *Brower Houses and David Brower Redwood*, July 2008, 17–18.

5. Joseph and Gayle Brower (Brower's youngest brother and Joseph's wife), interview by the author, Santa Rosa, Calif., March 26, 2013.

6. City of Berkeley Landmark Application, *Brower Houses*, 18–19; Brower interview by Turner, March 30, 1983.

7. Brower interview by Turner, March 30, 1983.

8. Joseph and Gayle Brower interview, March 26, 2013.

9. Brower, *For Earth's Sake*, 134.

10. Ross Brower to David Brower, n.d., Box 2, Folder 7, Brower Papers; David Brower to Ross Brower, August 21, 1935, Box 2, Folder 1, Brower Papers.

11. Michael McCloskey (former Sierra Club employee and executive director), interview by the author, Portland, Ore., January 22, 2013.

12. Brower, *For Earth's Sake*, 3.

13. Joseph Brower, telephone interview by the author, November 9, 2012; Brower, *For Earth's Sake*, 2.

14. Brower interview by Turner, March 30, 1983, 2; City of Berkeley Landmark Application, *Brower Houses*, 16–17; Joseph Brower, telephone interview, November 9, 2012.

15. Phyllis D. Knowles (former tenant of Ross Brower) to David Brower, July 4, 1995, Box 6, Folder 15, Brower Papers.

16. Edith Brower Grindle (David Brower's sister) to David Brower, n.d., Box 2, Folder 12, Brower Papers; Joseph Brower, telephone interview by the author, August 23, 2013.

17. City of Berkeley Landmark Application, *Brower Houses*, 23–24. The Haste Street apartments were sold several times. In the period from 2008 to 2013, historical preservationists complained about plans by the current owner to substantially rehabilitate the properties. Some changes were made, including additional protection for the redwood tree that towers over the street, and work on the apartments began in late 2012.

18. Teddie Allison (Edith Grindle's daughter), eulogy for Edith Ellen Brower Grindle, November 13, 1987, Box 2, Folder 12, Brower Papers; Joseph and Gayle Brower interview, March 26, 2013.

19. Brower, *For Earth's Sake*, 18–19.

20. David Brower, butterfly collection notes, 1925–1927, Carton 14, Folder 1, Brower Papers; Brower, For Earth's Sake, 17.

21. Brower interview by Turner, March 30, 1983.

22. David Brower to Edie Rubel (widow of Donald Rubel), October 7, 1987, Box 11, Folder 18, Brower Papers; Joseph Brower telephone interview, November 12, 2012.

23. Brower, *For Earth's Sake*, 25–26; John McPhee, *Encounters with the Archdruid* (New York: Farrar, Straus and Giroux, 1971), 34; David Brower, Yosemite trip diary, 1931–1933, Box 14, Folder 7, Brower Papers.

24. Brower to Edie Rubel, October 7, 1987; David Brower to A. F. Clark Jr. (deputy director, Historical Division, War Department), April 17 and 26, 1946, Box 11, Folder 6, Brower Papers; Anne Brower quoted in Kenneth Brower, "Climbing the Spiral Staircase," *California*, Spring 2013, 49.

25. Brower interview by Turner, March 30, 1983; Michael J. Ybarra, "Exhibition: A Solitary, Singular Life," *Wall Street Journal*, September 10, 2009.

26. David R. Brower, "Environmental Activist, Publicist, and Prophet," an oral history conducted in 1974–1978 by Susan Schrepfer, Regional Oral History Office, BL, 1979, 10.

27. Ibid.; David Brower, eulogy for Ansel Adams, n.d., Carton 5, Folder 67, Brower Papers.

28. Joseph E. Taylor III, *Pilgrims of the Vertical: Yosemite Rock Climbers and Nature at Risk* (Cambridge, Mass.: Harvard University Press, 2010), 110; McPhee, *Encounters with the Archdruid*, 232.

29. David Brower to Hervey Voge (Sierra Club friend), October 29, 1934, Box 10, Folder 13, Brower Papers.

30. Taylor, *Pilgrims of the Vertical*, 45–52; Brower interview by Turner, March 30, 1983.

31. Brower interview by Turner, March 30, 1983.

32. Hervey Voge to Virginia Ferguson (secretary, Sierra Club), November 9, 1934, Box 10, Folder 14, Brower Papers; David R. Brower, "Far from the Madding Mules: A Knapsacker's Retrospective," *Sierra Club Bulletin*, February 1935, 68–77.

33. "Killer Mountain Goal of Alpinists," *New York Times*, June 9, 1935.

34. Brower to Voge, June 14, 1935, Box 10, Folder 15, Brower Papers; Richard M. Leonard, Mount Waddington essay, n.d., Box 7, Folder 2, Brower Papers; Brower, *For Earth's Sake*, 44–57.

35. Taylor, *Pilgrims of the Vertical*, 44–47, 58–59; Brower, "Environmental Activist, Publicist, and Prophet," 33–34.

36. Taylor, *Pilgrims of the Vertical*, 59–61; Brower to Morgan Harris (mountain climber), November 13, 1936, Box 5, Folder 15, Brower Papers.

37. Brower to Richard Leonard, September 25, 1935, Box 7, Folder 2, Brower Papers.

38. Robert Ormes, "A Piece of Bent Iron," *Saturday Evening Post*, July 22, 1939, 13.

39. David R. Brower, "It Couldn't Be Climbed," *Saturday Evening Post*, February 3, 1940, 24–25, 70–75; Bestor Robinson, "The First Ascent of Shiprock," *Sierra Club Bulletin*, February 1940, 1–7; Underhill quoted in Brower, *For Earth's Sake*, 67.

3. The Club

Brower was told that the quotation in the epigraph had been chiseled on stone at the National Aquarium and that he was credited as the author, which surprised him. He eventually found the quote in an interview of him done at a North Carolina bar after he had had two martinis. He decided the words were too conservative. His preferred comment was, "We're not borrowing from our children, we're stealing from them—and it's not even considered to be a crime" (David Brower, with Steve Chapple, *Let the Mountains Talk, Let the Rivers Run: A Call to Those Who Would Save the Earth* [San Francisco: HarperCollins West, 1995], 1–2).

1. Ansel Adams, "A Conversation with Ansel Adams," an oral history conducted in 1972, 1974, and 1975 by Ruth Teiser and Catherine Harroun, Regional Oral History Office (ROHO), Bancroft Library (BL), University of California, Berkeley, 1978, 608–609; Richard Searle, "Grassroots Sierra Club Leader," an oral history conducted in 1976 by Paul Clark and included in "Southern Sierrans," ROHO, BL, 1976, 15; Richard Leonard, "Mountaineer, Lawyer, Environmentalist, Volume I," an oral history conducted in 1976 by Susan R. Schrepfer, ROHO, BL, 33; H. R. Swenson to Chairman, Membership Committee, August 28, 1946, Carton 8, Folder 9, Sierra Club Records, BANC MSS 71/103c, BL.

2. William B. Devall, "The Governing of a Voluntary Organization: Oligarchy and Democracy in the Sierra Club" (Ph.D. diss., University of Oregon, 1970), 94–106.

3. David Brower, *Grand Canyon Discussion at Silver Grotto and Marble Damsite May 31 to June 13, 1977*, audiocassette, recorded by Ron Hayes, Grand Canyon National Park Museum Collection, GRCA 40237, Grand Canyon National Park, Ariz.

4. Leonard, "Mountaineer, Lawyer, Environmentalist, Volume I," 158–159; "Handbook Edition," *Sierra Club Bulletin*, December 1967, 57.

5. Edgar Wayburn, "Sierra Club Statesman, Leader of the Parks and Wilderness Movement: Gaining Protection for Alaska, the Redwoods, and Golden Gate Parklands," an oral history conducted in 1976–1981 by Ann Lage and Susan Schrepfer, ROHO, BL, 1985, 41.

6. David R. Brower, "Environmental Activist, Publicist, and Prophet," an oral history conducted in 1974–1978 by Susan Schrepfer, ROHO, BL, 1979, 39.

7. Fernando Penalosa, *Yosemite in the 1930s: A Remembrance* (Rancho Palos Verdes, Calif.: Quaking Aspen Books, 2002), 161–174; Scott Gediman (assistant superintendent for public and legislative affairs for the National Parks Service at Yosemite National Park), interview by the author, Yosemite National Park, Calif., August 2, 2012.

8. For an example of Brower's work in promoting Yosemite, see his press release on trout fishing, May 20, 1937, Carton 119, Folder 2, David Ross Brower Papers, BANC MSS 79/9c, BL. Brower kept many of his press releases in storage before donating them to the Bancroft Library. Other subjects included the sighting of a new geyser, the arrival of a descendant of one of Yosemite's founders, the benefits of skiing at Yosemite, and even the birth of children and weddings at Yosemite.

9. Michael P. Cohen, *The History of the Sierra Club, 1892–1970* (San Francisco: Sierra Club Books, 1988), 69; Nancy Newhall, *Ansel Adams*, vol. 1, *The Eloquent Light* (San Francisco: Sierra Club Books, 1963), 22–98; Jonathan Spaulding, *Ansel Adams and the American Landscape: A Biography* (Berkeley: University of California Press, 1995), 56–160.

10. David Brower, "A Tribute to Ansel Adams," *Sierra*, July–August 1984, 32.

11. Ibid.; Michael Adams (Ansel's son), telephone interview by the author, October 4, 2012.

12. Ed J. Pyle Jr. (official, Stith-Noble Corp.) to David Brower, May 18, 1939, Carton 119, Folder 5, Brower Papers; Stith-Noble Corp. to Brower, December 31, 1938, Carton 119, Folder 5, Brower Papers; David R. Brower, *For Earth's Sake: The Life and Times of David R. Brower* (Salt Lake City: Peregrine Smith Books, 1990), 59.

13. David Brower, interview by Tom Turner, March 30, 1983, transcript, Carton 18, Folder 49, Brower Papers.

14. Cohen, *History of the Sierra Club*, 70. Brower would serve on the Sierra Club board of directors from 1941 to 1952, when he was named executive director, although he was on leave for military service between 1942 and 1945.

15. Brower interview by Turner, March 30, 1983.

16. Ibid.; Brower, *For Earth's Sake*, 141–142.

17. Kenneth Brower, "Climbing the Spiral Staircase," *California*, Spring 2013, 47–48.

18. Brower interview by Turner, March 30, 1983.

19. David Brower, Echo Lake Diary, June 10–August 11, 1930, Box 14, Folder 6, Brower Papers; Joseph and Gayle Brower, interview by the author, Santa Rosa, Calif., March 26, 2013; Sarr quoted in Kenneth Brower, *The Wildness Within: Remembering David Brower* (Berkeley, Calif.: Heyday, 2012), 146.

20. Quoted in K. Brower, *Wildness Within*, 283.

21. David R. Brower, "Reflections on the Sierra Club, Friends of the Earth, and Earth Island Institute," an oral history conducted in 1999 by Ann Lage, ROHO, BL, 2012, 111; Brower interview by Turner, March 30, 1983; Anne Brower to Marjorie, February 28, 1969, Carton 20, Folder 14, Brower Papers.

22. Trish Anderton, "Soldiers Who Loved to Ski," *Appalachia*, Summer–Fall 2014, 65.

23. Joseph E. Taylor III, *Pilgrims of the Vertical: Yosemite Rock Climbers and Nature at Risk* (Cambridge, Mass.: Harvard University Press, 2010), 92–93; K. Brower, "Climbing the Spiral Staircase," 48; Brower to Ross Brower (father), February 21, 1943, Box 2, Folder 7, Brower Papers; Brower to Ross Brower, November 24, 1942, Box 2, Folder 7, Brower Papers; Brower to parents, February 7, 1943, Box 2, Folder 2, Brower Papers; Brower to Richard Leonard, May 27, 1942, Box 7, Folder 5, Brower Papers.

24. Jeffrey Ingram (former Sierra Club employee), interview by the author, Tucson, Ariz., June 11, 2013.

25. John D'Emilio, "Capitalism and Gay Identity," in *The Material Queer: A LesBiGay Cultural Studies Reader*, ed. Donald Morton (Boulder, Colo.: Westview Press, 1995), 267–268.

26. Taylor, *Pilgrims of the Vertical*, 92–100.

27. Brower, *For Earth's Sake*, 144.

28. Brower, vignettes from ex-mountaineer speech, November 16, 1991, Carton 10, Folder 57, Brower Papers; "David Brower: Tireless Environmental Activist," *Mother Earth News*, May–June 1973, 3; David R. Brower, "Pursuit in the Alps," *Sierra Club Bulletin*, April 1946, 32–45.

29. Brower, *For Earth's Sake*, 128.

30. Robert Underhill (secretary and treasurer, University of California) to Brower, July 1, 1942, and June 5, 1946, Carton 119, Folder 8, Brower Papers.

31. August Fruge, "A Publisher's Career with the University of California Press, the Sierra Club, and the California Native Plant Society," an oral history conducted in 1997–1998 by Suzanne B. Riess, ROHO, BL, 2001, 66–69. Fruge recalled at one time placing Brower in charge of the production of all University of California Press books, which put Brower in the middle of an internal office fight between Fruge and the superintendent of the printing plant, a man named Tommasini. Tommasini could not criticize Fruge directly, so he instead used Brower's inexperience in printing to constantly complain that Brower was

incompetent. Those complaints, many of which Fruge did not think were warranted, forced Fruge to move Brower back to the editorial side of the operation.

32. Brower, *For Earth's Sake*, 145–151.

33. David Brower, "How To Kill a Wilderness," *Sierra Club Bulletin*, August 1945, 2–5.

34. Harold C. Bradley and David R. Brower, "Roads in the National Parks," *Sierra Club Bulletin*, June 1949, 31–54.

35. Tom Turner, *Sierra Club: 100 Years of Protecting Nature* (New York: Abrams, 1991), 128–129.

36. Brower, "Environmental Activist, Publicist, and Prophet," 20.

37. Leonard, "Mountaineer, Lawyer, Environmentalist, Volume I," 28–29.

38. Ethel Rose Taylor Horsehall, "On the Trail with the Sierra Club, 1920s–1960s," an oral history conducted in 1979 by George Baranoski and included in "Sierra Club Women III," ROHO, BL, 1982, 9.

39. Francis Farquhar, "Sierra Club Mountaineer and Leader," an oral history conducted in 1974 by Ann Lage and Ray Lage and included in "Sierra Club Reminisces," ROHO, BL, 1974, 53; Laura Hayden, "Q&A: Kenneth Brower Remembers His Father," *Sierra*, June 21, 2012, http://sierraclub.typepad.com/greenlife/2012/06/kenneth-brower-remembers-father-an-interview-with-david-browers-son.html (accessed October 15, 2012); K. Brower, *Wildness Within*, 6.

40. Wayburn, "Sierra Club Statesman," 88–89; David R. Brower, "Sierra High Trip," *National Geographic*, June 1954, 844–869; Lewis F. Clark, "Perdurable and Peripatetic Sierran: Club Officer and Outing Leader 1928–1984," an oral history conducted in 1975–1977 by Marshall Kuhn and included in "Sierra Club Reminisces III, 1910s–1970s," ROHO, BL, 1984, 45.

41. Hayden, "Q&A: Kenneth Brower"; K. Brower, *Wildness Within*, 6.

42. Phillip S. Berry, "Sierra Club Leader, 1960s–1980s: A Broadened Agenda, a Bold Approach," an oral history conducted in 1981 and 1984 by Ann Lage, ROHO, BL, 1988, 5.

43. Richard Sill, comments at Sierra Club board of directors meeting, October 3, 1968, transcript, Box 18, Sierra Club Records, 1957–1980, NC1260, Special Collections, University of Nevada, Reno. Brower's sexuality was well known by many. See Ingram interview, June 11, 2013; Michael McCloskey (former Sierra Club executive director), interview by the author, Portland, Ore., January 22, 2013; Gary Soucie (former Sierra Club employee), telephone interview by the author, August 26, 2013; and Phillip Berry (former Sierra Club officer and president), interview by the author, Lafayette, Calif., April 24, 2013.

44. Fruge, "Publisher's Career," 69–70; Ingram interview, June 11, 2013.

45. Clark, "Perdurable and Peripatetic Sierran," 43; Executive Director, December 5, 1952, Sierra Club Board of Directors Meeting Minutes, BANC

FILM 2945, BL, 1–2; Leonard, "Mountaineer, Lawyer, Environmentalist, Volume I," 159; Fruge, "Publisher's Career," 160; Harold E. Crowe, "Announcement from the President," *Sierra Club Bulletin*, January 1953, 3–4.

46. D'Emilio, "Capitalism and Gay Identity," 268.

47. Ingram interview, June 11, 2013.

4. The Lesson

1. David Brower, foreword to Eliot Porter, *The Place No One Knew: Glen Canyon on the Colorado*, ed. David Brower (San Francisco: Sierra Club Books, 1963), 7.

2. David R. Brower, "Environmental Activist, Publicist, and Prophet," an oral history conducted in 1974–1978 by Susan Schrepfer, Regional Oral History Office (ROHO), Bancroft Library (BL), University of California, Berkeley, 1979, 140.

3. Michael McCloskey (former Sierra Club executive director), interview by the author, Portland, Ore., January 22, 2013.

4. Sierra Club, "Dinosaur Trips Under Way," press release, June 9, 1954, Carton 8, Folder 20, Sierra Club Records, BANC MSS 71/103c, BL.

5. Brower, "Environmental Activist, Publicist, and Prophet," 126–128; Sierra Club, *Two Yosemites* (Sierra Club Films, 1955); Mark W. T. Harvey, *A Symbol of Wilderness: Echo Park and the American Conservation Movement* (Albuquerque: University of New Mexico Press, 1994), 271–272; Mark W. T. Harvey, *Wilderness Forever: Howard Zahniser and the Path to the Wilderness Act* (Seattle: University of Washington Press, 2005), 182.

6. Wallace Stegner, preface to *This Is Dinosaur: Echo Park Country and Its Magic River*, ed. Wallace Stegner (New York: Knopf, 1955), v.

7. According to Harvey, "*This Is Dinosaur* became the first book-length publication in conservation history that sought to publicize a park or wilderness preserve and to aid in a preservation campaign" (*Symbol of Wilderness*, 257); Brower, "Environmental Activist, Publicist, and Prophet," 127–128.

8. Brower, "Environmental Activist, Publicist, and Prophet," 125; "Echo Park Dam Not Needed," *New York Times*, June 16, 1955.

9. Harvey, *Symbol of Wilderness*, 266; Sharon Toney, "Conservationists' Role in the Echo Park Dispute," unpublished paper, n.d., Superintendent's Office, Dinosaur National Monument, Dinosaur, Colo., 18.

10. Richard Leonard to Portia Bradley, January 11, 1954, Carton 64, Folder 23, Sierra Club Records, BL; Toney, "Conservationists' Role in the Echo Park Dispute," 26.

11. C. Edward Graves (western representative, National Parks Association), "Opponent Addresses Memorandum," n.d., Superintendent's Office, Dinosaur National Monument.

12. Executive Director's Activities, August 16, 1955, Sierra Club Board of Directors Meeting Minutes, 1892–1995, BANC FILM 2945, BL.

13. Anne Brower to Marjorie, February 28, 1969, Carton 20, Folder 14, David Ross Brower Papers, BANC MSS 79/9c, BL.

14. Phillip Berry, interview by the author, Lafayette, Calif., April 24, 2013.

15. Joseph and Gayle Brower, interview by the author, Santa Rosa, Calif., March 26, 2013.

16. Grant McConnell, "About the Natural Resources and the Conservation Movement," essay, September 1954, Carton 19, Folder 2, Sierra Club Records, BL.

17. David Brower to Leonard, trip memorandum, March 12, 1953, Carton 2, Folder 2, Sierra Club Office of the Executive Director Records, BANC MSS 2002/230c, BL.

18. Brower to Sierra Club board of directors, December 23, 1955, Carton 1, Folder 38, Executive Director Records.

19. Brower, eastern trip report, n.d., Carton 2, Folder 3, Executive Director Records.

20. David R. Brower, "Executive Director Report to the Directors," May 2, 1957, Carton 8, Folder 27, Sierra Club Records, BL.

21. Brower, "Environmental Activist, Publicist, and Prophet," 120–121.

22. Jeffrey Ingram, telephone interview by the author, November 12, 2012.

23. Quoted in Jon M. Cosco, *Echo Park: Struggle for Preservation* (Boulder, Colo.: Johnson Books, 1995), 77.

24. David R. Brower, "Future of the Sierra Club," memorandum, October 5, 1953, Carton 9, Folder 4, Brower Papers.

25. Club Organization and the Future, October 17, 1953, Sierra Club Minutes; Francis Farquhar, "Sierra Club Mountaineer and Leader," an oral history conducted in 1974 by Ann Lage and Ray Lage and included in "Sierra Club Reminisces," ROHO, BL, 7.

26. "Handbook Edition," *Sierra Club Bulletin*, December 1967, 42.

27. Carl Pope (former Sierra Club executive director), telephone interview by the author, July 8, 2013.

28. For example, see Brower, eastern trip report, September 28–October 16, 1956, Carton 2, Folder 3, Executive Director Records.

29. Harold Bradley (Sierra Club president) to Ansel Adams, January 8, 1959, Carton 1, Folder 1, Sierra Club Members Papers, BANC MSS 71/295c, BL; Alexander Hildebrand (Sierra Club president) to Brower, December 9, 1955, Carton 1, Folder 26, Executive Director Records; Hildebrand to Brower, n.d., Carton 1, Folder 26, Executive Director Records (regarding the instructions not to go to a meeting in 1957).

30. Brower to Leonard, October 1, 1954, Carton 65, Folder 1, Sierra Club Records, BL.

31. Brower, "Environmental Activist, Publicist, and Prophet," 118.

32. David Brower, "Importance of Environment and Nuclear Energy" (speech presented to the Associated Sportsmen of California, Sacramento, September 17, 1955), transcript, Carton 8, Folder 22, Sierra Club Records, BL.

33. Brower, "Environmental Activist, Publicist, and Prophet," 119–122.

34. Ibid., 129; Harvey, *Symbol of Wilderness*, 277.

35. Harvey, *A Symbol of Wilderness*, 277; Cosco, *Echo Park*, 88; Brower, Dinosaur journal, part II, December 19, 1955, Carton 65, Folder 20, Sierra Club Records, BL.

36. Harvey, *Symbol of Wilderness*, 285.

37. Brower, "Environmental Activist, Publicist, and Prophet," 135.

38. David R. Brower, *For Earth's Sake: The Life and Times of David R. Brower* (Salt Lake City: Peregrine Smith Books, 1990), 341.

39. Dinosaur Strategy, December 27, 1955, Sierra Club Minutes; Dinosaur, January 7, 1956, Sierra Club Minutes.

40. Bestor Robinson, "Thoughts on Conservation and the Sierra Club," an oral history conducted in 1974 by Ann Lage and Ray Lage and included in "Sierra Club Reminisces," ROHO, BL, 1974, 23.

41. Richard Leonard, "Mountaineer, Lawyer, Environmentalist, Volume I," an oral history conducted in 1976 by Susan R. Schrepfer, ROHO, BL, 1976, 120–121.

42. John C. Miles, *Guardians of the Park: A History of the National Parks and Conservation Association* (Washington, D.C.: Taylor & Francis, 1995), 197. The National Parks Association changed its name in 1970 to the National Parks and Conservation Association.

43. Alice Joy Keith (Glen Canyon advocate) to 125 congressional representatives, February 22, 1956, Carton 65, Folder 21, Sierra Club Records, BL.

44. Russell Martin, *A Story That Stands Like a Dam: Glen Canyon and the Struggle for the Soul of the West* (New York: Holt, 1989), 80.

45. Ibid., 45–47.

46. Glen Canyon, June 20, 1954, Sierra Club Minutes.

47. Wallace Stegner, "Backroads River," *Atlantic Monthly*, January 1948, 63.

48. Charles Eggert (photographer) to Fred Packard (National Parks Association), August 2, 1955, Carton 65, Folder 19, Sierra Club Records, BL.

49. "Stop Glen Canyon Dam, Utahan Asks," *Salt Lake Tribune*, February 12, 1954; William Halliday (member, Utah Committee for a Glen Canyon National Park) to Josef Muench (noted nature photographer), August 22, 1954, Carton 14, Folder 2, Executive Director Records.

50. Brower to Malcolm S. Ellingson (member, Utah Committee for a Glen Canyon National Park), February 7, 1955, Carton 14, Folder 1, Executive Director Records; Howard Zahniser with Carl Gustafson (Conservation Council), telephone conversation, December 20, 1955, transcript, Carton 65, Folder 22, Sierra Club Records, BL; Ken Sleight, interview by Ken Verdoia, KUED, University of Utah, aired October 1999, http://www.kued.org/productions/glencanyon/interviews/sleight/html (accessed October 18, 2013).

51. Karl B. Brooks, *Public Power, Private Dams: The Hells Canyon High Dam Controversy* (Seattle: University of Washington Press, 2009), 176–216; David Brower, "Natural Resources Hells Canyon," memorandum, n.d., Carton 73, Folder 27, Sierra Club Records, BL. The clash over Hells Canyon was between public and private power developments as well as between the Roosevelt and Truman concept of government-funded projects and the Eisenhower philosophy of less government. The original plan called for a dam in North America's deepest river valley, which stretched 10 miles across and was 7,993 feet deep, and would have supplied cheap electricity to the Bonneville Power Authority and customers in Oregon, Washington, and Idaho. Idaho Power Company instead proposed and built three smaller dams within the canyon. Brower and the club opposed the three small dams because they flooded 90 miles of bottomlands that the nation's largest elk herd depended on. In another dam project, Montana Power Company proposed building several small dams on the Flathead River. Federal dam builders responded by proposing the Glacier View Dam, which environmentalists thought they had defeated in the late 1940s because it would have flooded portions of Glacier National Park. Brower and the Sierra Club called for an alternative, the Paradise Dam on the Clark fork of the Flathead. The Clark fork was, as Brower once put it, in "pretty country, but nothing scenically outstanding" ("A Case for a Dam," *Sierra Club Bulletin*, February 1957, 3). The dam project had first been scuttled in the early 1950s because of opposition from local farmers and because both a major rail line and a highway would have had to be moved.

52. Brower to the Sierra Club board of directors, memorandum, October 19, 1956, Carton 13, Folder 11, Executive Director Records; McCloskey interview, January 22, 2013.

53. David Brower, interview by Ken Verdoia, KUED, University of Utah, aired October 1999, http://www.kued.org productions/glencanyon/interviews/brower/html (accessed October 18, 2013).

54. Stegner, "Backroads River," 63.

55. Ibid., 64; Keith to 125 congressional representatives, February 22, 1956.

56. Mark W. T. Harvey, "Defending the Park System: The Controversy Over Rainbow Bridge," *New Mexico Historical Review*, January 1998, 59.

57. Harvey, *Symbol of Wilderness*, 287–301.

5. Wilderness

1. Weldon Heald, "San Gorgonio—Southern California's Rooftop," *Living Wilderness*, Summer 1963, 12–16; John Robinson, *San Gorgonio: A Wilderness Preserved* (San Bernardino, Calif.: San Gorgonio Volunteers Association, 1991), 98–99.

2. Michael McCloskey (former Sierra Club executive director), interview by the author, Portland, Ore., January 22, 2013.

3. For the Wilderness Conference volumes published during Brower's years as executive director, see David Brower, ed., *The Meaning of Wilderness to Science* (San Francisco: Sierra Club Books, 1960); David Brower, ed., *Wilderness: American's Living Heritage* (San Francisco: Sierra Club Books, 1961); François Leydet, ed., *Tomorrow's Wilderness* (San Francisco: Sierra Club Books, 1963); David Brower, ed., *Wildlands in Our Civilization* (San Francisco: Sierra Club Books, 1964); Bruce Kilgore, ed., *Wilderness in a Changing World* (San Francisco: Sierra Club Books, 1966); Maxine McCloskey and James P. Gilligan, eds., *Wilderness and the Quality of Life* (San Francisco: Sierra Club Books, 1969); Maxine McCloskey, ed., *Wilderness: The Edge of Knowledge* (San Francisco: Sierra Club Books, 1969). The volume published in 1964 covered the proceedings from 1949 to 1957. The editor of each volume was free to add material beyond what was presented at the conference. As the editor of the volume published in 1961, Brower added an essay written by his son Kenneth. At the time, Ken Brower was sixteen years old and a high school junior.

4. David Brower, foreword to Brower, *Wildlands in Our Civilization*, 15.

5. Philip Shabecoff, *A Fierce Green Fire: The American Environmental Movement* (New York: Hill and Wang, 1993), 3–76; Roderick Nash, *Wilderness and the American Mind* (New Haven, Conn.: Yale University Press, 1967), 67–160.

6. Doug Scott, *The Enduring Wilderness: Protecting Our Natural Heritage Through the Wilderness Act* (Golden, Colo.: Fulcrum, 2004), 24–25.

7. Char Miller, *Gifford Pinchot and the Making of Modern Environmentalism* (Washington, D.C.: Island Press, 2001), 121; Paul W. Hirt, *A Conspiracy of Optimism: Management of the National Forests Since World War Two* (Lincoln: University of Nebraska Press, 1994), 32.

8. Nash, *Wilderness and the American Mind*, 182–199; Scott, *Enduring Wilderness*, 29; Heald, "San Gorgonio," 12–16.

9. Nash, *Wilderness and the American Mind*, 200–208; Glen O. Robinson, *The Forest Service: A Study in Public Land Management* (Baltimore: Johns Hopkins University Press, 1975), 154–158; Scott, *Enduring Wilderness*, 30–32.

10. Mark W. T. Harvey, *Wilderness Forever: Howard Zahniser and the Path to the Wilderness Act* (Seattle: University of Washington Press, 2005), x–xi.

11. David R. Brower, *For Earth's Sake: The Life and Times of David R. Brower* (Salt Lake City: Peregrine Smith Books, 1990), 230; Stewart L. Udall, *The Quiet Crisis and the Next Generation* (Layton, Utah: Gibbs Smith, 1988), 214–215.

12. Brower, *For Earth's Sake*, 226–229.

13. McCloskey interview, January 22, 2013.

14. Richard E. McArdle, "Dr. Richard E. McArdle: An Interview with the Former Chief, U.S. Forest Service, 1952–1962," an oral history conducted in 1973–1974 by Elwood R. Maunder, Forest History Society, Santa Cruz, Calif., 1975, 157.

15. "U.S. Wilderness System Proposed," *Sierra Club Bulletin*, June 1956, 3; David Brower to unlisted recipients, memorandum, March 1, 1956, Carton 280, Folder 25, Sierra Club Members Papers, BANC MSS 71/295c, Bancroft Library (BL), University of California, Berkeley; Brower, *For Earth's Sake*, 228–230.

16. David R. Brower, "Environmental Activist, Publicist, and Prophet," an oral history conducted in 1974–1978 by Susan Schrepfer, Regional Oral History Office, BL, 1979, 80.

17. Wallace Stegner, introduction to Brower, *Wildlands in Our Civilization*, 37; "Silly Wilderness Preservation Proposal," *Salt Lake Tribune*, March 8, 1960.

18. Stegner, introduction to Brower, *Wildlands in Our Civilization*, 37–38; Ben Price, "Wilderness Bill Stirs Fuss," *Boston Herald*, December 21, 1958; David Brower, "The Wilderness Bill: Nobody Wants It but the People," *Sierra Club Bulletin*, March 1960, 2.

19. Quoted in "Play Area Tend to Wilds Is Cited," *New York Times*, May 17, 1957.

20. Brower to Sierra Club board members, memorandum, July 5, 1957, Carton 1, Folder 38, Sierra Club Office of Executive Director Records, BANC MSS 2002/230c, BL; Brower to Conrad Wirth, April 11, 1957, Carton 88, Folder 33, Sierra Club Records, BANC MSS 71/103c, BL.

21. William O. Douglas, *Of Men and Mountains* (New York: Harper, 1950), 255–273.

22. In Box 2, Folder 16, Executive Director Records: Brower to William O. Douglas, February 14, 1957; Douglas to Brower, December 1, 1958; Douglas to Richard Leonard, December 9, 1958; Howard Zahniser to Leonard, December 21, 1958. See also Paul L. Montgomery, "Tour of Hudson Led by Douglas," *New York Times*, March 7, 1966; William O. Douglas, *My Wilderness: The Pacific West* (New York: Doubleday, 1960). Justice Douglas took several celebrated hikes to promote trails and wilderness in the East. His most famous was a 189-mile hike in 1954 to promote saving the Chesapeake & Ohio Canal. Brower accompanied the justice on shorter hikes along the canal that drew public attention to it. He also hiked with Douglas in the Hudson Valley of New York on March 6, 1966, to promote removing barriers for hiking.

23. Harvey, *Wilderness Forever*, 205.

24. Proposed Scenic Resources Review, January 7–8, 1956, Sierra Club Board of Director Meeting Minutes, 1892–1995, BANC FILM 2945, BL, 7–8; David Brower, "Scenic Resources for the Future," *Sierra Club Bulletin*, December 1956, 1–10.

25. John F. Warth, "The ORRRC Study—Where Does It Leave Preservation?" report, n.d., Carton 7, Folder 25, Executive Director Records; McArdle, "Dr. Richard E. McArdle," 72–73.

26. David Pearlman, "Nation and State Take Inventory," *Sierra Club Bulletin*, January 1959, 3; Brower to Fred A. Seaton (secretary of the interior), August 26, 1958, Carton 88, Folder 5, Sierra Club Records, BL; Laurence S. Rockefeller (chairman, ORRRC) to Arthur B. Johnson (president, Federation of Western Outdoor Clubs), February 1, 1960, Carton 7, Folder 24, Executive Director Records.

27. Harvey, *Wilderness Forever*, 198.

28. Quoted in ibid., 195.

29. Grant McConnell (political science professor, University of Chicago) to Brower, November 3, 1960, Carton 9, Folder 11, Executive Director Records.

30. Watkins quoted in Jackson J. Benson, *Wallace Stegner: His Life and Work* (New York: Viking, 1996), 231; see also Kenneth Brower, *The Wildness Within: Remembering David Brower* (Berkeley, Calif.: Heyday, 2012), 64–65.

31. Wallace Stegner, "Saga of a Letter: The Geography of Hope," *Living Wilderness*, December 1980, 43; this essay includes the "Wilderness Letter," which Stegner wrote in December 1960, and a history of the letter.

32. Ibid., 42.

6. Forest

1. David R. Brower, "Environmental Activist, Publicist, and Prophet," an oral history conducted in 1974–1978 by Susan Schrepfer, Regional Oral History Office (ROHO), Bancroft Library (BL), University of California, Berkeley, 1979, 83.

2. David Brower, *Grand Canyon Discussion at Silver Grotto and Marble Damsite, May 31 to June 13, 1977*, audiocassette, recorded by Ron Hayes, Grand Canyon National Park Museum Collection, GRCA 40237, Grand Canyon National Park, Ariz.; David Kupfer, "Final Interview: David R. Brower," August 5, 2000, http://www.wildnesswithin.com/kupfery.html (accessed August 18, 2014).

3. Constance I. Millar (paleoecologist, U.S. Forest Service), interview by the author, Inyo National Forest, Calif., July 30, 2012.

4. Susan Schrepfer, "Establishing Administrative Standing: The Sierra Club and the Forest Service, 1897–1958," *Pacific Historical Review*, February 1989, 55–81; Greg Reis (member, Mono Lake Committee) to the author, e-mail, September 12, 2012.

5. Hal Roth, "Puzzling Story at Mammoth," n.d., Carton 76, Folder 4, Sierra Club Records, BANC MSS 71/103c, BL.

6. Barbara Haddaway (citizen opponent of logging in Inyo National Forest), chronology of communication with Forest Service, August 4, 1952–December 20, 1954, report, n.d., Carton 87, Folder 23, Sierra Club Members Papers, BANC MSS 71/295c, BL.

7. Dan Wyant, "Battle Against Pumice Mining Bitterly Recalled," *Eugene Register-Guard*, October 4, 1971; John and Barbara Haddaway, proposed Mono Crater National Monument map, n.d., Carton 76, Folder 5, Sierra Club Records, BL; Conrad Wirth (National Parks director) to Harold Bradley, August 7, 1958, Carton 88, Folder 13, Sierra Club Records, BL. The Mono Lakes and Inyo Crater areas had been repeatedly considered for recognition and protection under the national park system. The Haddaway plan included a detailed map. The area immediately north of Deadman Summit is now the Mono Basin National Scenic Area. Brower commented on the scenic nature of the area several times, but his most interesting comment came in 1958. Wirth told the Sierra Club that he had received a request to make the area a national park, but he did not want to send anyone from his staff there because it might upset the Forest Service. Brower wrote back that the area was geologically extremely interesting, but there was some development, including a ski resort at Mammoth. He suggested it could be part of a national recreation plan (David Brower to Wirth, August 26, 1958, Carton 88, Folder 13, Sierra Club Records, BL).

8. David Brower, "Chronology of Mammoth–Mono Timber Cutting," report, n.d, Carton 76, Folder 2, Sierra Club Records, BL.

9. Roth, "Puzzling Story at Mammoth"; David Brower, Glass Creek report, August 25–28, 1953, Carton 76, Folder 2, Sierra Club Records, BL.

10. Dick Leonard to W. S. Davis (supervisor, Inyo National Forest), August 16, 1954, Carton 76, Folder 2, Sierra Club Records, BL.

11. Roth, "Puzzling Story at Mammoth"; Brower, Glass Creek report.

12. John Haddaway to Brower, April 29, 1954, Carton 76, Folder 1, Sierra Club Records, BL.

13. Richard E. McArdle, "Dr. Richard E. McArdle: An Interview with the Former Chief, U.S. Forest Service, 1952–1962," an oral history conducted in 1973–1974 by Elwood R. Maunder, Forest History Society, Santa Cruz, Calif., 1975, vi–ix; Harold K. Steen, *The U.S. Forest Service: A History* (Seattle: University of Washington Press, 1976), 279; Brower, "Environmental Activist, Publicist, and Prophet," 80; Michael P. Cohen, *The History of the Sierra Club, 1892–1970* (San Francisco: Sierra Club Books, 1988), 195–199.

14. "National Forest Service Matters: A Report," October 15, 1955, Sierra Club Board of Director Meeting Minutes, 1892–1995, BANC FILM 2945, BL;

Richard McArdle (Forest Service director) to Brower, December 29,1954, Carton 76, Folder 2, Sierra Club Records, BL.

15. Edgar Wayburn, "Sierra Club Statesman, Leader of the Parks and Wilderness Movement: Gaining Protection for Alaska, the Redwoods, and Golden Gate Parklands," an oral history conducted in 1976–1981 by Ann Lage and Susan Schrepfer, ROHO, BL, 1985, 24–26.

16. Wyant, "Battle Against Pumice Mining"; Millar interview, July 30, 2012; Constance I. Millar, "Case Studies in Ecosystem Management: The Mammoth–June Ecosystem Management Project, Inyo National Forest," in *Sierra Nevada Ecosystem Project: Final Report to Congress*, vol. 1, *Assessment Summaries and Management Strategies*, Centers for Water and Wildland Resources, Report no. 37 (Davis: University of California, 1996), 146–150.

17. Brower to John and Barbara Haddaway, September 28, 1956, Carton 76, Folder 5, Sierra Club Records, BL.

18. Paul W. Hirt, *A Conspiracy of Optimism: Management of the National Forests Since World War Two* (Lincoln: University of Nebraska Press, 1994), xx; Char Miller and V. Alaric Sample, "Gifford Pinchot and the Conservation Spirit," introduction to Gifford Pinchot, *Gifford Pinchot: Breaking New Ground Commemorative Edition* (Washington, D.C.: Island Press, 1998), xi–xvii.

19. McArdle, "Dr. Richard E. McArdle," 68.

20. Glen O. Robinson, *The Forest Service: A Study in Public Land Management* (Baltimore: Johns Hopkins University Press, 1975), 14–15, 154–158; Steen, *U.S. Forest Service*, 283; Hirt, *Conspiracy of Optimism*, 131.

21. McArdle, "Dr. Richard E. McArdle," 68.

22. Brower to McArdle, draft, n.d., Carton 84, Folder 36, Sierra Club Records, BL.

23. Brower, "Environmental Activist, Publicist, and Prophet," 66.

24. Hirt, *Conspiracy of Optimism*, xxiii.

25. Tom McAllister (outdoor editor, *Oregon Journal*), "Waldo, the Wild One," n.d., Carton 89, Folder 9, Sierra Club Records, BL; Robert Aufderheide (forest supervisor, Willamette National Forest) to Ray Ramey (secretary-treasurer, Lane County Chamber of Commerce), August 1, 1958, Carton 89, Folder 9, Sierra Club Records, BL.

26. Brower to Ira Gabrielson (member, Citizens Committee on Natural Resources), May 29, 1959, Carton 87, Folder 29, Sierra Club Records, BL.

27. David R. Simons, "These Are the Shining Mountains," *Sierra Club Bulletin*, October 1959, 2.

28. Doug Scott, *The Enduring Wilderness: Protecting Our Natural Heritage Through the Wilderness Act* (Golden, Colo.: Fulcrum, 2004), 35.

29. "Three Sisters Decision, Last Word or Just the Beginning?" *Sierra Club Bulletin*, February 1957, 10; Brower, "Environmental Activist, Publicist, and Prophet," 65.

30. The impact of global climate change since the 1950s has likely significantly reduced the number of glaciers in the North Cascades. For the number of glaciers, see Simons, "These Are the Shining Mountains," 2; and Weldon Heald, "Cascade Holiday," in *The Cascades: Mountains of the Pacific Northwest*, ed. Roderick Peattie (New York: Vanguard, 1949), 134.

31. Brock Evans, "Showdown for the Wilderness Alps of Washington's North Cascades," *Sierra Club Bulletin*, April 1968, 7–16; Harvey Manning, *The Wild Cascades: Forgotten Parkland* (San Francisco: Sierra Club Books, 1965), 3–117; Heald, "Cascade Holiday," 134.

32. "The NCCC Proposes: A North Cascades National Park," *Sierra Club Bulletin*, October 1963, 7–13; Evans, "Showdown for the Wilderness Alps," 9; U.S. Forest Service, "Glacier Peak Land Management Study," February 7, 1957, Carton 12, Folder 19, Sierra Club Office of Executive Director Records, BANC MSS 2002/230c, BL, 3; Manning, *Wild Cascades*, 114–115.

33. Wayburn, "Sierra Club Statesman," 32; Grant McConnell, "Conservation and Politics in the North Cascades," an oral history conducted in 1983 by Rod Holmgren and included in "Sierra Club Nationwide I," ROHO, BL, 1983, 32–34.

34. McConnell, "Conservation and Politics in the North Cascades," 33–34; Abigail Avery (financial donor, North Cascades film) to Brower, May 1, 1957, and Brower to Avery, May 10, 1957, Carton 23, Folder 33, Executive Director Records.

35. Sierra Club, *Wilderness Alps of Stehekin* (Sierra Club Films, 1957), http://content.sierraclub.org/brower/video (accessed October 18, 2013).

36. McConnell, "Conservation and Politics in the North Cascades," 34. Brower attempted to interest President Eisenhower in watching the North Cascades film. He wrote to the president that some critics said the film had an almost religious quality about it and that if that was true, it was because the mountains were so beautiful (Brower to Dwight D. Eisenhower, January 7, 1958, Carton 12, Folder 12, Executive Director Records).

37. David Brower, national forest speech, U.S. Forest Service Supervisors Meeting, Portland, Ore., April 5, 1957, transcript, Carton 9, Folder 14, David Ross Brower Papers, BANC MSS 79/9c, BL.

38. Both Harrison and Cliff quoted in Hirt, *Conspiracy of Optimism*, 131–132, 164.

39. National Forest Policy, September 18, 1960, Sierra Club Minutes, 2–7.

40. McArdle to Lewis Clark (president, Sierra Club board), April 6, 1961, Carton 8, Folder 28, Executive Director Records.

41. W. F. McCulloch (dean, School of Forestry, Oregon State College), resource-managers speech, U.S. Forest Service Supervisors Meeting, Portland, Ore., March 28, 1956, transcript, Carton 87, Folder 35, Sierra Club Records, BL.

42. Brower, "Environmental Activist, Publicist, and Prophet," 50.

43. Karl Onthank, "Not Yet a Lost Cause: Report on the Three Sisters," *Sierra Club Bulletin*, January 1958, 16–17; "Three Sisters Wilderness," Wilderness.net, n.d., http://www.wilderness.net/NWPS/wildView?WID=602 (accessed January 9, 2014). The Three Sisters Wilderness, which encompassed 281,190 acres, was established in 1964 when Congress passed the Wilderness Act.

44. "Three Sisters Decision, Last Word or Just the Beginning?" 10.

45. Manning, *Wild Cascades*, 115–116.

46. *Sierra Club Outdoor Newsletter Number 6*, August 22, 1960, Carton 6, Folder 27, Brower Papers.

47. Brower, "Environmental Activist, Publicist, and Prophet," 91.

48. David Brower, notes on Forest Tactics memorandum to Sierra Club board of directors, October 12, 1960, Carton 8, Folder 2, Sierra Club Records, BL; McArdle to Clark, November 15, 1960, Carton 8, Folder 27, Executive Director Records; Henry Clapper (executive secretary, Council of American Foresters) to Clark, November 18, 1960, Carton 8, Folder 27, Executive Director Records; Brower to Clapper, November 28, 1960, Carton 8, Folder 27, Executive Director Records.

49. Lewis Clark, "Perdurable and Peripatetic Sierran: Club Officer and Outing Leader 1928–1984," an oral history conducted in 1975–1977 by Marshall Kuhn and included in "Sierra Club Reminisces III, 1910s–1970s," ROHO, BL, 123; Clark to McArdle, December 30, 1960, Carton 8, Folder 27, Sierra Club Records, BL.

50. Richard Leonard, "Mountaineer, Lawyer, Environmentalist, Volume II," an oral history conducted in 1976 by Susan R. Schrepfer, ROHO, BL, 1976, 375–376.

51. Ibid.

52. Ibid., 339–340.

53. Brower, "Environmental Activist, Publicist, and Prophet," 93.

54. Brower to Sierra Club board, memorandum, October 12, 1960, Carton 8, Folder 2, Sierra Club Records, BL.

55. Brower to anonymous recipient in Portland, Ore., February 23, 1961, Carton 9, Folder 13, Executive Director Records.

56. Brower to William O. Douglas (justice, U.S. Supreme Court), November 12, 1961, Carton 87, Folder 29, Sierra Club Records, BL.

7. Parks

1. Walter Starr, *Guide to the John Muir Trail and the High Sierra Region* (San Francisco: Sierra Club Books, 1943), 17–19; Ansel Adams, "Tenaya Tragedy," *Sierra Club Bulletin*, November 1958, 1–13.

2. Fernando Penalosa, *Yosemite in the 1930s: A Remembrance* (Rancho Palos Verdes, Calif.: Quaking Aspen Book, 2002), 15–16; Tioga Road at Tenaya Lake, August 31, 1952, Sierra Club Board of Directors Meeting Minutes, 1892–1995, BANC FILM 2945, Bancroft Library (BL), University of California, Berkeley, 5–7.

3. David Brower, "Tioga Protest: What Happened Below Tenaya," *Sierra Club Bulletin*, October 1958, 1–7; Adams, "Tenaya Tragedy."

4. Ansel Adams to Harold Bradley, July 12, 1958, Carton 11, Folder 6, Sierra Club Office of Executive Director Records, BANC MSS 2002/230c, BL.

5. Brower, "Tioga Protest," 4–5; Conrad L. Wirth, *Parks, Politics, and the People* (Norman: University of Oklahoma Press, 1980), 359.

6. Brower, "Tioga Protest," 6.

7. Conrad L. Wirth to Sierra Club officials, memorandum, November 24, 1958, Carton 86, Folder 3, Sierra Club Records, BANC MSS 71/103c, BL; Wirth, *Parks, Politics, and the People*, 360.

8. David R. Brower, "Environmental Activist, Publicist, and Prophet," an oral history conducted in 1974–1978 by Susan Schrepfer, Regional Oral History Office (ROHO), BL, 1979, 57; Glen Binford (reporter, *Los Angeles Times*), Lake Tenaya story, n.d., Carton 11, Folder 12, Executive Director Records; David Brower to Stewart Udall (secretary of the interior), May 18, 1961, Carton 8, Folder 4, Sierra Club Records, BL.

9. Richard West Sellars, *Preserving Nature in the National Parks: A History* (New Haven, Conn.: Yale University Press, 1997), 58–59.

10. Brower, "Tioga Protest," 6.

11. Sellars, *Preserving Nature in the National Parks*, 36.

12. Susan Schrepfer, *The Fight to Save the Redwoods: A History of Environmental Reform, 1917–1978* (Madison: University of Wisconsin Press, 1983), xiii–xiv; Peggy Wayburn, "The Tragedy of Bull Creek," *Sierra Club Bulletin*, January 1960, 10–11.

13. Raymond F. Desmann, *The Destruction of California* (New York: Macmillan, 1965), 88–90; Schrepfer, *Fight to Save the Redwoods*, 130–162; Claire Reynolds, exec. prod., and Sam Greene, prod. and dir., *Redwood National Park: Preserving Ancient Forests*, DVD (KEET-TV, 2009).

14. David R. Simons, "These Are the Shining Mountains," *Sierra Club Bulletin*, October 1959, 2.

15. David Simons (college student), prospectus of proposed trip memorandum, Carton 72, Folder 19, Sierra Club Records, BL; Richard Tablor (chairman, Sierra Club Natural Science Committee) to Mr. and Mrs. Ralph Simons, May 31, 1956, Carton 72, Folder 19, Sierra Club Records, BL; Michael Cohen, *The History of the Sierra Club, 1892–1970* (San Francisco: Sierra Club Books, 1988), 220–228; "The NCCC Proposes: A North Cascades National Park," *Sierra Club Bulletin,* October 1963, 7–13.

16. David Brower, "David Ralph Simons 1936–1960," *Sierra Club Bulletin,* November 1960, 25

17. Michael McCloskey, *In the Thick of It: My Life in the Sierra Club* (Washington, D.C.: Island Press, 2005), 40–41.

18. John B. Oakes, "Conservation: Fight for Parks," *New York Times,* October 2, 1960.

19. Wolfgang Saxon, "Conrad L. Wirth, 93; Led National Parks Service," *New York Times,* July 28, 1993; Brower, "Environmental Activist, Publicist, and Prophet," 58; Wirth, *Parks, Politics, and the People,* 4–6.

20. Carsten Lien, *Olympic Battleground: The Power Politics of Timber Preservation* (San Francisco: Sierra Club Books, 1991), 268–298; Brower, "Environmental Activist, Publicist, and Prophet," 48–50; Brower, statement to Governor Langlie's Committee on Olympic National Park, October 26, 1953, Carton 12, Folder 26, Executive Director Records. Brower was very familiar with Olympic Park and its history, having worked to defeat an effort in 1953 to reduce the park's boundaries to open up more land for logging.

21. Brower to Sierra Club board of directors, memorandum on proposed hotel at Wonder Lake, October 21, 1961, Carton 8, Folder 6, Sierra Club Records, BL; David Brower, statement to U.S. Senate Committee on Interior and Insular Affairs, February 26, 1955, Carton 82, Folder 21, Sierra Club Records, BL; Ansel Adams, "A Conversation with Ansel Adams," an oral history conducted in 1972, 1974, and 1975 by Ruth Teiser and Catherine Harroun, ROHO, BL, 1978, 600; Arthur E. Harrison, "Mt. Rainier Tramway Plan Defeated," *Sierra Club Bulletin,* January 1955, 6; David Brower, statement to U.S. Senate Committee on Interior and Insular Affairs, October 15, 1956, Carton 12, Folder 11, Executive Director Records.

22. Bernard DeVoto, "Let's Close the National Parks," *Harper's,* October 1953, 52; Sellars, *Preserving Nature in the National Parks,* 182–183.

23. Sellars, *Preserving Nature in the National Parks,* 183–184; U.S. Department of the Interior, "Mission 66 Program," press release, February 2, 1956, Carton 88, Folder 24, Sierra Club Records, BL.

24. Wirth, *Parks, Politics, and the People,* 237–284; Brower, "Environmental Activist, Publicist, and Prophet," 54.

25. Brower, Mission 66 notes, December 23, 1955, Carton 8, Folder 10, Executive Director Records; Brower, "Environmental Activist, Publicist, and Prophet," 54.

26. Brower to Wirth, April 11, 1957, Carton 88, Folder 33, Sierra Club Records, BL.

27. Wirth to Alexander Hildebrand, telegram, May 2, 1957, and Hildebrand to Wirth, May 3, 1957, Carton 88, Folder 33, Sierra Club Records, BL.

28. Adams to Brower, Dick Leonard, and Harold Bradley, July 27, 1957, Carton 88, Folder 34, Sierra Club Records, BL.

29. David Brower, "Mission 66 Is Proposed by Reviewer of Park Service's New Brochure on Wilderness," *National Parks Magazine*, January–March 1958, 1–4, 45–47.

30. Harold Crowe (board member and former president, Sierra Club) to Wirth, January 12, 1958, Carton 1, Folder 5, Executive Director Records; Wirth to Spencer M. Smith (secretary, Citizens Committee on Natural Resources), March 26, 1958, Carton 88, Folder 34, Sierra Club Records, BL.

31. Wirth to Smith, March 26, 1958; Horace Albright (former director, National Park Service), comments on Brower and Mission 66, April 4, 1958, Carton 90, Folder 30, Sierra Club Records, BL.

32. Harold E. Crowe, "Sierra Club Physician, Baron, and President," an oral history conducted in 1973 by Richard Searle and included in "Sierra Club Reminisces II," ROHO, BL, 1975, 10–11.

33. Adams to Stewart Udall, June 12, 1961, Carton 93, Folder 29, Sierra Club Records, BL; Edgar Wayburn, Tioga Road dedication report, June 24, 1961, Carton 93, Folder 29, Sierra Club Records, BL; Wirth, *Parks, Politics, and the People*, 302–303.

8. Glen Canyon

1. Russell Martin, *A Story That Stands Like a Dam: Glen Canyon and the Struggle for the Soul of the West* (New York: Holt, 1989), 86–87.

2. Jedediah Rogers, "Glen Canyon Unit," U.S. Bureau of Reclamation, 2006, http://www.usbr.gov/projects/ImageServer?imgName=Doc_1232657383034.pdf, 21–22 (accessed September 27, 2014); Jack Goodman, "Taming the Colorado," *Saturday Evening Post*, September 15, 1962, 26–31.

3. Rogers, "Glen Canyon Unit," 23.

4. David Brower to William Halliday, March 13, 1958, Carton 81, Folder 17, Sierra Club Records, BANC MSS 71/103c, Bancroft Library (BL), University of California, Berkeley; Utah Committee for a Glen Canyon National Park, report, February 1958, Carton 72, Folder 26, Sierra Club Records, BL; Brower to Howard Zahniser, letter and report, July 5, 1955, Carton 65, Folder 19, Sierra

Club Records, BL; Martin, *Story That Stands Like a Dam*, 96; James Lawrence Powell, *Dead Pool, Lake Powell, Global Warming, and the Future of Water in the West* (Berkeley: University of California Press, 2008), 158.

5. David R. Brower, "Environmental Activist, Publicist, and Prophet," an oral history conducted in 1974–1978 by Susan Schrepfer, Regional Oral History Office (ROHO), BL, 1979, 141.

6. Powell, *Dead Pool*, 159–160.

7. Brower to Halliday, March 13, 1958.

8. Floyd Dominy, interview by Ken Verdoia, KUED, University of Utah, aired October 1999, http://www.kued.org/productions/glencanyon/interview/Dominy/html (accessed October 18, 2013).

9. Dominy interview by Verdoia; Marc Reisner, *Cadillac Desert: The American West and Its Disappearing Water* (New York: Penguin Books, 1986), 218; John McPhee, *Encounters with the Archdruid* (New York: Farrar, Straus and Giroux, 1971), 153–158.

10. Reisner, *Cadillac Desert*, 231.

11. Quoted in ibid., 242.

12. Quoted in McPhee, *Encounters with the Archdruid*, 168.

13. Reisner, *Cadillac Desert*, 216–217.

14. "Protective Measures for Rainbow Bridge," *Sierra Club Bulletin*, February 1957, 14.

15. Floyd Dominy, "Oral History Interviews," conducted April 6, 1984, and April 8, 1986, by Brit Allan Story, U.S. National Archives and Records Administration, College Park, Md., 205–206; Powell, *Dead Pool*, 131.

16. David R. Brower, "Rainbow Promise Breaking: New Threat at Echo Park," *Sierra Club Bulletin*, April–May 1960, 16.

17. Rainbow Bridge National Monument, February 6, 1960, Sierra Club Board of Directors Meeting Minutes, 1892–1995, BANC FILM 2945, BL, 4–5, and January 22, 1961, 8; Brower to Luna Leopold, May 26, 1961, Carton 14, Folder 12, Sierra Club Office of Executive Director Records, BANC MSS 2002/230c, BL. For Brower, the situation became even worse when the prestigious journal *Science* in February 1961 published an article by Angus Woodbury, a biology professor at the University of Utah, questioning the need for the dam to protect Rainbow Bridge (Angus Woodbury, "Protecting Rainbow Bridge," *Science*, August 26, 1960, 519–528). Brower called him "Black Woodbury" and complained that *Science* was allowing itself to be used.

18. Brower, "Environmental Activist, Publicist, and Prophet," 144.

19. Keith Schneider and Cornelia Dean, "Stewart L. Udall, Conservationist in Kennedy and Johnson Cabinets, Dies at 90," *New York Times*, March 20, 2010.

20. Brower, "Environmental Activist, Publicist, and Prophet," 145.

21. Schneider and Dean, "Stewart L. Udall"; Alfred A. Knopf (publisher) to Brower, January 14, 1963, Carton 14, Folder 18, Executive Director Records.

22. Stewart Udall to Wayne Aspinall, August 27, 1960, Carton 14, Folder 10, Executive Director Records; Mark W. T. Harvey, *A Symbol of Wilderness: Echo Park and the American Conservation Movement* (Albuquerque: University of New Mexico Press, 1994), 55–56.

23. Natural Resources Council of America, "Group Meets with Udall on Rainbow Bridge," press release, February 22, 1961, Carton 81, Folder 19, Sierra Club Records, BL.

24. Dominy, "Oral History Interviews," 207.

25. Ibid.

26. John O'Reilly, "Udall at the Bridge," *Sports Illustrated*, May 15, 1961, 26–27.

27. Martin, *Story That Stands Like a Dam*, 226–227. See also Bruce Kilgore, "Rainbow Bridge: Final Act," *Sierra Club Bulletin*, June 1961, 8–9.

28. David Brower, "Rainbow Bridge and the Quicksands of Time," *Sierra Club Bulletin*, June 1961, 2.

29. David Brower, "Please Keep Those Glen Canyon Tunnels Open Until Rainbow Bridge Protection Is Certain," *Sierra Club Bulletin*, March–April 1962, 2–3.

30. Brower to S. Udall, March 18, 1962, Carton 14, Folder 14, Executive Director Records; Martin, *Story That Stands Like a Dam*, 233.

31. Arthur Johnson (consulting engineer) to Brower, May 3, 1962, Carton 14, Folder 13, Executive Director Records; Randall Henderson (editor, *Desert* magazine) to Brower, June 18, 1958, Carton 14, Folder 7, Executive Director Records. *Desert* suggested that the reservoir should be raised high enough so that boaters could glide beneath it, a proposition that horrified many conservationists. Randall Henderson, the magazine editor, told Brower that the blasting necessary for the barrier dam construction would likely be far more destructive than the reservoir.

32. Rainbow Bridge, July 22, 1962, Sierra Club Minutes, 6–7.

33. Lawrence E. Davies, "Udall Is Urged to 'Obey the Law,'" *New York Times*, January 12, 1963; Sierra Club, "Sierra Club Charges Reclamation Betrayal of National Park System," press release, n.d., Carton 81, Folder 17, Sierra Club Records, BL. Lawsuits such as the one about Rainbow Bridge prompted William O. Douglas to resign from the Sierra Club board. Despite their differences, Udall would call on Brower for help, sometimes on important issues. In 1967, he appealed to Brower for assistance on a speech about population issues. "Scribble in the margins and be ruthless," Udall wrote to Brower. "I want it to be the most important statement of my political career" (S. Udall to Brower, January 31, 1967, Carton 7, Folder 20, Executive Director Records).

34. David Brower, "Glen Canyon: The Year of the Last Look," *Sierra Club Bulletin*, June 1962, 7.

35. David R. Brower, "Reflections on the Sierra Club, Friends of the Earth, and Earth Island Institute," an oral history conducted in 1999 by Ann Lage, ROHO, BL, 2012, 54–55; David Brower, "Pennington Glen Canyon Film Release," *Sierra Club Bulletin*, June, 1965, 19.

36. Eliot Porter, *The Place No One Knew: Glen Canyon on the Colorado*, ed. David Brower (San Francisco: Sierra Club Books, 1963); Michael P. Cohen, *The History of the Sierra Club, 1892–1970* (San Francisco: Sierra Club Books, 1988), 319.

37. Brower to S. Udall, June 22, 1963, Carton 63, Folder 1, Sierra Club Records, BL.

38. Harold Gilliam, "The Sierra Club's First Century," *San Francisco Chronicle*, April 26, 1992.

39. Martin, *Story That Stands Like a Dam*, 221.

40. Barry Goldwater to Brower, July 22, 1963, Aspinall to Brower, July 23, 1963, and Ernest Gruening to Brower, July 18, 1963, Carton 21, Folder 52, Executive Director Records.

41. Edward Weeks, "The Peripatetic Reviewer," *Atlantic Monthly*, May 1964, 134. Dominy countered Brower's book and film by producing his own versions touting Lake Powell. In 1965, he convinced Senator Carl Hayden of Arizona to use his influence with the Government Printing Office to produce *Jewel of the Colorado*. The 75-cent book and a similar film extolled the glories of Lake Powell. Dominy said they were produced "to answer the Sierra Club nonsense about how we've destroyed that area" (Dominy, "Oral History Interviews," 212; see also Martin, *Story That Stands Like a Dam*, 244–245).

42. David Brower, interview by Ken Verdoia, KUED, University of Utah, aired October 1999, http://www.kued.org/productions/glencanyon/interviews/brower/html.

43. David Brower, "Lake Powell and the Canyon That Was," *Sierra Club Bulletin*, April–May 1963, 2.

44. Jack Goodman, "How Wahweap Creek Became Wahweap Marina," *New York Times*, July 9, 1967.

45. Wallace Stegner, "Lake Powell," *Holiday*, May 1966, 64–68, 148–151.

46. Dominy interview by Verdoia.

47. Mark W. T. Harvey, "Defending the Park System: The Controversy over Rainbow Bridge," *New Mexico Historical Review*, January 1998, 62.

48. Scott Miller, "Undamming Glen Canyon: Lunacy, Rationality, or Prophecy?" *Stanford Environmental Law Journal*, January 2000, 121–207; Sandra Blakeslee, "Drought Unearths a Buried Treasure, *New York Times*, November 2, 2004. Lake Powell has proved difficult to manage. In 1983, the reservoir was so full that engineers worried that the spillways would crumble and give way, creating a major tidal wave. A few years later a drought was so severe that many of the side canyons, including the vast Cathedral Canyon, began to reemerge.

49. Polly Dyer (Sierra Club leader) to Brower, August 22, 1999, Box 4, Folder 24, David Ross Brower Papers, BANC MSS 79/9c, BL.

50. David Brower, *Grand Canyon Discussion at Silver Grotto and Marble Damsite, May 31 to June 13, 1977*, audiocassette, recorded by Ron Hayes, Grand Canyon National Park Museum Collection, GRCA 40237, Grand Canyon National Park, Ariz.

51. Brower interview by Verdoia.

52. Martin, *Story That Stands Like a Dam*, 312–313; Edward Abbey, *The Monkey Wrench Gang* (1975; reprint, New York: Perennial, 2000); Dave Foreman, *Confessions of an Eco-Warrior* (New York: Harmony Books, 1991), iv.

53. Gilliam, "Sierra Club's First Century."

54. Martin, *Story That Stands Like a Dam*, 11–12; David Brower, journal, January 21, 1963, Box 18, Folder 11, Brower Papers.

9. Progress

1. Stewart Udall, "Stewart L. Udall Oral History Interview V," an oral history interview conducted on October 31, 1969, by Joe B. Frantz, Lyndon Baines Johnson Library, Austin, Tex., 8.

2. Mark Harvey, *Wilderness Forever: Howard Zahniser and the Path to the Wilderness Act* (Seattle: University of Washington Press, 2005), 229.

3. Doug Scott, *The Enduring Wilderness: Protecting Our Natural Heritage Through the Wilderness Act* (Golden, Colo.: Fulcrum, 2004), 51; Paul W. Hirt, *A Conspiracy of Optimism: Management of the National Forests Since World War Two* (Lincoln: University of Nebraska Press, 1994), 231.

4. U.S. Department of the Interior, "Three New National Seashores," press release, November 11, 1962, Carton 10, Folder 14, Sierra Club Office of Executive Director Records, BANC MSS 2002/230c, Bancroft Library (BL), University of California, Berkeley; Richard West Sellars, *Preserving Nature in the National Parks: A History* (New Haven, Conn.: Yale University Press, 1997), 205–206.

5. Sierra Club, *An Island in Time*, brochure, n.d., Carton 10, Folder 12, Executive Director Records; David Brower to Representatives John Kyl, Glenn Cunningham, and John Baldwin, telegram, March 6, 1962, Carton 10, Folder 12, Executive Director Records.

6. Sellars, *Preserving Nature in the National Parks*, 206.

7. Ibid., 191–192; Conrad L. Wirth, *Parks, Politics, and the People* (Norman: University of Oklahoma Press, 1980), 361.

8. Harold K. Steen, *The U.S. Forest Service: A History* (Seattle: University of Washington Press, 1976), 297–305; Hirt, *Conspiracy of Optimism*, 171–190.

9. Brower to Horace Albright and twenty-eight others, memorandum, June 3, 1960, Carton 9, Folder 9, Executive Director Records.

10. Multiple Use Bill, May 7, 1960, Sierra Club Board of Directors Meeting Minutes, 1892–1995, BANC FILM 2945, BL, 2; Edgar Wayburn, "Sierra Club Statesman, Leader of the Parks and Wilderness Movement; Gaining Protection for Alaska, the Redwoods and Golden Gate Parklands," an oral history conducted in 1976–1981 by Ann Lage and Susan Schrepfer, Regional Oral History Office (ROHO), BL, 1985. 28.

11. Brower to Howard Zahniser, June 3 , 1960, and Zahniser to Brower, June 15, 1960, Carton 9, Folder 9, Executive Director Records; David R. Brower, "Environmental Activist, Publicist, and Prophet," an oral history conducted in 1974–1978 by Susan Schrepfer, ROHO, BL, 1979, 230.

12. John P. Saylor, "Wilderness: The Outlook from Capitol Hill," in *Wilderness, America's Living Heritage*, ed. David Brower (San Francisco: Sierra Club Books, 1961), 147–151; Harvey, *Wilderness Forever*, 222.

13. Harvey, *Wilderness Forever*, 227–228; Scott, *Enduring Wilderness*, 52; David Brower, testimony, U.S. House of Representatives, Subcommittee on Public Lands, *National Wilderness Preservation Act: Hearing on S. 174, H.R.293, H.R.299, H.R. 496, H.R. 776, H.R. 1762, H.R. 1925, H.R. 2008, and H.R.8237*, 87th Cong., 1st sess., November 6, 1961, 891.

14. Brower, testimony, U.S. House of Representatives, Subcommittee on Public Lands, *National Wilderness Preservation Act*, November 6, 1961, 891

15. David Brower, *For Earth's Sake: The Life and Times of David R. Brower* (Salt Lake City: Peregrine Smith Books, 1990), 398; U.S. Department of the Interior, "Secretary Udall Pledges Support of ORRRC Proposals," press release, February 1, 1962, Carton 7, Folder 25, Executive Director Records; John F. Warth, "The ORRRC Study—Where Does It Leave Preservation?" report, n.d., Carton 7, Folder 25, Executive Director Records; John F. Warth to Brower, n.d., Carton 7, Folder 25, Executive Director Records.

16. Craig W. Allin, *The Politics of Wilderness Preservation* (Westport, Conn.: Greenwood Press, 1982), 125–126.

17. William O. Douglas to Brower, March 28, 1962, Carton 90, Folder 27, Sierra Club Records, BANC MSS 71/103c, BL.

18. Quoted in "Aspinall Gives Views on Wilderness Bill," *Grand Junction Daily Sentinel*, June 3, 1963; see also Harvey, *Wilderness Forever*, 235.

19. Stephen C. Sturgeon, *The Politics of Western Water: The Congressional Career of Wayne Aspinall* (Tucson: University of Arizona Press, 2002), 16–55; Jeffrey Ingram, telephone interview by the author, November 12, 2012; "Aspinall Raps Riots, Enlivens Meeting," *Denver Post*, n.d., clipping, Box 190, Wayne Aspinall Papers, Penrose Special Collections and Archives, University of Denver.

20. Sturgeon, *Politics of Western Water*, 63–64; William M. Blair, "Feud Threatens Wilderness Bill," *New York Times*, March 1, 1963. In 1961, Aspinall blocked the Senate version of the bill, and in 1962 he had his committee pass legislation that opened up wilderness for development rather than protecting it. After that, the parliamentary gamesmanship turned exotic. Aspinall insisted that his bill could come up for a vote on the House floor only if no amendments were added. In a very rare congressional rebuke, the ploy failed, and the issue was dead for the year. In 1963, mining interests began questioning whether the president had the legal authority to transfer lands into national monument protection, a policy that had been implemented in the administration of Theodore Roosevelt. This questioning of the law was another stalling ploy by Aspinall, who announced he would not consider further wilderness legislation and perhaps any conservation-related issues in his committee until the issue was resolved.

21. Paul Brooks, "Congressman Aspinall vs. the People of the United States," *Harper's*, March 1963, 60–63.

22. David Brower, testimony, U.S. House of Representatives, Subcommittee on Public Lands, *National Wilderness Preservation Act: Hearing on S. 174, H.R. 293, H.R. 299, H.R. 496, H.R. 776, H.R. 1762, H.R. 1925, H.R. 2008, and H.R. 8237*, 87th Cong., 2nd sess., May 9, 1962, 1366–1374.

23. Julius Duscha, "Wilderness Bill Not for Aspinall," *Washington Post*, April 12, 1963.

24. Brower to Ike Livermore (Brower's friend), April 4, 1958, Carton 90, Folder 26, Sierra Club Records, BL.

25. Michael McCloskey, *In the Thick of It: My Life in the Sierra Club* (Washington, D.C.: Island Press, 2005), 38–41. McCloskey said the one place he had trouble raising support for the wilderness bill within Aspinall's district during the campaign was near Durango, Colorado. When he was being introduced to people who might be able to help the campaign, he met a rancher who owned a dance hall and bar. When McCloskey finished his spiel, the man said, "There was a shooting here last week, and there's just about to be another." McCloskey rapidly retreated and never knew how serious the threat might have been.

26. Andrew N. Smith (wilderness bill advocate) to Brower, January 13, 1964, Carton 90, Folder 27, Sierra Club Records, BL. Brower and McCloskey exchanged numerous letters about the planning for these hearings and their growing optimism that the bill would prevail in the House.

27. Sturgeon, *Politics of Western Water*, 64–65.

28. Mike and Brandy to Dick Leonard, telegram, May 5, 1964, Carton 90, Folder 29, Sierra Club Records, BL.

29. Brower, *For Earth's Sake*, 230; David Brower, "Wilderness and the Constant Companion," *Sierra Club Bulletin*, September 1964, 3.

30. Conrad Wirth to Wayne Aspinall, August 13, 1964, Box 148, Aspinall Papers.

31. Scott, *Enduring Wilderness*, 40, 54 (Sundquist quote); Allin, *Politics of Wilderness Preservation*, 135; David Brower, "Wilderness and the Constant Advocate," *Living Wilderness*, Spring–Summer 1964, 43.

32. Harvey, *Wilderness Forever*, 248–249.

33. Weldon Heald, "San Gorgonio—Southern California's Rooftop," *Living Wilderness*, Summer 1963, 12–16.

34. John Robinson, *San Gorgonio: A Wilderness Preserved* (San Bernardino, Calif.: San Gorgonio Volunteers Association, 1991), 107–128.

35. A list of wilderness area names is given in "Creation and Growth of the National Wilderness Preservation System," Wilderness.net, n.d., http://www.wilderness.net/NWPS/AtoZ (accessed August 17, 2013).

10. Books

1. Sierra Club, "LeConte Memorial Lodge," n.d., http://www.sierraclub.org.education/leconte/ (accessed July 17, 2012).

2. Ansel Adams, "A Conversation with Ansel Adams," an oral history conducted in 1972, 1974, and 1975 by Ruth Teiser and Catherine Harroun, Regional Oral History Office (ROHO), Bancroft Library (BL), University of California, Berkeley, 1978, 462–464.

3. J. Donald Adams, "Speaking of Books," *New York Times*, May 15, 1960; David Brower to Ansel Adams, September 27, 1956, Carton 93, Folder 4, Sierra Club Records, BANC MSS 71/103c, BL.

4. Brower to Nancy Newhall (book editor, Sierra Club), November 29, 1958, Carton 21, Folder 42, Sierra Club Office of the Executive Director Records, BANC MSS 2002/230c, BL.

5. Ansel Adams, *This Is the American Earth* (San Francisco: Sierra Club Books, 1960); "This Is the American Earth," *Sierra Club Bulletin*, October 1959, 1; David R. Brower, "Reflections on the Sierra Club, Friends of the Earth, and Earth Island Institute," an oral history conducted in 1999 by Ann Lage, ROHO, BL, 2012, 19.

6. "Fairchild Award Given to Writer," *Rochester Times-Union*, November 10, 1960; Charles Poore, "Books of the Times," *New York Times*, July 9, 1960. There were a few exceptions to the admiration critics expressed about the Sierra Club books. A review in *Science* praised the photos but panned the text, commenting, "Any purchaser who can get past the book's repellant title will probably not even notice the text and the pictures are magnificent" (Edward S. Deevey, "Review," *Science*, December 9, 1960, 1759). Brower replied in a letter to the *Science* editor December 20, 1960, complaining that the review could severely impair the book's success and that the reviewer

should have read it carefully instead of skimming it. He said that the *Science* reviewer had missed Newhall's thesis by a mile (Brower to *Science* editor, December 20, 1960, Carton 21, Folder 42, Executive Director Records).

7. Brower to Sierra Club board of directors, publishing proposal memorandum, September 9, 1960, Box 15, Sierra Club Records, 1957–1980, NC1260, Special Collections, University of Nevada, Reno.

8. David R. Brower, "Environmental Activist, Publicist, and Prophet," an oral history conducted in 1974–1978 by Susan Schrepfer, ROHO, BL, 1979, 212.

9. August Fruge, "A Publisher's Career with the University of California Press, the Sierra Club, and the California Native Plant Society," an oral history conducted in 1997–1998 by Suzanne B. Riess, ROHO, BL, 2001, 92.

10. Ibid., 2–7; Mary Rourke, "August Fruge, 94; Publisher Transformed UC Press," *Los Angeles Times*, July 18, 2004.

11. Brower, "Environmental Activist, Publicist, and Prophet," 213.

12. Nancy Newhall, *Ansel Adams*, vol. 1, *The Eloquent Light* (San Francisco: Sierra Club Books, 1963); Michael McCloskey (former executive director, Sierra Club), interview by the author, Portland, Ore., January 22, 2013. McCloskey said that after he succeeded Brower as executive director, he elected to change the typeface in the book because it was "too feminine," and he preferred a more masculine look.

13. Jacob Deshin, "Viewpoint: Eliot Porter, Medication to Conservation," *Popular Photography*, n.d., clipping, Carton 22, Folder 21, Executive Director Records.

14. Gene Thornton, "Reappraising Eliot Porter," *New York Times*, December 16, 1979; William H. Honan, "Eliot Porter, Photographer, Is Dead at 88," *New York Times*, November 2, 1990.

15. Deshin, "Viewpoint"; Brower to Eliot Porter, February 3, 1961, Box 7, Folder 14, Executive Director Records

16. David Brower, "The Story Behind It: The Most Beautiful Book of Its Kind Ever Produced," *Sierra Club Bulletin*, October 1962, 6–7.

17. David Brower, journal, various dates, Boxes 17–21, David Ross Brower Papers, BANC MSS 79/9c, BL.

18. Brower, "The Story Behind It." Ian Ballantine was very impressed by how thoroughly Brower supervised the final production of his books at Barnes Press. Ballantine said that Brower liked to have absolute control. He would approve the color form by form and get each pressman personally involved. He said it was a more productive system than approving sample sheets back in the office. Ballantine said that Brower had photographers watch the press run to better understand it. In addition, he said Brower ordered engravings of more subjects than the book could handle. Some inevitably came out as mediocre products, but Brower always had extras. "There was never, ever any compromise with quality,"

Ballantine commented (Ian Ballantine [book publisher], introduction essay, n.d., Carton 64, Folder 2, Brower Papers). Ansel Adams told a much different story in relationship to the first book published, *This Is the American Earth*. That book was produced by a different company in New York, and Adams just happened to drop by before the book went to press. He said Brower and others had not proofed any of the pages, and he began to see glaring errors. "I ordered a stop to everything, and we went carefully thorough it," he said. "You never saw such a mess in your life, and every time you changed a detail, you had to re-make a whole page" (Adams, "Conversation with Ansel Adams," 474).

19. Henry David Thoreau, with photographs by Eliot Porter, *In Wildness Is the Preservation of the World* (San Francisco: Sierra Club Books, 1962); Harry C. Kenney, "From the Bookshelf," *Christian Science Monitor*, November 29, 1962; Ronald Singer, "Reviews," *Natural History*, May 1963, 8.

20. "Battle of the Wilderness," *Newsweek*, October 3, 1966, 108.

21. John G. Mitchell, "On Environmental Publishing," *Not Man Apart*, November 1973, 3.

22. "Handbook Edition," *Sierra Club Bulletin*, December 1967, 49–55.

23. Brower, "Environmental Activist, Publicist, and Prophet," 210.

24. Bruce Kilgore (secretary, Sierra Club Publications Committee), Publications Committee meeting minutes, November 2, 1962, Carton 20, Folder 2, Executive Director Records; Eliot Porter, *The Place No One Knew: Glen Canyon on the Colorado*, ed. David Brower (San Francisco: Sierra Club Books, 1963).

25. August Fruge, publications program memorandum, August 13, 1963, Carton 20, Folder 6, Executive Director Records.

26. Brower to August Fruge, August 23, 1963, Carton 20, Folder 6, Executive Director Records; David Brower, "The High Cost of Consensus," memorandum to the Sierra Club Publications Committee, October 9, 1963, Carton 20, Folder 7, Executive Director Records; David Brower, "Why the Club Should Publish," memorandum to the Publications Committee, October 14, 1963, Carton 20, Folder 7, Executive Director Records.

27. Brower, journal, various dates, Boxes 18–20, Brower Papers.

28. Annual audits, 1954–1989, Carton 25, Folders 1–9, Sierra Club Records, BL; budgets, 1956–1978, Carton 25, Folders 15–19, Sierra Club Records, BL; financial statements, Carton 25, Folders 32–40, Sierra Club Records, BL; financial statements, Carton 26, Folders 1–10, Sierra Club Records, BL; Bruce Kilgore, Publications Committee meeting minutes, June 24, 1963, Carton 49, Folder 7, Sierra Club Records, BL.

29. Brower, "Environmental Activist, Publicist, and Prophet," 210; Richard Leonard, "Mountaineer, Lawyer, Environmentalist, Volume I," an oral history conducted in 1976 by Susan R. Schrepfer, ROHO, BL, 1976, 365; Edgar Wayburn, "Sierra

Club Statesman, Leader of the Parks and Wilderness Movement: Gaining Protection for Alaska, the Redwoods, and Golden Gate Parklands," an oral history conducted in 1976–1981 by Ann Lage and Susan Schrepfer, ROHO, BL, 1985, 225, 206.

30. Brower to Glen Canyon book supporters, July 25, 1963, Carton 21, Folder 50, Executive Director Records; Brower to Hugh Barnes (owner, Barnes Press), January 7, 1963, Carton 21, Folder 47, Executive Director Records. After moving on from the Sierra Club, Brower would continue to raise money by seeking contributions and loans, especially when he was in charge of Friends of the Earth between 1969 and 1986. He would solicit and receive hundreds of thousands of dollars to keep the organization running but still would eventually create a debt of several hundred thousand dollars. Friends of the Earth defaulted on many of those loans.

31. Sierra Club Executive Committee, directive on publishing, memorandum, November 23, 1963, Carton 49, Folder 7, Sierra Club Records, BL; Bruce Kilgore, Publications Committee meeting minutes, November 25, 1963, Carton 49, Folder 7, Sierra Club Records, BL.

32. Philip Hyde and François Leydet, *The Last Redwoods* (San Francisco: Sierra Club Books, 1964).

33. Emanuel Fritz, "Book Reviews," *Journal of Forestry*, September, 1964, 641–643; E. H. Linford, "Belated Crusade to Save the Redwoods," *Salt Lake Tribune*, February 15, 1964; Edward Weeks, "The Peripatetic Reader," *Atlantic Monthly*, May 1964, 134.

34. Brower to George Dusheck (reporter, *San Francisco News-Call Bulletin*), October 19, 1962, Carton 21, Folder 53, Executive Director Records.

35. Brower to Adams, February 14, 1964, Carton 20, Folder 9, Executive Director Records.

36. Ansel Adams to Brower, February 11, 1964; Brower to Adams, February 14, 1964; and Adams to Brower, February 15, 1964, all in Carton 20, Folder 9, Executive Director Records. The earlier letter is Adams to Brower, May 6, 1963, Box 1, Folder 3, Sierra Club Members Papers, BANC MSS 71/295c, BL.

37. Porter to Brower, September 9, 1964, Box 7, Folder 16, Executive Director Records.

38. Brower, "Environmental Activist, Publicist, and Prophet," 213.

39. Leonard, "Mountaineer, Lawyer, Environmentalist, Volume I," 365.

40. François Leydet, *Time and the River Flowing: Grand Canyon*, ed. David Brower (San Francisco: Sierra Club Books, 1964).

41. Quoted in Harry Gilroy, "Prize Is Awarded Crusading Book," *New York Times*, April 23, 1965.

42. Brower to Norman G. Dyhronfurth (leader, American Mount Everest Expedition), December 17, 1964, Carton 22, Folder 14, Executive Director

Records; Naomi Bliven, "Books," *New Yorker*, March 12, 1966, 173; Thomas F. Hornbein, preface, in Thomas F. Hornbein and Norman G. Dyhrenfurst, *Everest: The West Ridge* (San Francisco: Sierra Club Books, 1965), 12.

43. Fruge, "Publisher's Career," 76; Adams, "Conversation with Ansel Adams," 559; Wallace Stegner to Adams, December 29, 1964, Carton 2, Folder 26, Sierra Club Members Papers.

44. Michael P. Cohen, *The History of the Sierra Club, 1892–1970* (San Francisco: Sierra Club Books, 1988), 347–348. Brower was extremely upset about some portions of Cohen's history of the Sierra Club, including the reference in the text to the criticism of his son's editing of the Big Sur book. At the conclusion of the paragraph, Cohen added this sentence: "Kenneth Brower, beyond the fray, went on to a distinguished literary career." For the book in question, see David Brower, ed., *Not Man Apart* (San Francisco: Sierra Club Books, 1965).

45. Adams to Will Siri, August 12, 1965, Carton 10, Folder 4, Sierra Club Records, BL.

46. Stegner to Adams, December 29, 1964; annual audits, 1954–1989, Carton 25, Folders 1–9, Sierra Club Records, BL; budgets, 1956–1978, Carton 25, Folders 15–19, Sierra Club Records, BL; financial statements, 1956–1978, Carton 25, Folders 32–40, Sierra Club Records, BL; financial statements, 1963–1971, Carton 26, Folders 1–10, Sierra Club Records, BL.

47. Hugh Nash (secretary), Publications Committee meeting minutes, November 6, 1967, Carton 20, Folder 2, Executive Director Records.

48. Brower, "Environmental Activist, Publicist, and Prophet," v.

49. Renny Russell, *Rock Me on the Water: A Life on the Loose* (Questa, N.M.: Animist Press, 2007), 164–167.

50. Terry Russell and Renny Russell, *On the Loose* (San Francisco: Sierra Club Books, 1967); Nash, Publications Committee meeting minutes, November 6, 1967; Brower, "Reflections on the Sierra Club," 25–26; Russell, *Rock Me on the Water*, 176–182. Terry Russell would never live to see the finished book. In July 1965, the two Russell brothers took a trip down the Green River. They came to the Steer Ridge Rapid, the largest they had seen on the trip. After scouting it, they rowed into the river without life jackets and quickly capsized. Renny Russell grabbed onto a floating object and survived. Terry Russell drowned.

11. Escalating the Risks

1. Relation of Sierra Club to Public Land Administering Agencies, July 7, 1959, Sierra Club Board of Directors Meeting Minutes, 1892–1995, BANC FILM 2945, Bancroft Library (BL), University of California, Berkeley; Relations with Government Agencies, December 6, 1959, Sierra Club Minutes.

2. David R. Brower, "Environmental Activist, Publicist, and Prophet," an oral history conducted in 1974–1978 by Susan Schrepfer, Regional Oral History Office (ROHO), BL, 1979, 90, 208, 211; Hugh Nash to Edgar Wayburn, April 28, 1964, Box 7, Folder 9, Sierra Club Office of the Executive Director Records, BANC MSS 2002/230c, BL.

3. Effectiveness of Communication, October 14, 1962, Sierra Club Minutes.

4. Edgar Wayburn and Allison Alsup, *Your Land and Mine: Evolution of a Conservationist* (San Francisco: Sierra Club Books, 2004), 98.

5. Brower, "Environmental Activist, Publicist, and Prophet," 211.

6. David Brower to Stewart Udall, January 13, 1961, Carton 7, Folder 16, Executive Director Records.

7. Ansel Adams to S. Udall, December 15, 1960, Carton 190, Stewart L. Udall Papers, AZ 372, University of Arizona Libraries, Special Collections, University of Arizona, Tucson.

8. Barbara Brower, interview by the author, Portland, Ore., May 24, 2013.

9. Gary Soucie, telephone interview by the author, August 26, 2013.

10. Barbara Brower interview, May 24, 2013.

11. Kenneth Brower, *The Wildness Within: Remembering David Brower* (Berkeley, Calif.: Heyday, 2012), 7–8.

12. Barbara Brower interview, May 24, 2013.

13. Joseph Brower and Gayle Brower, interview by the author, Santa Rosa, Calif., March 26, 2013; David Brower, "Strawberry Waffles," n.d., Carton 6, Folder 40, David Ross Brower Papers, BANC MSS 79/9c, BL; Rex Burnham (president, Bay Side Press) to Brower, July 10, 1995, Carton 6, Folder 40, Brower Papers.

14. David R. Brower, *For Earth's Sake: The Life and Times of David R. Brower* (Salt Lake City: Peregrine Smith Books, 1990), 159–165. Isabelle was a macaque monkey originally from Indonesia that the Brower family inherited in 1962 from an anthropologist when he left the University of California at Berkeley. Twelve-year-old Barbara Brower volunteered to take care of the monkey as a family pet, and Isabelle would live with the Brower family for another twenty-two years. Brower recounted the antics of living with a monkey, who was not house trained and who would often be tethered to a rope that reached to the roof, where she often perched unless she was escaping to a neighbor's house.

15. Barbara Brower interview, May 24, 2013; Gayle Brower interview, March 26, 2013.

16. Michael McCloskey (former executive director, Sierra Club), interview by the author, Portland, Ore., January 22, 2013.

17. Barbara Brower interview, May 24, 2013.

18. Brower to William O. Douglas, March 29, 1962, Box 2, Folder 17, Executive Director Records; Curt Meine, Michael Soule, and Reed F. Noss,

"A Mission-Driven Discipline: The Growth of Conservation Biology," *Conservation Biology*, June 2006, 631–651.

19. Rachel Carson to Brower, telegram, October 11, 1963, Box 2, Folder 6, Executive Director Records; Barbara Brower interview, May 24, 2013; Brower to Shirley Briggs (secretary, Carson Trust), September 21, 1967, Box 8, Folder 6, Executive Director Records.

20. Brower to Conrad Wirth, May 27, 1953, Carton 80, Folder 10, Sierra Club Records, BANC MSS 71/103c, BL; Use of Pesticides at Tuolumne Meadows, July 4, 1959, Sierra Club Minutes, 3; David Brower, letter to the editor, *New York Times*, August 28, 1963, clipping, Carton 280, Folder 8, Sierra Club Members Papers, BANC MSS 71/295c, BL; Brower to Sierra Club board of directors, pesticides memorandum, August 17, 1964, Carton 191, Folder 2, Sierra Club Members Papers.

21. "Handbook Edition," *Sierra Club Bulletin*, December 1967, 45–49; David Brower and David Sive, "Two Davids, One Goliath," *Sierra Club Bulletin*, February 1966, 2; David Brower, "The Need for Conservation," speech, February 3, 1966, transcript, Carton 10, Folder 10, Sierra Club Records, BL; David Brower, journal, January 23, 1965, Carton 19, Folder 9, Brower Papers.

22. Barbara Brower interview, May 24, 2013.

23. Brower to C. W. Richard Atkins (member, Fresno Chamber of Commerce), September 26, 1959, Carton 86, Folder 28, Sierra Club Records, BL; Eunice Elton (Trans Sierra Highway opponent), to Brower, November 29, 1964, and Brower to Elton, n.d., Carton 86, Folder 25, Sierra Club Records, BL.

24. Michael McCloskey, *In the Thick of It: My Life in the Sierra Club* (Washington, D.C.: Island Press, 2005), 57; Brower to Michael McCloskey, December 21, 1964, Carton 10, Folder 16, Executive Director Records.

25. Mineral King, May 1–2, 1965, Sierra Club Minutes, 11–15.

26. David Brower, job descriptions memorandum, October 30, 1959, Carton 24, Folder 7, Executive Director Records; "Directors Launch Campaign for Expanded Grand Canyon Protection," *Sierra Club Bulletin*, June 1963, 6–7; budget, 1964, Carton 36, Folder 5, Sierra Club Members Papers.

27. Michael McCloskey, interview by the author, May 24, 2013, Portland, Ore.

28. Brower, "Environmental Activist, Publicist, and Prophet," 209–210; Washington Office, October 23, 1963, Sierra Club Board of Directors Meeting Minutes, 1892–1995, BL, 5; Budget 1964—Washington Representative, Sierra Club Board of Director Meeting Minutes, 10–12.

29. Soucie interview, October 26, 2012.

30. Staff Organization, June 9, 1963, Sierra Club Minutes; budget, 1969, Carton 25, Folder 6, Sierra Club Records, BL; Berry interview, April 24, 2013.

31. Stephanie Mills, *Whatever Happened to Ecology?* (San Francisco: Sierra Club Books, 1989), 112.

32. Brock Evans, "Environmental Campaigner, from the Northwest Forests to the Halls of Congress," an oral history conducted in 1982 and 1984 by Ann Lage and included in "Building the Sierra Club's National Lobbying Program, 1967–1981," ROHO, BL, 1985, 36–37.

33. Brock Evans (Northwest representative, Sierra Club), telephone interview by the author, September 11, 2012.

34. Soucie interview, October 26, 2012.

35. Will Siri to Brower, November 24, 1964, Carton 12, Folder 5, Executive Director Records.

36. McCloskey interview, May 24, 2013; McCloskey, *In the Thick of It*, 48–49.

37. McCloskey interview, January 22, 2013.

38. Gary Soucie to Brower, McCloskey, Wayburn, memoranda, May 23, 1967, May 31, July 5, and August 18, 1967, as well as February 18, March 29, and October 31, 1968, Carton 13, Folder 14, Executive Director Records; Soucie to McCloskey, memorandum, March 5, 1969, Carton 13, Folder 14, Executive Director Records; Soucie to McCloskey and Brower, memorandum, April 16, 1969, Carton 13, Folder 14, Executive Director Records. Soucie's memos were extremely detailed, describing trips and activities he took on behalf of the club. In one, he described a very hectic schedule with much travel and an office with five employees, two secretaries, and an art gallery, where the phone was constantly ringing.

39. Soucie interview, October 26, 2012.

40. McCloskey interview, January 22, 2013.

41. Michael McCloskey, telephone interview by the author, July 10, 2014; Jeffrey Ingram, telephone interview by the author, July 8, 2014.

42. McCloskey interview, May 24, 2013.

43. Audit statements, 1963–1969, Carton 25, Folders 1–9, Sierra Club Records, BL; Brower, "Environmental Activist, Publicist, and Prophet," 334–335. The club's revenue and year-end surplus or deficit by year were: 1963, $908,000/-$93,000; 1964, $1,340,000/$193,000; 1965, $1,480,000/-$28,000; 1966, $1,813,000/$56,000; 1967, $2,321,000/-$65,000; 1968, $3,111,000/-$158,000; 1969, $2,048,000/-$119,000. For accounting and budgetary reasons, the year 1969 was only nine months. The publication revenue and year-end surplus or deficit by year were: 1963, $381,000/-$14,000; 1964, $516,000/-$17,000; 1965, $697,000/-$15,000; 1966, $604,000/-$119,000; 1967, $1,077,000/-$63,000; 1968, $1,558,000/-$36,000; 1969, $481,000/-$237,000. Brower argued that the publication figures were misleading because in his opinion overhead expenses charged to the publication program were too high. However, he often minimized the expenses of the publication program while emphasizing its benefits.

44. Brower to Sierra Club board of directors, memorandum, September 5, 1966, Carton 10, Folder 13, Sierra Club Records, BL; Brower to the Sierra Club board of directors, memorandum, September 15, 1966, Carton 10, Folder 13, Sierra Club Records, BL; Brower, Sierra Club financial memorandum, March 22, 1965, Carton 20, Folder 7, Executive Director Records; Brent Blackwelder, telephone interview by the author, January 3, 2013; Avis Ogilvy Moore, telephone interview by the author, February 19, 2013. On March 1, 1965, Brower headed back to New York with the understanding that the club might finish with a very small budget deficit. When he returned twelve days later, he was told that the club was in the red by $99,000, with $65,000 of the deficit in the books program. He was shocked. Yet he was able to work with the club's financial manager, Clifford Rudden, and together they found mistakes and adjustments that dramatically reduced the budget deficit (Brower, Sierra Club financial memorandum, March 22, 1965). A year later William Siri and George Marshall, who was replacing Siri as president, were so worried about the financial picture that they asked Rudden to draw up a series of recommendations on how to cut costs (Brower to Sierra Club board of directors, memoranda, September 5 and 15, 1966). Brower was bitterly opposed to the reductions and offended that neither Siri nor Rudden had consulted him. In later years, he would express less concern about financial problems. He would become known for a particular aphorism: if a group was not spending itself into the red, it was not doing its job to protect the planet. The aphorism was particularly appealing to many of Brower's young followers. But for the people who were trying to keep an organization alive—Friend of the Earth at the time he made this comment—it grated.

45. Lewis F. Clark, "Perdurable and Peripatetic Sierran: Club Officer and Outing Leader 1928–1984," an oral history conducted in 1975–1977 by Marshall Kuhn and included in "Sierra Club Reminisces III, 1910s–1970s," ROHO, BL, 1984, 43.

46. Evans telephone interview, September 11, 2012.

47. Ansel Adams, "A Conversation with Ansel Adams," an oral history conducted in 1972, 1974, and 1975 by Ruth Teiser and Catherine Harroun, ROHO, BL, 1978, 677.

48. Edgar Wayburn, "Sierra Club Statesman, Leader of the Parks and Wilderness Movement: Gaining Protection for Alaska, the Redwoods, and Golden Gate Parklands," an oral history conducted in 1976–1981 by Ann Lage and Susan Schrepfer, ROHO, BL, 1985, 182–186.

49. Edgar Wayburn and Allison Alsup, "Dr. Edgar Wayburn, MD: 1906–2010," Sierra Club, n.d., http://www.sierraclub.org/history/wayburn/ (accessed September 21, 2013).

50. Wayburn, "Sierra Club Statesman," 186.

51. Wayburn and Alsup, *Your Land and Mine*, 106.

52. Glen Dawson, "Pioneer Rock Climber and Ski Mountaineer," an oral history conducted in 1975 by Richard Searle and included in "Sierra Club Reminiscences II," ROHO, BL, 1975, 28.

53. Phillip S. Berry, "Sierra Club Leader, 1960s–1980s: A Broadened Agenda, a Bold Approach," an oral history conducted in 1981 and 1984 by Ann Lage, ROHO, BL, 1988, 22.

54. See, for example, John C. Miles, *Guardians of the Park: A History of the National Parks and Conservation Foundation* (Washington, D.C.: Taylor & Francis, 1995), 213–214.

55. David R. Brower, "Reflections on the Sierra Club, Friends of the Earth, and Earth Island Institute," an oral history conducted in 1999 by Ann Lage, ROHO, BL, 2012, 122.

56. Brower to Paul Brooks, February 10, 1965, Box 1, Folder 21, Sierra Club Members Papers.

57. "Wilderness Bill of Rights Called for by Justice Douglas," August 19, 1961, Box 2, Folder 17, Executive Director Records; Douglas to Brower, December 12, 1961, Box 2, Folder 17, Executive Director Records; Douglas to Charles A. Reith (Yale University Law School professor), March 3, 1962, Box 2, Folder 17, Executive Director Records; Douglas to Sierra Club board of directors, October 1, 1962, Box 2, Folder 17, Executive Director Records. Brower recruited Douglas in 1960 to run for the board, but Douglas's tenure was short. In October 1962, he left because the Sierra Club was becoming more litigious, and he feared that this approach would complicate his job as a Supreme Court justice. Douglas would write several books about the wilderness, including one calling for a wilderness bill of rights, and for years he continued to counsel Brower.

58. Adams, "Conversation with Ansel Adams," 676.

59. Joel Hildebrand, "Sierra Club Leader and Ski Mountaineer," an oral history conducted in 1974 by Ann Lage and included in "Sierra Club Reminiscences I, 1900–1960," ROHO, BL, 1974, 29; Joel Hildebrand to Sierra Club board of directors, May 11, 1965, Carton 1, Folder 48, Sierra Club Members Papers; Wallace Stegner to Brower, May 31, 1965, Carton 10, Folder 6, Sierra Club Records, BL; Adams to Siri, May 16, 1965, Carton 1, Folder 48, Sierra Club Members Papers.

60. Richard Leonard, "Mountaineer, Lawyer, Environmentalist Volume I," an oral history conducted in 1976 by Susan R. Schrepfer, ROHO, BL, 1976, 366–367.

61. Ibid., 366–367, 382; Adams's comments on the rift are in Adams, "Conversation with Ansel Adams," 675.

62. Brower, "Environmental Activist, Publicist, and Prophet," 217–218.

63. Ibid., 216; Brower, "Reflections on the Sierra Club," 23.

64. Adams to George Marshall (member, Sierra Club board), September 4, 1959, Carton 1, Folder 21, Executive Director Records.

65. Adams to Brower, September 14, 1966, Box 1, Folder 4, Executive Director Records.

66. Adams to Brower, November 9, 1966, Carton 1, Folder 6, Sierra Club Members Papers.

67. Brower, "Environmental Activist, Publicist, and Prophet," 218.

68. Wayburn, "Sierra Club Statesman," 216.

12. Grand Canyon

1. Quoted in Grant McConnell, "Conservation and Politics in the North Cascades," an oral history conducted in 1983 by Rod Holmgren and included in "Sierra Club Nationwide I," Regional Oral History Office (ROHO), Bancroft Library (BL), University of California, Berkeley, 1983, 34.

2. Rich Johnson, *The Central Arizona Project, 1918–1968* (Tucson: University of Arizona Press, 1977), 6–12; Byron Pearson, *Still the Wild River Runs: Congress, the Sierra Club, and the Fight to Save Grand Canyon* (Tucson: University of Arizona Press, 2002), 6–10.

3. Horace Albright to Elwood Mead (commissioner, U.S. Bureau of Reclamation), January 11, 1933, Carton 10, Folder 12, Sierra Club Records, BANC MSS 71/103c, BL.

4. Bestor Robinson, "Thoughts on Conservation and the Sierra Club," an oral history conducted in 1974 by Ann Lage and Ray Lage and included in "Sierra Club Reminisces," ROHO, BL, 1974, 24; J. F. Carithers (lobbyist) to Fred Packard (National Parks Association), May 5, 1956, Carton 72, Folder 26, Sierra Club Records, BL.

5. Conrad Wirth to Stewart Udall, n.d., Box 166, Stewart Udall Papers, AZ 372, University of Arizona Libraries, Special Collections, University of Arizona, Tucson.

6. Carithers to Packard, May 5, 1956; Harrison Humphries, "Park Service Head Opposes Colorado River Power Plan," *Washington Post*, December 7, 1961.

7. Bridge Canyon Dam, November 12, 1949, Sierra Club Board of Directors Meeting Minutes, 1892–1995, BANC FILM 2945, BL, 2–3.

8. Robinson, "Thoughts on Conservation and the Sierra Club," 24–28.

9. Quoted in Pearson, *Still the Wild River Runs*, 16.

10. Dams in National Parks, National Monuments, and Wilderness Areas, September 3, 1950, Sierra Club Minutes, 2.

11. William M. Blair, "Vast Water Plan Outlined by U.S.," *New York Times*, January 22, 1963; Pearson, *Still the Wild River Runs*, 31–33; U.S. Department of the Interior, "Udall Announces New Water Plan," press release, January 22, 1963, Box 166, S. Udall Papers.

12. Wallace Stegner, "The Artist as Environmental Activist," an oral history conducted in 1982 by Ann Lage, ROHO, BL, 1983, 15.

13. Pearson, *Still the Wild River Runs*, 45; Colorado River Canyon, May 4, 1963, Sierra Club Minutes, 7–8.

14. Martin Litton, interview by the author, May 26, 2013, Portola Valley, Calif.

15. Kevin Fedarko, "Ain't It Just Grand," *Outside*, May 31, 2005, http://www.outsideonline.com/templates/Outside_Print_Template?content=123280513 (accessed October 2, 2012).

16. Martin Litton, telephone interview by the author, October 8, 2012.

17. On engineers already boring holes in mid-1963, see anonymous to Edgar Wayburn, July 14, 1963, Carton 17, Folder 4, Sierra Club Office of the Executive Director Records, BANC MSS 2002/230c, BL. The anonymous letter writer began, "It is later than we think": he had flown over the Buckfarm Canyon area and found a trailer and work crews doing fieldwork. He estimated that they had been there since January that year.

18. François Leydet, *Time and the River Flowing: Grand Canyon*, ed. David Brower (San Francisco: Sierra Club, 1964), 50–52, Powell quoted on 52.

19. Ibid., 118–121, Powell quoted on 121.

20. L. W. Lane Jr., *The Sun Never Sets: Reflections on a Western Life* (Palo Alto, Calif.: Stanford University Press, 2013), 102–104.

21. Colorado River Canyon, May 4, 1963, Sierra Club Minutes, 7–8; Litton telephone interview, October 8, 2012.

22. Edgar Wayburn, "Sierra Club Statesman, Leader of the Parks and Wilderness Movement: Gaining Protection for Alaska, the Redwoods, and Golden Gate Parklands," an oral history conducted in 1976–1981 by Ann Lage and Susan Schrepfer, ROHO, BL, 1985, 197.

23. David R. Brower, "Environmental Activist, Publicist, and Prophet," an oral history conducted in 1974–1978 by Susan Schrepfer, ROHO, BL, 1979, 143.

24. "Audubon Group Hears Brower," *Arizona Daily Star* (Tucson), November 10, 1964.

25. Stewart Udall and Floyd Dominy, testimony, in U.S. House of Representatives, Subcommittee on Irrigation and Reclamation, *Lower Colorado River Basin Project: Hearing on H.R. 4671, 4672–4706, 9248*, 89th Cong., 1st sess., August 8–13, 1965, and September 1, 1965, 100–237.

26. Dominy, testimony in ibid., 105.

27. Pearson, *Still the Wild River Runs*, 81.

28. Wayne Aspinall to Mr. and Mrs. Charles Worth (Aspen-based constituents), November 19, 1964, Box 147, Wayne Aspinall Papers, Penrose Special Collections and Archives, University of Denver.

29. Floyd Dominy, "Oral History Interviews," conducted on April 6, 1984, and April 8, 1986, by Brit Allan Story, National Archives and Records Administration, College Park, Md., 85.

30. Ibid., 85–86; U.S. Department of the Interior, "Bridge and Marble Canyon Dams and Their Relationship to Grand Canyon National Park and Monument," n.d., Carton 17, Folder 12, Executive Director Records; Pearson, *Still the Wild River Runs*, 107.

31. David Brower to Hugh Barnes, September 9, 1964, Carton 22, Folder 4, Executive Director Records. *The Last Redwoods* (San Francisco: Sierra Club Books, 1964), by Philip Hyde and François Leydet, encountered numerous technical problems at the printer, and Brower complained in this letter to Hugh Barnes that it seemed that Barnes Press was beginning to tire from the many demands made by the Sierra Club and Dave Brower.

32. Morris Udall and David Brower, testimony, U.S. House, Subcommittee on Irrigation and Reclamation, *Lower Colorado River Basin Project*, August 8–13, 1965, and September 1, 1965, 802–809. Brower worried that the hydroelectricity was going to be used for "peaking power," meaning that it would be generated only during peak demand, usually in the morning or late afternoon, which would disrupt the flow of the Colorado within the Grand Canyon.

33. Brower, note on copy of May 5, 1965, *Salt Lake Tribune* story, May 18, 1965, Carton 10, Folder 9, Sierra Club Records, BL.

34. S. Udall, testimony, U.S. House, Subcommittee on Irrigation and Reclamation, *Lower Colorado River Basin Project*, August 8–13, 1965, and September 1, 1965, 147; Pearson, *Still the Wild River Runs*, 102–103.

35. Brower to members of the Grand Canyon Task Force, May 13, 1965, Carton 15, Folder 26, Executive Director Records. See also Carton 15, Folders 23–27, and Carton 16, Folders 1–15, Executive Director Records, which contain hundreds of communications from citizens opposed to the two Grand Canyon dams. At first, Brower and other staffers attempted to get back to the writers when they could. Later, after the task force was created with the people who wrote these letters, a more concerted set of mailings went out. In the mailing of May 13, 1965, Brower sent one hundred reprints of an article, "The Chips Are Down," to each task force member. Task force members were also asked to contribute donations that varied from $4 to $1,000.

36. Pearson, *Still the Wild River Runs*, 102.

37. "Grand Canyon 'Cash Registers,'" *Life*, May 7, 1965, 4; Pearson, *Still the Wild River Runs*, 93.

38. Pearson, *Still the Wild River Runs*, 91–92.

39. Alan Carlin (economist, RAND Corp.) to Brower, March 8, 1965, Carton 15, Folder 23, Executive Director Records.

40. Laurence Moss, testimony, U.S. House of Representatives, Subcommittee on Irrigation and Reclamation, *Lower Colorado River Basin Project: Hearing on H.R. 4671 and Similar Bills*, 89th Cong., 2nd sess., May 9–13 and 18, 1966, 1540–1547.

41. Jeffrey Ingram, telephone interview by the author, November 12, 2012.

42. Ibid.

43. Richard C. Bradley, "Ruin for the Grand Canyon?" *Audubon*, January–February 1966, 34–41.

44. Kate Stone Lombardi, "Recalling the Glory Days of *Reader's Digest*," *New York Times*, October 1, 2010.

45. Floyd Dominy to Harriet C. Young (editor, *Reader's Digest*), February 2, 1966, Carton 16, Folder 5, Executive Director Records; Richard Bradley to Brower, February 11, 1966, Carton 15, Folder 3, Executive Director Records.

46. Brower to Robert Kellogg (Grand Canyon dams opponent), April 28, 1966, Carton 16, Folder 16, Executive Director Records.

47. Pearson, *Still the Wild River Runs*, 123.

48. Litton interview, October 8, 2012.

49. Brower to Kellogg, April 28, 1966.

50. Ben Avery, "Magazine's Motives Cause for Alarm," *Arizona Republic*, March 27, 1966; Walter W. Meeks, "Sierra Club Prepares a 'Low Blow' at CAP," unidentified newspaper, n.d., clipping, Box 1, GRCA 32364, Grand Canyon National Park Museum Collection, Grand Canyon National Park, Ariz.

51. Ingram telephone interview, November 12, 2012.

52. Brower to Kellogg, April 28, 1966.

53. Pearson, *Still the Wild River Runs*, 120–121.

54. Brower to Kellogg, April 28, 1966.

55. Charles H. Callison (Audubon Society) to Brower, April 3, 1966, Carton 15, Folder 3, Executive Director Records; William W. Meek, "Barry, Swinging Late, Hits Hardest at Canyon Forum," *Arizona Republic*, April 1, 1966. The stories that resulted from the journalists' tour of the Grand Canyon during the conference included William Logan, "Grand Canyon Battle in Depth," *Rocky Mountain News* (Denver), April 4, 1966; Kimmish Hendrick, "The Battle of Grand Canyon," *Christian Science Monitor*, May 19, 1966; "Controversy in the Grand Canyon: Beauty of Colorado River Dams?" *National Observer*, April 25, 1966; Robert Sylvester, "The Battle for the Grand Canyon," *Daily News* (New York), May 8, 1966; Bert Hanna, "Rivals Vow Congress Fight," *Denver Post*, April 3, 1966; Dean Kearsh, "Debate at Rim on Grand Canyon," *Kansas City Star*, April 6, 1966; and United Press International, "Grand Debate: Flood Canyon?" *Deseret News* (Salt Lake City), April 1, 1966.

56. Ingram telephone interview, November 12, 2012; Johnson, *Central Arizona Project*, 11; Alan Carlin, testimony, U.S. House, Subcommittee on Irrigation and Reclamation, *Lower Colorado River Basin Project*, May 9–13 and 18, 1966, including comments by Representatives Craig Hosmer and John Saylor, 1513–1514.

57. Sierra Club, "Government Secrecy," press release, May 12, 1966, Box 211, Aspinall Papers; United Press International, "Opponent of Dam Charges Suppression," *Dallas Chronicle*, May 13, 1966. Brower and George Hartzog detested each other. Brower on Hartzog: "He was always a hale and hearty man. He had a powerful handshake. He'd look you in the eye—and forget about wilderness" (quoted in Pearson, *Still the Wild River Runs*, 67). Hartzog on Brower: "You were just a no good son of a bitch if you didn't agree with everything he said" (quoted in Pearson, *Still the Wild River Runs*, 67).

58. Howard B. Stricklin (superintendent, Grand Canyon National Park) to George Hartzog, April 1, 1966, Box 1, GRCA 32364, Grand Canyon National Park Museum Collection; S. Udall to Brower, May 9, 1966, Box 1, GRCA 32364, Grand Canyon National Park Museum Collection.

59. Morris Udall to Les Alexander, memorandum, April 26, 1966, Box 476, Morris Udall Papers, MS 35, University of Arizona Libraries, Special Collections, University of Arizona, Tucson; Henry F. Dobyns, testimony, U.S. House, Subcommittee on Irrigation and Reclamation, *Lower Colorado River Basin Project*, May 9–13 and 18, 1966, 1577–1579.

60. Bruce Pearson, "We Have Almost Forgotten How to Hope: The Hualapai, the Navajo, and the Fight for the Central Arizona Project, 1944–1968," *Western Historical Quarterly*, Autumn 2000, 297–316. Dam proponents went so far as to draft a letter in the name of George Rocha, the Hualapai chairman, lamenting how the tribe had "almost forgotten how to hope for a better life." The letter was actually written by Rich Johnson, lawyer for the Central Arizona Project, and released to the press.

61. Henry Dobyns and Stephen Jett, testimony, U.S. House, Subcommittee on Irrigation and Reclamation, *Lower Colorado River Basin Project*, May 9–13 and 18, 1966, 1581–1587; Pearson, "We Have Almost Forgotten How to Hope," 310–312.

62. Brower, testimony, U.S. House, Subcommittee on Irrigation and Reclamation, *Lower Colorado River Basin Project*, 1486.

63. "Dam the Canyon?" *Newsweek*, May 30, 1966, 27.

64. Brower to Grand Canyon Task Force, urgent memorandum, April 13, 1966, Carton 10, Folder 12, Sierra Club Records, BL; "Grand Canyon," *Sierra Club Bulletin*, May 1966, 1–16; Jeff Ingram, "Ingram Events Journal, 1966–68, Part 1, Added to 6/3/11," http://gcfutures.blogspot.com/2011/06/ingram-events-journal-1966-8-part-1.html (accessed September 19, 2012).

13. Losing While Winning

1. Michael P. Cohen, *The History of the Sierra Club, 1892–1970* (San Francisco: Sierra Club Books, 1988), 352–355.

2. David Brower, "Grand Canyon Battle Ads," in *Grand Canyon of the Living Colorado*, ed. Roderick Nash (San Francisco: Sierra Club and Ballantine Books, 1970), 130.

3. Sierra Club, "An Open Letter to President Johnson on the Last Chance *Really* to Save the Redwoods," newspaper advertisement, Box 5, Redwood National Park Establishment Papers Series, Redwood National Park, Orick, Calif. Other redwoods park ads included "History Will Think It Most Strange That America Could Afford the Moon and $4 Billion Airplanes While a Patch of Primeval Redwoods—Not Too Big for a Man to Walk Through in a Day—Was Considered Beyond Its Means" and "Legislation by Chain-Saw?"

4. Richard Leonard, "Mountaineer, Lawyer, Environmentalist, Volume I," an oral history conducted in 1976 by Susan R. Schrepfer, Regional Oral History Office (ROHO), Bancroft Library (BL), University of California, Berkeley, 1976, 170; David Brower to Stewart Udall, January 6, 1966, and S. Udall to Brower, January 10, 1966, Carton 10, Box 18, Sierra Club Office of the Executive Director Records, BANC MSS 2002/230c, BL; Susan Schrepfer, *The Fight to Save the Redwoods: A History of Environmental Reform, 1917–1978* (Madison: University of Wisconsin Press, 1983), 139.

5. Brower to John Oakes, March 28, 1966, Carton 10, Folder 19, Executive Director Records; response to Redwood ad, memorandum, March 31, 1966, Carton 10, Folder 19, Executive Director Records; David R. Brower, "Environmental Activist, Publicist, and Prophet," an oral history conducted in 1974–1978 by Susan Schrepfer, ROHO, BL, 1979, 362.

6. Brower, "Environmental Activist, Publicist, and Prophet," 147.

7. Donald W. Carson and James W. Johnson, *Mo: The Life and Times of Morris K. Udall* (Tucson: University of Arizona Press, 2001), 124; Bryon Pearson, *Still the Wild River Runs: Congress, the Sierra Club, and the Fight to Save Grand Canyon* (Tucson: University of Arizona Press, 2002), 142–143.

8. J. M. Cullen (district director, IRS) to Sierra Club, June 10, 1966, Carton 3, Folder 20, Executive Director Records.

9. Brower, "Environmental Activist, Publicist, and Prophet," 147.

10. "Free Speech Apparently Does Not Apply If Objecting to Action of Great Society," *Fort Lauderdale News*, June 28, 1966; "The Smell of Retribution," *Wichita Eagle*, June 28, 1966; unidentified newspaper clipping, *Albuquerque Tribune*, Carton 3, Folder 27, Executive Director Records; "Scorecard," *Sports Illustrated*, July 18, 1966; "IRS and the Grand Canyon," *New York Times*, June 17, 1966.

11. Quoted in "IRS Threatens the Sierra Club," *New York Times*, June 12, 1966.

12. Brower to unidentified recipients, confidential memorandum, June 24, 1966, Carton 2, Folder 21, Executive Director Records; Brower, "Environmental Activist, Publicist, and Prophet," 151.

13. "No Barr Involved in Tax Wrangle," *Arizona Republic*, June 16, 1966; Pearson, *Still the Wild River Runs*, 144.

14. Edgar Wayburn, "Sierra Club Statesman, Leader of the Parks and Wilderness Movement: Gaining Protection for Alaska, the Redwoods, and Golden Gate Parklands," an oral history conducted in 1976–1981 by Ann Lage and Susan Schrepfer, ROHO, BL, 1985, 293; Brower, "Environmental Activist, Publicist, and Prophet," 151.

15. Pearson, *Still the Wild River Runs*, 146–147; "Sierra Club Gains in Fight on Taxes," *New York Times*, August 7, 1966.

16. Brower, "Environmental Activist, Publicist, and Prophet," 148; newspaper advertisements memorandum, June 9, 1966–January 14, 1969, Carton 115, Folder 15, Sierra Club Members Papers, BANC MSS 71/295c, BL. Between 1966 and 1969, the Sierra Club spent $143,000 on newspaper advertisements to advance its conservation agenda. The most well-known ad, "SHOULD WE ALSO FLOOD THE SISTINE CHAPEL SO TOURISTS CAN GET NEARER THE CEILING?" ran in six different publications on various dates, including *Wall Street Journal*, August 23, 1966; *San Francisco Chronicle*, August 28, 1966; *Saturday Review*, September 10, 1966; *National Review*, September 20, 1966; *Harper's*, October 1966; and *Ramparts*, October 1966 (noted in the ad memorandum, June 9–January 14, 1969).

17. Cohen, *History of the Sierra Club*, 163.

18. A. Lincoln Green (tax expert) to Brower, June 16, 1966, Carton 3, Folder 20, Executive Director Records.

19. "We Defend the Parks," *Sierra Club Bulletin*, January 1955, 3–5.

20. "The Sierra Club Foundation," *Sierra Club Bulletin*, December 1968, 12.

21. William Zimmerman (Sierra Club lobbyist) to Brower, November 17, 1965, Carton 3, Folder 19, Executive Director Records.

22. Michael McCloskey to Brower, memorandum, June 27, 1966, Carton 3, Folder 20, Executive Director Records; Edward J. Doyle (agent, IRS) to Sierra Club, June 21, 1966, Carton 3, Folder 21, Executive Director Records; David Brower, statement at press conference, July 22, 1966, Carton 16, Folder 18, Executive Director Records; "Sierra Club Gains in Fight on Taxes"; Avis Ogilvy Moore, telephone interview by the author, February 19, 2013.

23. Quoted in Donovan Bess, "A New Attack by Sierra Club," *San Francisco Chronicle*, July 2, 1966.

24. Morris Udall to Walter Reuther (president, United Auto Workers), July 19, 1966, Box 476, Morris Udall Papers, University of Arizona Libraries, Special Collections, University of Arizona, Tucson; M. Udall to John Oakes, July 22, 1966, Box 476, M. Udall Papers; Paul A. McKalip (editor, *Tucson Citizen*) to M. Udall, July 13, 1966, Box 476, M. Udall Papers; Pearson, *Still the Wild River Runs*, 149.

25. "Aspinall Hits Sierra Club Lobby Tactics," *Grand Junction (Colo.) Daily Sentinel*, July 6, 1966; "Dam Foes Twist Truth, Moss Says," *Phoenix Gazette*, June 22, 1966; Barry Goldwater, remarks at press conference, July 27, 1966, transcript, Box 476, M. Udall Papers; Dominy quoted in "Dam Builder Raps the Conservationists," *Louisville Times*, n.d., clipping, Carton 58, Folder 21, Sierra Club Members Papers.

26. L. M. Alexander, Central Arizona Project memorandum, September 1, 1966, Box 476, M. Udall Papers.

27. Wayne Aspinall, speech to the Colorado State Grange, October 1, 1966, transcript, Box 190, Wayne Aspinall Papers, Penrose Special Collections and Archives, University of Denver; Jeff Ingram, "Ingram Events Journal, 1966–68, Part 1, Added to 6/3/11," http://gcfutures.blogspot.com/2011/06/ingram-events-journal-1966-8-part-2.html (accessed September 19, 2012).

28. Pearson, *Still the Wild River Runs*, 158.

29. Beaty quoted in ibid., 159–160; David Brower, *Grand Canyon Discussion at Silver Grotto and Marble Damsite, May 31 to June 13, 1977*, audiocassette, recorded by Ron Hayes, Grand Canyon National Park Museum Collection, GRCA 40237, Grand Canyon National Park, Ariz.

30. Roderick Nash, *Wilderness and the American Mind* (New Haven, Conn.: Yale University Press, 1967), 229–236; Marc Reisner, *Cadillac Desert: The American West and Its Disappearing Water* (New York: Viking Penguin, 1986), 287.

31. Aspinall, speech to the Colorado State Grange.

32. Brower, *Grand Canyon Discussion at Silver Grotto and Marble Damsite May 31 to June 13, 1977*.

33. Bob Beier, "Aspinall Raps Opposition to Project," *Albuquerque Tribune*, November 18, 1966.

34. U.S. Bureau of Reclamation employee and reports quoted in Reisner, *Cadillac Desert*, 288.

35. Cohen, *History of the Sierra Club*, 364; Leonard, "Mountaineer, Lawyer, Environmentalist, Volume I," 174.

36. Stewart Udall, Southwest water plan memorandum, January 28, 1967, Box 169, Stewart L. Udall Papers, AZ 372, University of Arizona Libraries, Special Collections, University of Arizona, Tucson; Pearson, *Still the Wild River Runs*, 164.

Reisner reports that Udall was so concerned about Dominy's reaction to dropping the two dams that he waited until Dominy had left the country to introduce the legislation (*Cadillac Desert*, 289–290).

37. "Dominy Says River Project in Mess," *Farmington (N.M.) Times*, October 26, 1966.

38. David Brower, testimony, U.S. House of Representatives, Subcommittee on Irrigation and Reclamation, *Colorado River Basin Project: Hearing on H.R. 3300 and Similar Bills*, 90th Cong., 1st sess., March 13–17, 1967, 466.

39. Brower, *Grand Canyon Discussion at Silver Grotto and Marble Damsite May 31 to June 13, 1977*.

40. Associated Press, "EPA Is Said to Plan a Curb on Dirty Air at Grand Canyon," *New York Times*, February 1, 1991.

41. Wallace Turner, "Sierra Club Loses Exemption on Tax," *New York Times*, December 21, 1966.

42. David Brower, "Sierra Club Attacks IRS Decision," press release, December 20, 1966, Carton 10, Folder 16, Sierra Club Records, BANC MSS 71/103c, BL.

43. David Brower, Sierra Club appealing IRS decision, memorandum, December 22, 1966, Carton 2, Folder 22, Executive Director Records.

44. Gary Torre, "Labor and Tax Attorney, 1949–1982; Sierra Club Foundation Trustee, 1968–1981, 1994–1998," an oral history conducted in 1998 by Carl Williams, ROHO, BL, 1999, 135.

45. Cohen, *History of the Sierra Club*, 381–383.

46. David R. Brower, "Reflections on the Sierra Club, Friends of the Earth, and Earth Island Institute," an oral history conducted in 1999 by Ann Lage, ROHO, BL, 2012, 46.

47. Torre, "Labor and Tax Attorney," 135; Wayburn, "Sierra Club Statesman," 242–243.

48. Brower to Henry H. Fowler (secretary of the U.S. Treasury), night of May 20, 1968, Carton 1, Folder 33, Executive Director Records; Torre, "Labor and Tax Attorney," 134.

49. Torre, "Labor and Tax Attorney," 134; Wayburn, "Sierra Club Statesman," 242.

50. David Brower, defense of IRS actions, memorandum, March 18, 1969, Carton 57, Folder 11, Sierra Club Members Papers.

51. Michael McCloskey, "Sierra Club Executive Director: The Evolving Club and the Environmental Movement, 1961–1981," an oral history conducted in 1981 by Susan Schrepfer, ROHO, BL, 1983, 137.

52. Wayburn, "Sierra Club Statesman," 242–244.

14. Diablo and Galápagos

1. John Wills, *Conservation Fallout: Nuclear Protest at Diablo Canyon* (Reno: University of Nevada Press, 2006), 90–91, 103; Mark Evanoff, "Boondoggle at Diablo: Saga of Greed, Deception, Ineptitude—and Opposition," *Not Man Apart*, September 1981, D1–D12.

2. David R. Brower, "Environmental Activist, Publicist, and Prophet," an oral history conducted in 1974–1978 by Susan Schrepfer, Regional Oral History Office (ROHO), Bancroft Library (BL), University of California, Berkeley, 1979, 220.

3. Evanoff, "Boondoggle at Diablo," D1; Wills, *Conservation Fallout*, 39.

4. Kenneth Brower, *The Wildness Within: Remembering David Brower* (Berkeley, Calif.: Heyday, 2012), 71.

5. Bodega Head Atomic Park, May 4, 1963, Sierra Club Board of Directors Meeting Minutes, 1892–1995, BANC FILM 2945, BL, 11; Bodega Head, June 9, 1963, Sierra Club Minutes, 4.

6. Brower, "Environmental Activist, Publicist, and Prophet," 198.

7. Wills, *Conservation Fallout*, 39–40.

8. Virginia Cornell, *Defender of the Dunes: The Kathleen Goddard Jones Story* (Carpinteria, Calif.: Manifest, 2001), 91–94. Given the name Kathleen Goddard at birth in 1907, Jones was married three times and thus had varying names, including Kathleen or Kathy Goddard, Shiraz, Jackson, and finally Jones.

9. Edgar Wayburn and Allison Alsup, *Your Land and Mine: Evolution of a Conservationist* (San Francisco: Sierra Club Books, 2004), 112–113.

10. Kathleen Goddard Jones [Kathy Jackson], "Defender of California's Nipomo Dunes, Steadfast Sierra Club Volunteer," an oral history conducted in 1983 by Anne Van Tyne and included in "Sierra Club Nationwide II," ROHO, BL, 1984, 28–29.

11. Edgar Wayburn, "Sierra Club Statesman, Leader of the Parks and Wilderness Movement: Gaining Protection for Alaska, the Redwoods, and Golden Gate Parklands," an oral history conducted in 1976–1981 by Ann Lage and Susan Schrepfer, ROHO, BL, 1985, 231.

12. Nipomo Dunes, May 7–8, 1966, Sierra Club Minutes; Wayburn, "Sierra Club Statesman," 231; Martin Litton to Sherman L. Sibley (president, PG&E), June 13, 1966, Carton 81, Folder 18, Sierra Club Members Papers, BANC MSS 71/295c, BL.

13. David Brower, journal, May 24, 1963, Box 20, Folder 4, David Ross Brower Papers, BANC MSS 79/9c, BL.

14. Litton to Sibley, June 13, 1966; Ansel Adams to Litton, July 12, 1966, Carton 81, Folder 18, Sierra Club Members Papers; Litton to Sierra Club board members, September 9, 1966, Carton 81, Folder 18, Sierra Club Members Papers.

15. Michael McCloskey, "Sierra Club Executive Director: The Evolving Club and the Environmental Movement, 1961–1981," an oral history conducted in 1981 by Susan Schrepfer, ROHO, BL, 1983, 93; Wayburn, "Sierra Club Statesman," 56; Jones, "Defender of California's Nipomo Dunes," 29. Although Wayburn was satisfied at the time that Diablo was a suitable site, he changed his mind later. He returned to the site, saw the extent of development in the canyon and the power lines that extended well beyond the canyon, and felt that he had been deceived by PG&E on that first visit.

16. Brower, "Environmental Activist, Publicist, and Prophet," 223; Diablo Canyon, September 17, 1966, Sierra Club Minutes; Wayburn, "Sierra Club Statesman," 234.

17. McCloskey, "Sierra Club Executive Director," 93.

18. Quoted in Wills, *Conservation Fallout*, 60.

19. Wayburn, "Sierra Club Statesman," 234.

20. Martin Litton, interview by the author, May 26, 2013, Portola Valley, Calif.; Phillip Berry, interview by the author, Lafayette, Calif., April 24, 2013.

21. Richard Sill (Sierra Club Council) to Fred Eissler and Litton, February 7, 1967, Carton 10, Folder 4, Sierra Club Office of the Executive Director Records, BANC MSS 2002/230c, BL.

22. Brower to Sill, February 13, 1967, Carton 10, Folder 4, Executive Director Records.

23. Phillip S. Berry, "Sierra Club Leader, 1960s–1980s: A Broadened Agenda, a Bold Approach," an oral history conducted in 1981 and 1984 by Ann Lage, ROHO, BL, 1988, 25–26.

24. Berry interview, April 24, 2013.

25. "Atomic Plant Site Stirs Wide Coast Conservation Controversy," *New York Times*, April 4, 1967; Cornell, *Defender of the Dunes*, 127; Wills, *Conservation Fallout*, 42–48.

26. Evanoff, "Boondoggle at Diablo," D4; Berry, "Sierra Club Leader," 27.

27. Diablo Canyon Petition, January 7–8, 1967, Sierra Club Minutes, 3–10; "Background on Nipomo Dunes–Diablo Canyon Issue," *Sierra Club Bulletin*, February 1967, 12.

28. William Siri to George Marshall, January 11, 1967, Carton 10, Folder 3, Executive Director Records; Brower to Marshall, March 6, 1967, Carton 10, Folder 5, Executive Director Records; Hugh Nash, "PG&E Difficulties with Existing Plants," memorandum, July 1969, Carton 19, Folder 39, Brower Papers.

29. Hugh Nash to Litton, January 22, 1967, Carton 10, Folder 3, Executive Director Records.

30. Nash, "PG&E Difficulties."

31. Richard Searle, "Grassroots Sierra Club Leader," an oral history conducted in 1976 by Paul Clark and included in "Southern Sierrans," ROHO, BL, 1976, 33.

32. Marshall to Sierra Club board of directors, March 6, 1967, Carton 10, Folder 4, Executive Director Records; Nash, "PG&E Difficulties."

33. "Background on Nipomo Dunes–Diablo Canyon Issue," *Sierra Club Bulletin*, February 1967, 12; Sierra Club Atlantic Chapter, "Diablo Canyon Referendum," press release, March 8, 1967, copy supplied to the author by George Alderson; Wills, *Conservation Fallout*, 60–61.

34. Nash, "PG&E Difficulties"; *This Is the Issue*, Carton 10, Folder 6, Executive Director Records; Stewart Ogilvy (chairman, Atlantic chapter) to Nash and Litton, April 22, 1967, Carton 87, Folder 28, Sierra Club Members Papers.

35. Results of Election, May 6, 1967, Sierra Club Minutes.

36. Eliot Porter to Marshall, January 20, 1967, Carton 1, Folder 49, Serra Club Members Papers.

37. David Brower, "Excerpts from Publications Committee Minutes," memorandum, November 6, 1967, Carton 22, Folder 23, Executive Director Records. Brower in this memo cited minutes back to 1963 that he believed supported his role in developing the Galápagos book.

38. Ibid.

39. David Brower, foreword to *Galápagos*, vol. 1, *Discovery*, ed. Kenneth Brower, with photographs by Eliot Porter (London: Sierra Club Books, 1968), 24–26.

40. Brower, journal, February 15–19, 1966, Box 20, Folder 3, Brower Papers; Bob Golden to Porter, November 3, 1965, Carton 22, Folder 23, Executive Director Records; Brower to U.S. Selective Service, November 23, 1965, Carton 22, Folder 23, Executive Director Records. The islands' isolation created the unusual biological diversity that was so attractive to scientists. But with only limited air and sea traffic, getting all of the equipment there for a three-month stay proved challenging. There were also the odd tasks, as when Brower, a retired army officer, had to write to Stephen Porter's draft board. Vietnam was heating up, and Brower urged the board not to change young Porter's draft status while he took three months off from college to assist his father.

41. Brower to Edgar Wayburn, November 27, 1967, Carton 235, Folder 5, Sierra Club Members Papers.

42. Porter to Clifford Rudden, July 3, 1966, Carton 22, Folder 23, Executive Director Records; Marshall to Brower, September 3, 1966, Carton 22, Folder 24, Executive Director Records; Porter to Marshall, September 5, 1966, Carton 22, Folder 24, Executive Director Records.

43. Russell Train (president, Conservation Foundation) to Brower, February 15, 1966, Carton 20, Folder 24, Executive Director Records.

44. Brower, "Excerpts from Publications Committee Minutes."

45. Porter to Marshall, January 20, 1967.

46. Ansel Adams to Porter, February 1, 1967, Carton 1, Folder 49, Sierra Club Members Papers; Adams, statement to the Sierra Club board of directors, February 15, 1967, Carton 1, Folder 49, Sierra Club Members Papers.

47. Brower, "Environmental Activist, Publicist, and Prophet," 217.

48. Richard Leonard, "Mountaineer, Lawyer, Environmentalist, Volume II," an oral history conducted in 1976 by Susan R. Schrepfer, ROHO, BL, 1976, 345–346.

49. Brower, "Environmental Activist, Publicist, and Prophet," 239; Leonard, "Mountaineer, Lawyer, Environmentalist, Volume II," 345–346; Brower to Wayburn, June 29, 1967, Carton 1, Folder 33, Executive Director Records.

50. Herb Caen, "In One Ear," *San Francisco Chronicle*, May 1, 1967; Brower, journal, May 2, 1967, Box 20, Folder 6, Brower Papers.

51. Letters, telegrams, postcards, Carton 10, Folder, Sierra Club Records, BANC MSS 71/103c, BL. The folder contains about sixty-five letters, telegrams, postcards, a petition signed by thirty-three congressional representatives (primarily Democrats but also including a few Republicans), and testimony in the *Congressional Record* supporting Brower and protesting his firing.

52. Scott Thurber, "Sierra Club Battle over Brower's Job," *San Francisco Chronicle*, May 6, 1967; "Sierra Club Keeps Brower," *Oakland Tribune*, May 7, 1966; George Dusheck, "Sierra's Chief Wins," *San Francisco Examiner*, May 7, 1967; Scott Thurber, "Ouster Plan That Failed," *San Francisco Chronicle*, May 8, 1967.

15. Conflict

1. Stanley A. Cain, "Address at the Diamond Jubilee Banquet," *Sierra Club Bulletin*, January, 1968, 7–13; Edgar Wayburn, "A President's Message," *Sierra Club Bulletin*, April–May 1967, 4.

2. Exhibit B, September 14–15, 1967, Sierra Club Board of Directors Meeting Minutes, 1892–1995, BANC FILM 2945, Bancroft Library (BL), University of California, Berkeley, 43–47; David Brower to Edgar Wayburn, June 6, 1967, Carton 19, Folder 42, David Ross Brower Papers, BANC MSS 79/9c, BL; William Siri, "Reorganization of Club Publishing Activities," memorandum to Sierra Club board of directors, May 1, 1967, Carton 5, Folder 6, Sierra Club Office of the Executive Director Records, BANC MSS 2002/230c, BL; Sierra Club, "Publishing Reorganization Plan," press release, May 9, 1967, Carton 5, Folder 6, Executive Director Records.

3. Edgar Wayburn, "Sierra Club Statesman, Leader of the Parks and Wilderness Movement: Gaining Protection for Alaska, the Redwoods, and Golden Gate Parklands," an oral history conducted in 1976–1981 by Ann Lage and Susan Schrepfer, Regional Oral History Office (ROHO), BL, 1985, 237–238.

4. "Agreement on Redwoods Park Plan," *San Francisco Chronicle*, October 20, 1967; Sierra Club, "Georgia-Pacific Cutting Redwoods in Emerald Mile," press release, December 15, 1967, Anthrop Series 1, Carton 1, Redwood National Park Establishment Papers Series, Redwood National Park, Orick, Calif.

5. Stewart Udall, "Stewart L. Udall Oral History Interview V," an oral history interview conducted on October 31, 1969, by Joe B. Frantz, Lyndon Baines Johnson Library, Austin, Tex., 15; "North Cascades National Park," *Seattle Times*, January 30, 1967.

6. Mark Evanoff, "Boondogle at Diablo: Saga of Greed, Deception, Ineptitude—and Opposition," *Not Man Apart*, September 1981, D4; Hugh Nash, "PG&E Difficulties with Existing Plants," memorandum, July 1969, Carton 19, Folder 39, Brower Papers; Jack Hope, "The King Besieged," *Natural History*, November 1968, 52–56, 72–82. The other major Sierra Club conservation issue of the late 1960s was Mineral King, and it affected the redwoods park campaign. Some politicians tried to broker a deal in which if northern California gave up valuable redwood timber property, Mineral King would in exchange be developed as a ski resort in southern California. The argument was that if there were an economic loss in the North, it should be offset by an economic gain in the South. The Mineral King fight would fester for years, and it was complicated by the fact that although the national club opposed Disney's planned ski resort, the local chapter favored it.

7. Brower to Wayburn, "The Next Five Years," memorandum, November 21, 1967, Carton 10, Folder 21, Sierra Club Records, BANC MSS 71/103c, BL.

8. Max Linn (John Muir Institute) to Brower, memorandum, November 1967, Carton 19, Folder 6, Executive Director Records; Articles of Incorporation, John Muir Institute, January 5, 1968, Carton 19, Folder 6, Executive Director Records; Brower to Clifford Rudden (Sierra Club controller), memorandum, December 26, 1967, Carton 19, Folder 6, Executive Director Records. Brower planned to make a second payment to the John Muir Institute from his discretionary fund in 1968, but there is no record that this payment occurred.

9. David Brower, "The Earth's Wild Places," memorandum, n.d., Carton 19, Folder 42, Brower Papers.

10. Financial Report, December 9–10, 1967, Sierra Club Minutes, 13–18.

11. David Brower, London office chronology memorandum, May 1967–June 8, 1968, Carton 20, Folder 14, Executive Director Records; Sally Walker (Sierra Club donor) to Brower, November 14, 1967, Carton 20, Folder 19, Executive Director Records; David Brower, journal, September 29, 1967–January 23, 1968, Box 20, Folder 7, Brower Papers.

12. Ansel Adams, "A Conversation with Ansel Adams," an oral history conducted in 1972, 1974, and 1975 by Ruth Teiser and Catherine Harroun, ROHO, BL, 1978, 681.

13. David R. Brower, "Environmental Activist, Publicist, and Prophet," an oral history conducted in 1974–1978 by Susan Schrepfer, ROHO, BL, 1979, 227.

14. Executive Director's Discretionary Fund, February 3–4, 1968, Sierra Club Minutes, 2–3.

15. Brower, journal, April 1, 1968–May 1, 1968, Box 21, Folder 1, Brower Papers.

16. Ibid., March 21, 1968 (including mention of Caen). The report that Brower was about to be fired was a false alarm.

17. Brower to Norman G. Dyhrenfurth (leader, Everest Expedition), May 26, 1966, and Dyhrenfurth to Brower, June 4, 1966, Carton 22, Folder 14, Executive Director Records.

18. Brower to Phillip Berry, March 1, 1969, Carton 10, Folder 24, Sierra Club Records, BL; Brower to Wayburn, publishing contracts memorandum, August 29, 1968, Carton 49, Folder 4, Sierra Club Records, BL.

19. Phillip Berry, reply to charges in the executive director's memorandum to the Sierra Club board of directors, January 20, 1969 Carton 11, Folder 3, Sierra Club Records, BL.

20. Phillip Berry, interview by the author, Lafayette, Calif., April 24, 2013.

21. Brower, "Environmental Activist, Publicist, and Prophet," 241.

22. Berry, reply to charges.

23. Richard Leonard, transcript of Executive Committee meeting, June 10, 1968, Carton 19, Folder 48, Brower Papers.

24. Berry interview, April 24, 2013.

25. Executive Director's Discretionary Fund, 1967, Carton 25, Folder 22, Sierra Club Records, BL; Rudden to Berry, October 21, 1968, Carton 25, Folder 22, Sierra Club Records, BL. Even after Brower was supposed to get the Sierra Club president's approval to spend money from the discretionary fund and the amount in the fund was capped at $25,000, he continued to spend more than he was allotted, and in the first nine months after the change the fund was over budget by nearly $11,000. That was about the same amount that went for renting and equipping the club's New York office. More than $9,000 went into salaries for Allan Horlin and J. Robert Horlin, who were employees in the London office, which was supposed to be financed by Walker's gift.

26. Brower to Sierra Club Executive Committee, memorandum, March 18, 1968, Carton 10, Folder 24, Sierra Club Records, BL.

27. Financial reports, May 3, 1968, Sierra Club Minutes, 2–6.

28. Richard Sill to Orrin H. Bonney (leader, Texas chapter), November 5, 1967, Box 18, Sierra Club Records, 1957–1980, NC1260, Special Collections, University of Nevada, Reno; Wayburn, "Sierra Club Statesman," 229.

29. Brower to Norman Garrod (principal, Garrod & Lofthouse), August 3, 1968, Carton 21, Folder 4, Executive Director Records; Brower to Berry, n.d., Carton 10, Folder 26, Sierra Club Records, BL; Brower, journal, July 7–23, 1968, Box 21, Folder 1, Brower Papers.

30. David Hales (employee, Sierra Club) to Brower, December 17, 1968, Carton 20, Folder 25, Executive Director Records; Brower, journal, July 30–August 21, 1968, Box 21, Folder 1, Brower Papers; Brower to Garrod, August 3, 1968.

31. Publications Committee, September 14–15, 1968, Sierra Club Minutes, 7. See also Kenneth Brower, ed., and Eliot Porter, *Galápagos*, vol. 1, *Discovery* (San Francisco: Sierra Club Books, 1968), and *Galápagos*, vol. 2, *Prospect* (San Francisco: Sierra Club Books, 1968).

32. Harvard Gunn (public-relations specialist) to Brower, November 18, 1968, Carton 21, Folder 8, Executive Director Records.

33. Carroll Harris (president, MacKenzie & Harris) to Joe Linn (employee, Sierra Club), July 29, 1968, Carton 49, Folder 4, Sierra Club Records, BL; Hugh Nash, Publications Committee meeting minutes, December 13, 1968, Carton 26, Folder 38, Sierra Club Records, BL.

34. Brower to Allan Horlin and Robert Horlin, October 23, 1968, Carton 20, Folder 25, Executive Director Records; Clifford Rudden, "Mismanagement of English Gift," memorandum to Sierra Club board of directors, November 21, 1968, Carton 235, Folder 4, Sierra Club Members Papers,.

35. Brower to Sherman Sibley, June 11, 1968, Carton 19, Folder 37, Brower Papers; Berry, Frederick Eissler, Patrick Goldsworthy, Luna Leopold, Litton, Laurence I. Moss, John Oakes, and Eliot Porter to Sibley, June 27, 1968, Carton 19, Folder 37, Brower Papers.

36. Sibley to Wayburn, July 2, 1968, Carton 19, Folder 37, Brower Papers; Wayburn to Sierra Club board of directors, memorandum, July 9, 1968, Carton 19, Folder 37, Brower Papers; "Inconsistent Club," *San Francisco Examiner*, July 3, 1968; David Brower, "Editor's Mail Box, Diablo Power Site," *San Francisco Examiner*, July 9, 1968.

37. Diablo Canyon, September 14–15, 1968, Sierra Club Minutes, 21–30; Evanoff, "Boondogle at Diablo," D4; Brower, "Environmental Activist, Publicist, and Prophet," 223; Phillip S. Berry, "Sierra Club Leader, 1960s–1980s: A Broadened Agenda, a Bold Approach," an oral history conducted in 1981 and 1984 by Ann Lage, ROHO, BL, 1988, 29.

38. Diablo Canyon Policy Review, December 14–15, 1968, Sierra Club Minutes, 8–10.

39. Publications Reorganization Committee, Internal Procedures and Retention of Executive Director, September 14–15, 1968, Sierra Club Minutes, 8–9, 32–41, 11–13

40. "Adams Asks Ouster of Chief," *Monterey Peninsula Herald* (Monterey, Calif.), September 25, 1968; Alan Cline, "Turmoil Over Sierra Club Leadership," *San Francisco Examiner & Chronicle*, October 6, 1968.

41. Inquiry into Allegations [by] Directors Sill, Leonard, and Adams, October 19, 1968, Sierra Club Minutes, 1–7; meeting transcript, October 19, 1968, Carton 1, Folder 18, Executive Director Records (including Sive's comments and thoughts).

42. Brower to Sierra Club board of directors, memorandum, November 29, 1968, Carton 19, Folder 44, Brower Papers; Executive Director's Report in Reply to Charges, December 14–15, 1968, Sierra Club Minutes.

43. James W. Moorman, "Attorney for the Environment 1966–1981, Center for Law and Social Policy, Sierra Club Legal Defense Fund, Department of Justice, Division of Lands and Natural Resources," an oral history conducted in 1984 by Ted Hudson, ROHO, BL, 1994, 32; Gordon Robinson (employee, Sierra Club) to Brower, December 3, 1968, Carton 124, Folder 24, Brower Papers; James Elliott Bryant (chairman, Rocky Mountain chapter) to Brower and Sierra Club board of directors, September 11, 1968, Carton 1, Folder 50, Sierra Club Members Records.

44. Polly Dyer, "Preserving Washington's Parklands and Wilderness," an oral history conducted in 1983 by Susan Schrepfer and included in "Pacific Northwest Conservationists," ROHO, BL, 1986, 106; Ansel Adams to the Sierra Club board of directors and board president, September 3, 1967, Carton 1, Folder 21, Executive Director Records; Berry, "Sierra Club Leader, 1960s–1980s," 30.

45. Michael McCloskey, "Four Major New Conservation Laws: A Review and a Preview," *Sierra Club Bulletin*, November 1968, 4–10; Patrick Goldsworthy to Mr. and Mrs. Ralph Simons, October 10, 1968, Box 8, Folder 9, Executive Director Records.

46. McCloskey, "Four Major New Conservation Laws," 4; Brower, "Environmental Activist, Publicist, and Prophet," 161.

47. Susan Schrepfer, *The Fight to Save the Redwoods: A History of Environmental Reform, 1917–1978* (Madison: University of Wisconsin Press, 1983), 226; Claire Reynolds, exec. prod., and Sam Greene, prod. and dir., *Redwood National Park: Preserving Ancient Forests*, DVD (KEET-TV, 2009).

16. Campaign

1. "Brower May Quit Post in Sierra Club," *San Francisco Examiner*, January 6, 1969.

2. David R. Brower, "Environmental Activist, Publicist, and Prophet," an oral history conducted in 1974–1978 by Susan Schrepfer, Regional Oral History Office (ROHO), Bancroft Library (BL), University of California, Berkeley, 1979, 240.

3. David Brower to Clifford Rudden, London account memorandum, November 27, 1968, Carton 20, Folder 14, Sierra Club Office of the Executive Director Records, BANC MSS 2002/230c, BL.

4. *Sierra Club Explorer*, Fall 1968, Carton 19, Folder 41, David Ross Brower Papers, BANC MSS 79/9c, BL; Brower to Rudden, *Explorer* memorandum, December 11, 1968, Carton 49, Folder 4, Sierra Club Records, BANC MSS 71/103c, BL; Brower to Phillip Berry, December 10, 1968, Carton 11, Folder 1, Sierra Club Records, BL.

5. Edgar Wayburn, "Sierra Club Statesman, Leader of the Parks and Wilderness Movement: Gaining Protection for Alaska, the Redwoods, and Golden Gate Parklands," an oral history conducted in 1976–1981 by Ann Lage and Susan Schrepfer, ROHO, BL, 1985, 243; *Sierra Club Explorer*, January 11, 1969, Sierra Club Board of Directors Meeting Minutes, 1892–1995, BANC FILM 2945, BL, 4–5.

6. Brower to Mrs. Henry S. Francis Jr., October 30, 1968, Carton 19, Folder 42, Brower Papers; Brower to Mrs. Lyndon B. Johnson, November 16, 1968, Carton 19, Folder 42, Brower Papers; Elizabeth Carpenter (press secretary to Mrs. Johnson) to Brower, November 21, 1968, Carton 19, Folder 42, Brower Papers.

7. "Actions of Board at December Meeting," *Sierra Club Bulletin*, January 1969, 5.

8. International Book and Television Series, December 14–15, 1969, Sierra Club Minutes, 3.

9. Stewart Ogilvy, "Sierra Club Expansion and Evolution: The Atlantic Chapter, 1957–1969," an oral history conducted in 1978 by Jeri Nunn and included in "The Sierra Club Nationwide I," ROHO, BL, 1982, 12; Michael McCloskey, interview by the author, Portland, Ore., May 24, 2013.

10. "Actions of Board at December Meeting," 5.

11. Publications Committee meeting transcript, January 10, 1969, Carton 20, Folder 3, Executive Director Records.

12. Proposed International Book Series, February 8–9, 1969, Sierra Club Minutes, 8.

13. Michael McCloskey, *In the Thick of It: My Life in the Sierra Club* (Washington, D.C.: Island Press, 2005), 95.

14. Wayburn, "Sierra Club Statesman," 240.

15. "Earth National Park," advertisement, Carton 2, Folder 22, Executive Director Records.

16. Wayburn, "Sierra Club Statesman," 241.

17. Ibid.; David Brower, journal, January 14, 1969, Box 21, Folder 3, Brower Papers. Brower was in New York the morning of January 14, appearing on NBC's *Today Show* to speak not about the Earth Island ad but about Alaska governor Walter Hickel. President-elect Richard Nixon had nominated Hickel to succeed

Stewart Udall as interior secretary. The Sierra Club might be divided on other issues, but it was united in opposition to Hickel. It feared that Big Oil would control Hickel and that he would open up federal lands to oil drilling, especially in Alaska. Two days later, Brower testified in the Senate that the Sierra Club had never before opposed a presidential nomination but that now the stakes were too high. Despite the opposition, Hickel was eventually sworn in. See Drew Pearson, "Big Oil Man Visits the President-Elect," *San Francisco Chronicle*, December 23, 1968; "Hickel Receives Approval from Senate," *Sierra Club Bulletin*, February 1969, 3; Walter Rugaber, "Hickel Sworn In at White House Ceremony and Hailed by Nixon," *New York Times*, January 25, 1969; Brock Evans, telephone interview by the author, September 11, 2012.

18. Brower, "Environmental Activist, Publicist, and Prophet," 230–231.

19. Scott Thurber, "Brower Cut Off from Funds," *San Francisco Chronicle*, January 31, 1969; Ansel Adams to William Siri, January 12, 1969, Carton 124, Folder 26, Brower Papers; "Brower Called a Dictator," *San Francisco Examiner*, February 5, 1969; Richard Leonard, "Mountaineer, Lawyer, Environmentalist, Volume II," an oral history conducted in 1976 by Susan R. Schrepfer, ROHO, BL, 1976, 386.

20. Organizational and Internal Affairs, February 8–9, 1969, Sierra Club Minutes, 1–8.

21. Brower, journal, February 8–9, 1969, Box 21, Folder 3, Brower Papers.

22. Michael McCloskey, "Sierra Club Executive Director: The Evolving Club and the Environmental Movement, 1961–1981," an oral history conducted in 1981 by Susan Schrepfer, ROHO, BL, 1983, 90.

23. Organizational and Internal Affairs, February 8–9, 1969, Sierra Club Minutes, 1–8.

24. Peggy Wayburn, "Author and Environmental Activist," an oral history conducted in 1990 by Ann Lage, ROHO, BL, 1992, 97; Committee for an Active Bold Constructive Sierra Club, "Shall the Sierra Club Revert to Its Days as a Society of Companions on the Trail?" n.d., copy supplied to the author by George Alderson; "Executive Committee Recommendations," *Bonanza*, March 1969, copy supplied to the author by George Alderson. Some board members such as Richard Leonard and Philip Berry were not up for reelection because terms were staggered. The stakes were so high that several other candidates who were neutral or not part of the slates were asked to withdraw so that voters could have a clear choice.

25. Committee for an Active Bold Constructive Sierra Club, "Shall the Sierra Club Revert to Its Days as a Society of Companions on the Trail?"; CMC, "All Living Past Presidents of the Sierra Club Join with These Dedicated Members," n.d., copy supplied to the author by George Alderson.

26. Ansel Adams, "A Conversation with Ansel Adams," an oral history conducted in 1972, 1974, and 1975 by Ruth Teiser and Catherine Harroun, ROHO, BL, 1978, 687.

27. Eliot Porter to Edgar Wayburn, January 23, 1969, Carton 124, Folder 26, Brower Papers.

28. Brower to Henry Siegel (Brower's lawyer), January 20, 1969, Carton 20, Folder 26, Sierra Club Records, BL; Siegel to Ed Daubs, James McCracken, Maynard Munger, Jean Searle, Peggy Singer, Bob Van Allen, Anne Van Tyne, Al Donner, LaVerne Ireland, Mark Massie, Bill Olmsted, Jack G. Roof, Gene Turney, and Raymond J. Sherwin, January 22, 1969, Carton 20, Folder 26, Brower Papers; Richard Leonard to CMC, January 26, 1969, Carton 2, Folder 38, Sierra Club Members Papers, BANC MSS 71/295c, BL.

29. United Press International, "2 Sierra Club Factions Seek Control in Bitter Row," February 8, 1969, clipping, Carton 20, Folder 34, Brower Papers; "World Conservation Stressed by Author," *Seattle Times*, March 8, 1969; Norman Sklarewitz, "Conservationists Turn from Guarding Nature to Internal Warfare," *Wall Street Journal*, February 13, 1969.

30. Brower, journal, February 21–April 5, 1969, Box 21, Folder 3, Brower Papers. Brower campaign venues ranged from Fort Collins, Colorado, to Oakridge, Tennessee; Seattle; Eugene, Oregon; Chicago; San Diego, Los Angeles, Palo Alto, San Francisco, and Carmel, California; Poughkeepsie, New York; and Amherst, Massachusetts. The audience was often receptive, although groups such as the Lone Star chapter in Texas pointedly asked both camps to stay away because they had already heard enough.

31. Brower to the Sierra Club and the IRS, memorandum, March 19, 1969, Carton 20, Folder 13, Brower Papers; David Brower, "The CMC Christmas Party Charge," memorandum, March 19, 1969, Carton 20, Folder 13, Brower Papers; David Brower, response to Berry charge, memorandum, March 19, 1969, Carton 20, Folder 13, Brower Papers; David Brower, "Effort to Prevent Oil Disaster," press release, n.d., Carton 100, Folder 15, Sierra Club Members Papers.

32. Adams to Wayburn, January 12, 1969, Carton 124, Folder 26, Brower Papers; Bruce Kennedy to Sill, February 7, 1969, NC1260, Box 5, Sierra Club Records, 1957–1980, NC1260, Special Collections, University of Nevada, Reno (UNR); Thomas H. Jukes to CMC Steering Committee, n.d., Carton 83, Folder 21, Sierra Club Members Papers.

33. Richard Searle, "Grassroots Sierra Club Leader," an oral history conducted in 1976 by Paul Clark and included in "Southern Sierrans," ROHO, BL, 1976, 39.

34. Brower, "CMC Christmas Party Charge"; Sierra Club Board of Inquiry, transcript, October 19, 1968, Carton 19, Folder 44, Sierra Club Records, BL; Raffi Bedayn to concerned Sierra Club members, memorandum, February 28,

1969, Carton 83, Folder 17, Sierra Club Members Papers; CMC, "All Living Past Presidents of the Sierra Club Join with These Dedicated Members"; audit statements, 1963–1969, Carton 25, Folders 1–9, Sierra Club Records, BL. According to audit reports, the year-end deficit over the five years before the election in 1969, from 1963 to 1968, was $198,000, and the deficit for 1968 was $36,000.

35. Phillip Berry, interview by the author, Lafayette, Calif., April 24, 2013; the Great Hero Club newsletter, n.d., Carton 196, Folder 33, Sierra Club Members Papers.

36. Brock Evans, "Environmental Campaigner, from the Northwest Forests to the Halls of Congress," an oral history conducted in 1982 and 1984 by Ann Lage and included in "Building the Sierra Club's National Lobbying Program, 1967–1981," ROHO, BL, 1985, 87.

37. Patrick Goldsworthy to Wayburn, March 11, 1969, Box 5, Sierra Club Records, UNR; CMC, "Hugh Nash, *Sierra Club Bulletin* Editor Suspended," press release, March 17, 1969, Box 5, Sierra Club Records, UNR; "Nash Raps Sierra Club Critic," *San Francisco Examiner*, March 21, 1969; McCloskey interview, May 24, 2013.

38. Jackson J. Benson, *Wallace Stegner: His Life and Work* (New York: Viking, 1996), 334–335; Wallace Stegner, "Bitten by the Worm of Power," letter to the editor, *Palo Alto Times*, n.d., clipping, Carton 11, Folder 4, Sierra Club Records, BL. After Stegner calmed down, he regretted what he had written because he had great admiration for Brower's accomplishments. "But it was a question there where it did seem that the actual existence of the Sierra Club was at hazard, and I didn't think Dave had the temperance, as it were, to back off enough to save it" (Wallace Stegner, "The Artist an Environmental Advocate," an oral history conducted in 1982 by Ann Lage, ROHO, BL, 1983, 27).

39. Anne Brower, "Worm of Power Not Visible," letter to the editor, *Palo Alto Times*, February 27, 1969.

40. Kenneth Brower to Wallace Stegner, February 16, 1969, Carton 20, Folder 15, Brower Papers; Kenneth Brower, letter to the editor, *San Francisco Chronicle*, n.d., clipping, Carton 20, Folder 31, Brower Papers.

41. Anne Brower to Harriet, February 6, 1969, Carton 20, Folder 14, Brower Papers.

42. Anne Brower to Leonard, October 8, 1968, Carton 20, Folder 14, Brower Papers.

43. Anne Brower to Argonauts chapter, November 8, 1968, Carton 20, Folder 14, Brower Papers; Nancy Newhall to Anne Brower, November 18, 1968, Carton 20, Folder 14, Brower Papers.

44. Anne Brower to Marjorie, February 28, 1969, Carton 20, Folder 14, Brower Papers.

45. Barbara Brower, interview by the author, Portland, Ore., May 24, 2013; Leonard, "Mountaineer, Lawyer, Environmentalist, Volume II," 394.

46. David Brower, "Note," April 15, 1969, Carton 20, Folder 26, Brower Papers; Brower, comment on Sierra Club election, 1969, n.d. Carton 20, Folder 29, Brower Papers.

47. Kenneth F. Dickey Jr. (chairman, Judges of Election) to Berry, April 22, 1969, Carton 11, Folder 4, Sierra Club Records, BL; Brower, journal, April 16, 1969, Box 21, Folder 3, Brower Papers.

48. John Flannery, "Sierra Club Post," n.d., Carton 20, Folder 29, Brower Papers.

49. Paul and Elizabeth Wilson to Brower, April 17, 1969, Carton 20, Folder 14, Brower Papers; Dale Jones to Brower, n.d., Carton 20, Folder 4, Brower Papers; Walt Woodward, "Sierra Club Will Survive War," *Seattle Times*, April 18, 1969; "Sierra Elections Should Not Signal Retreat, Discord," *Louisville Courier-Journal*, April 21, 1969; Harold Gilliam, "The Sierra Club: Time for a Reappraisal," *San Francisco Chronicle-Examiner*, April 27, 1969.

50. Brower to Sierra Club Publications Reorganization Committee and board of directors, April 25, 1969, Carton 19, Folder 42, Brower Papers.

51. Leonard to CMC slate of Sierra Club board of directors, April 20, 1969, Carton 83, Folder 21, Sierra Club Members Papers; Leonard, "Mountaineer, Lawyer, Environmentalist, Volume II," 350–352.

52. Brower to Ken Brower, April 25, 1969, Carton 20, Folder 29, Brower Papers.

53. John McPhee, *Encounters with the Archdruid* (New York: Farrar, Straus and Giroux, 1971), 209–218; James W. Moorman, "Attorney for the Environment 1966–1981, Center for Law and Social Policy, Sierra Club Legal Defense Fund, Department of Justice, Division of Lands and Natural Resources," an oral history conducted in 1984 by Ted Hudson, ROHO, BL, 1994, 34.

54. Quoted in McPhee, *Encounters with the Archdruid*, 215.

55. Appendix D, David Brower's Statement, May 3–4, 1969, Sierra Club Minutes, 45–46.

56. Richard Leonard, "Mountaineer, Lawyer, Environmentalist, Volume I," an oral history conducted in 1976 by Susan R. Schrepfer, ROHO, BL, 1976, 154; Philip Hager, "Executive Head of Sierra Club Resigns Post," *Los Angeles Times*, May 4, 1969; Ed Arnow, "David Brower Farewell Speech at Sierra Club," KPIX, San Francisco, May 3, 1969, http://diva.sfsu.edu/collections/sfbatv/bundles/189379 (accessed December 10, 2013).

57. Executive Director's Report and Resignation, May 3–4, 1969, Sierra Club Minutes, 6–9; Hager, "Executive Head of Sierra Club Resigns Post"; "1969 Annual Banquet, Claremont Hotel, Berkeley, May 3," *Sierra Club Bulletin*, May 1969, 4.

58. Gary Soucie (former employee, Sierra Club), interview by the author, Williamstown, Mass., October 26, 2012; Evans telephone interview, September 11, 2012; Evans, "Environmental Campaigner," 88.

59. Rudden to Brower, July 30, 1969, Carton 20, Folder 1, Brower Papers; August Fruge, "London Office of the Sierra Club," memorandum to Executive Committee, Publications Committee, Ted Wilentz (publications director, Sierra Club), and Rudden, May 13, 1969, Carton 49, Folder 4, Sierra Club Records, BL.; Soucie interview, October 26, 2012; Jeffrey Ingram, telephone interview by the author, November 12, 2012; audit statements, 1963–1989, Carton 25, Folders 1–9, Sierra Club Records, BL. The Sierra Club finances began to improve in 1972, according to the audit reports showing end-of-year debits or credits: 1969, –$119,000; 1970, $84,000; 1971, –$469,000; 1972, $89,000.

60. Evans, "Environmental Campaigner," 88.

61. Kelly Duane, *Monumental: David Brower's Fight for Wild America*, DVD (KEET-TV, First Run Features, 2004); Christopher Reed, "David Brower, Environmental Champion Whose Passion Founded Friends of the Earth," *Guardian* (London), November 7, 2000; Richard Severo, "David Brower, an Aggressive Champion of U.S. Environmentalism, Is Dead at 88," *New York Times*, November 7, 2000; Adam Bernstein, "David Brower Dies; Transformed Sierra Club into Powerful Force," *Washington Post*, November 7, 2000; "Friend of the Earth," *St. Louis Post-Dispatch*, November 9, 2000; David Balzar, "David Brower, Crusader for the Environment, Dies at 88," *Los Angeles Times*, November 7, 2000. Some reports on Brower did attempt to provide a more balanced account of his dismissal, such as the *Los Angeles Times* report.

62. Ann Brower to Marjorie, February 28, 1969, Carton 20, Folder 14, Brower Papers.

63. Michael P. Cohen, *The History of the Sierra Club, 1892–1970* (San Francisco: Sierra Club Books, 1988), 435–459.

64. Brower, "Environmental Activist, Publicist, and Prophet," 249.

65. David Brower, with Steve Chapple, *Let the Mountains Talk, Let the Rivers Run: A Call to Those Who Would Save the Earth* (San Francisco: HarperCollins West, 1995), 193.

17. Echoes

1. David R. Brower, "Environmental Activist, Publicist, and Prophet," an oral history conducted in 1974–1978 by Susan Schrepfer, Regional Oral History Office (ROHO), Bancroft Library (BL), University of California, Berkeley, 1979, 263.

2. David Brower, comment on Joe Browder memorandum to the Friends of the Earth board of directors and Advisory Council, March 7, 1972, Carton 67, Folder 3,

David Ross Brower Papers, BANC MSS 79/9c, BL; Joseph Browder, telephone interview by the author, November 28, 2012; Brent Blackwelder, telephone interview by the author, January 3, 2013; David Brower, note, September 11, 1976, Carton 86, Folder 24, Brower Papers. Two significant quarrels in 1972 should have served as a warning. One was between Brower and Joe Browder (not to be confused with David's brother Joseph) and much of the Washington, D.C., staff. It was nasty. Brower accused Browder of insurrection and disloyalty. Browder accused Brower of borrowing from a young widow who worked in the Washington office and refusing to repay the money that had come from her late husband's insurance policy. Browder eventually took the entire staff and created a rival environmental lobbying organization, the Environmental Policy Center. The split with Max Linn and the John Muir Institute was even worse, not just because of the vicious exchanges but because the money was so intertwined. How could anyone conceive of Brower sharing power with someone else? Although the arrangement collapsed in 1972, the dispute lingered for years as the two sides fought over money.

3. Quoted in Lawrence E. Davies, "Naturalists Get a Political Arm," *New York Times*, September 17, 1969.

4. David R. Brower, *For Earth's Sake: The Life and Times of David R. Brower* (Salt Lake City: Peregrine Smith Books, 1990), 487; Frank Graham, "David Brower: Last of the Optimists?" *Audubon*, September 1982, 62–73. Brower had hoped to obtain more funding from Anderson. After Friends of the Earth (FOE) and other environmentalists opposed the Alaska Pipeline, however, that did not happen because the Atlantic Richfield Company (ARCO) was a major principal in the pipeline project.

5. Kenneth Brower, *The Wildness Within: Remembering David Brower* (Berkeley, Calif.: Heyday, 2012), 122; Bill Devall, "David Brower," *Environmental Review*, Autumn 1985, 243–244; David Brower to Daniel Patrick Moynihan (White House aide), August 12, 1969, Carton 86, Folder 26, Brower Papers.

6. Angela Taylor, "Fur Coats: Facing Extinction at Conservationists' Hand?" *New York Times*, December 30, 1969; Tom Brown, "Alaskan Interview," *Not Man Apart*, January 1971, 12–14; David Brower, "Should Atomic Energy Go Back to the Drawing Board?" *Not Man Apart*, August 1972, 2–8; David Brower and Jim Harding, "Horsepower Sense," *Not Man Apart*, June 1974, 1–7; Catherine Smith, "Carter's Alaska Move: 100 Million Acres Saved," *Not Man Apart*, January–February 1979, 1; Jerry Emory, "Condor Recovery: Hands-on or Hands-off?" *Not Man Apart*, November 1980, 15–17; "Update: Dave Brower, the Environmentalist and the Bomb," *Mother Earth News*, September–October 1982, 94–96; David R. Brower, "It's Healing Time on Earth," *Earth Island Journal*, Winter 1980, 51–52; Stephanie Mills, *Whatever Happened to Ecology?* (San Francisco: Sierra Club

Books, 1989), 111. The pages of the FOE publication *Not Man Apart* provide a detailed summary of Brower's activities and concerns in the 1970s and early 1980s.

7. Brower, "Environmental Activist, Publicist, and Prophet," 289.

8. "An ECO from the Third World," *Not Man Apart*, May 1974, 7.

9. Edwin Matthews, "Dreams Can Begin It," *Not Man Apart*, July 1979, 8–9, 14–15; David Chatfield, "That's FOE All Over," *Not Man Apart*, January–February 1979, 8–9.

10. Steve Leahy (Canadian journalist), telephone interview by the author, December 18, 2012; Mark Tamhane (Australian journalist), e-mail correspondence with the author, January 5, 2013; *Friends of the Earth International at 40* (Amsterdam: Friends of the Earth International, 2011), http://www.foei.org/wp-content/uploads/2011/06/10-foei-at-40-lr.pdf (accessed October 6, 2015).

11. Blackwelder telephone interview, January 3, 2013.

12. John McPhee, *Encounters with the Archdruid* (New York: Farrar, Straus and Giroux, 1971), 79–84.

13. K. Brower, *Wildness Within*, 160–163.

14. "FOE's Branches," *Not Man Apart*, December 1984, 6; Ed Dobson, telephone interview by the author, June 14, 2010; Boyd Norton, telephone interview by the author, January 16, 2013; Mills, *Whatever Happened to Ecology?* 111.

15. Avis Ogilvy Moore, telephone interview by the author, February 19, 2013.

16. Brower, "Environmental Activist, Publicist, and Prophet," 328–333; William Hogan, "What Makes David Run?" n.d., Carton 64, Folder 40, Brower Papers; "FOE Book May Backfire on Maui," *Not Man Apart*, February 1979, 4; David R. Brower, "Headlands," *Not Man Apart*, September 1976, 7–10. See also Garrett DeBell, ed., *The Environmental Handbook* (New York: Ballantine and Friends of the Earth Books, 1970); Kenneth Brower and Robert Wenkam, *Micronesia: Island Wilderness* (San Francisco: Friends of the Earth and Seabury Press, 1973); Robert Wenkam, *Maui: The Last Hawaiian Place* (San Francisco: Friends of the Earth, 1971); Kenneth Brower, with photographs by William R. Curtsinger, *The Wake of the Whale* (New York: Dutton, 1979); and Kenneth Brower, ed., *Guale, the Golden Coast of Georgia* (San Francisco: Friends of the Earth, 1974).

17. Joe Kane, "The Friends of David Brower," *Outside*, December 1984, 39.

18. Brower, FOE publications program memorandum, n.d., Carton 64, Folder 6, Brower Papers; Kane, "The Friends of David Brower," 40; summary of FOE's financial history, memorandum, n.d., Carton 76, Folder 36, Brower Papers.

19. Moore interview, February 19, 2013.

20. Dan Gabel, telephone interview by the author, January 10, 2013.

21. Jeff Knight (FOE employee), "Loans as a Source of Cash," memorandum, June 5, 1984, Carton 76, Folder 29, Brower Papers; Brower to Maitland A. Edey (FOE lender), January 9, 1974, Carton 76, Folder 29, Brower Papers; FOE loans,

n.d., Carton 76, Folder 29, Brower Papers. The spreadsheet of FOE loans listed 270 that were outstanding, including 15 that were for $5,000 or more, one of them for $10,000, and another for $20,000. It indicated that Anne H. Brower was owed $1,500, but this item was crossed out, and the amount was reduced to $450. Knight's June 1984 memorandum indicates that loans were a funding mainstay and that the organization was staying afloat financially because appeals to the membership to renew their loans usually had an 80 percent success rate.

22. Blackwelder telephone interview, January 3, 2013.

23. Wayne Steiger (president, Flow Pay Inc.), telephone interview by the author, March 5, 2013; David Stolow (faculty director, Public and Nonprofit Management Program, Boston University), e-mail correspondence with the author, March 1, 2013.

24. Brower to FOE Foundation board, November 31, 1984 [sic], Carton 76, Folder 1, Brower Papers. Brower estimated the amount he had personally contributed since 1975 to be $460,000, including $210,000 in honoraria and royalties, $175,000 in nonreimbursed expenses, $35,000 in losses related to the John Muir Institute, and $39,000 in other contributions.

25. Moore interview, February 19, 2013.

26. Tom Turner and David Chatfield, "Brower Still on Board," *Not Man Apart*, July–August 1984, 7; Philip Shabecoff, "Conservation Figure Ousted for Resisting Orders to Cut Staff," *New York Times*, July 7, 1984.

27. Paul Rauber, "With Friends Like These," *Mother Jones*, November 1986, 35–37, 47–49; Harold Gilliam, "David Brower, Down but Not Out," n.d., Brower File, Yosemite Park Museum and Library, Yosemite National Park, El Portal, Calif.

28. Tom Turner, "From Heresy to Conventional Wisdom at Blinding Speed: A History of Earth Island Institute's 25 Years," *Earth Island Journal*, Spring 2008, http://www.earthisland.org/journal/index.php/eij/article/from_heresy_to_conventional_wisdom_at_blinding_speed/ (accessed January 27, 2010); Norton telephone interview, January 16, 2013; David R. Brower, *Work in Progress* (Salt Lake City: Peregrine Smith Books, 1991), 340–343.

29. Quoted in Ellen Chrismer, "Esteemed Leader Has Quit," *Modesto Bee*, May 20, 2000.

30. Bruce Hamilton, interview by the author, San Francisco, August 7, 2012; Brock Evans, telephone interview by the author, September 11, 2012.

31. David Brower, press release, n.d., Carton 23, Folder 12, Brower Papers; Michael McCloskey, *In the Thick of It: My Life in the Sierra Club* (Washington, D.C.: Island Press, 2005), 348–349; Eric Brazil, "Sierra Club's Squabble over Nuclear Arms," *San Francisco Examiner*, July 14, 1986; Michelle Perrault to M. Stong, D. Willigan, S. Rauh, D. Moses, and D. Brower, memorandum, February 7, 1986, Carton 22, Folder 2, Brower Papers.

32. David Brower, "Let the River Run Through It," *Sierra*, March–April 1997, 42–44; Hamilton interview, August 7, 2012; Carl Pope (former executive director, Sierra Club), telephone interview by the author, July 8, 2013.

33. John Woolefenden, "Brower on Population Bomb," *Monterey Peninsula Herald* (Monterey, Calif.), October 15, 1968; Paul Rogers, "Border Control Proposal Rejected," *San Jose Mercury News*, April 26, 1998 (quote from Brower); Alan Kruper (affirmation coordinator, Sierra Club) to Chris Rankin (Sierra Club), memorandum, November 1, 1997, Carton 23, Folder 32, Brower Papers; Carl Pope to Brower, November 18, 1997, Carton 23, Folder 32, Brower Papers.

34. Brower to Perry Knowlton (president, Curtis-Brown Literary Agency), January 13, 1982, Carton 1, Folder 3, Brower Papers; Thomas A. Bisdale (general counsel, Hearst Corp.) to Knowlton, March 4, 1985, Carton 1, Folder 3, Brower Papers; Madge Baird (editorial director, Peregrine Smith Books) to Brower, March 3, 1987, Carton 1, Folder 4, Brower Papers.

35. Roger B. Swain, "The Archdruid Himself," *New York Times*, March 20, 1990.

36. Michael Fischer (executive director, Sierra Club), "Publication of the History of the Sierra Club," memorandum to the Sierra Club board of directors, August 24, 1988, Carton 24, Folder 30, Brower Papers; Ken Brower to Jim Cohee (senior editor, Sierra Club Books), September 3, 1988, Carton 24, Folder 31, Brower Papers; Richard A. Cellarius (president, Sierra Club board of directors) to Brower, September 17, 1988, Carton 24, Folder 31, Brower Papers; David R. Brower, "Reflections on the Sierra Club, Friends of the Earth, and Earth Island Institute," an oral history conducted in 1999 by Ann Lage, ROHO, BL, 2012, 153–154. Fischer's memo to the board attached a three-page letter from Cohen explaining why he was not making further changes to the book and seventeen pages of notes from Brower alleging errors in the draft. Ken Brower's letter contained another eight pages of notes alleging errors.

37. McPhee, *Encounters with the Archdruid*, 216–217; Brower, "Environmental Activist, Publicist, and Prophet," 241; Daniel Coyle, "The High Cost of Being David Brower," *Outside*, December 1995, http://outsideonline.com/outdoor-adventure/The-High-Cost-of-Being-David-Brower.html?page=all (accessed January 9, 2013).

38. Brower, "Reflections on the Sierra Club," 71.

39. Ansel Adams to Brower, July 28, 1975, Box 3, Folder 4, Brower Papers; Brower, "Environmental Activist, Publicist, and Prophet," 192; Coyle, "High Cost of Being David Brower."

40. Brower, "Environmental Activist, Publicist, and Prophet," 283–284; Barbara Brower, interview by the author, Portland, Ore., May 24, 2013.

41. Quoted in Coyle, "High Cost of Being David Brower."

42. Barbara Brower to Karin Marie Kunoichi, e-mail, September 13, 1999, Carton 125, Folder 15, Brower Papers; "Environmental Pioneer Hit by Heart Trouble," *San Francisco Examiner*, April 26, 1995; Earth Island Institute, "David R. Brower and Prof. Paul Ehrlich Jointly Nominated for 1998 Nobel Peace Prize," press release, February 7, 1998, Carton 125, Folder 13, Brower Papers; Asahi Glass Foundation, "1998 Blue Planet Prize," press release, n.d., http://www.af-info.or.jp/en/blueplanet/list.html (accessed December 12, 2013); Brower, note, November 11, 1986, Carton 22, Folder 5, Brower Papers.

43. Quoted in Mikhail Davis, "Obituary: Anne Hus Brower (1913–2001)," *Berkeley Daily Planet*, November 28, 2001.

44. Mikhail Davis, e-mail, July 20, 2000, Carton 125, Folder 24, Brower Papers; Davis, e-mail, August 20, 2000, Carton 125, Folder 25, Brower Papers; Davis, e-mail, July 4, 2000, Carton 125, Folder 27, Brower Papers; Eric Brazil, "Environment Pioneer Brower Energizes S.F. State Students," *San Francisco Examiner*, April 28, 2000; Glen Martin, "Longtime Sierra Club Leader Quits," *San Francisco Chronicle*, May 19, 2000; Davis to Stephanie Milburn, e-mail, August 2, 2000, Carton 125, Folder 25, Brower Papers.

45. David Balzar, "David Brower, Crusader for the Environment, Dies at 88," *Los Angeles Times*, November 7, 2000; Mark Dowie, *Losing Ground: American Environmentalism at the Close of the Twentieth Century* (Cambridge, Mass.: MIT Press, 1995), 105.

46. Tom Turner, "David R. Brower, July 1, 1912–November 5, 2000," copy supplied to the author by Stephanie Mills.

47. David R. Brower, "Brower on the Yosemite Valley Plan," *San Francisco Chronicle*, November 20, 2002.

Epilogue

1. Mark Dowie, "I Intend to Keep Shouting," *Men's Journal*, February 2001, 80–81.

2. Benjamin Kline, *First Along the River: A Brief History of the U.S. Environmental Movement* (Lanham, Md.: Rowman & Littlefield, 2011), 89–96.

3. G. Jon Roush, "Conservation's Hour—Is Leadership Ready?" in *Voices from the Environmental Movement: Perspectives for a New Era*, ed. Donald Snow (Washington, D.C.: Island Press, 1992), 25; Katherine Bagley, "Field Guide to the Modern U.S. Environmental Movement," *Inside Climate Change*, April 7, 2014, http://insideclimatenews.org (accessed March 12, 2015); National Center for Charities Statistics, "Public Charities: NTEE = C (Environment)," 2013, http://nccs.urban.org/database/overview.cfm (accessed March 12, 2015).

4. Gallup, "Climate Change, Environment" (1989–2015), n.d., http://gallup. com/pll/1615/environment.aspx (accessed May 29, 2015); Justin McCarthy, "About Half in U.S. Say Environmental Protection Falls Short," *Gallup Poll Briefing*, April 9, 2015, 1; Kline, *First Along the River*, 180–183; Roush, "Conservation's Hour," 22–23; "United States: Endangered Species, the Environmental Movement," *Economist*, February 18, 2006, 50.

5. Roush, "Conservation's Hour," 26.

6. Cited in Donald Snow, *Inside the Environmental Movement: Meeting the Leadership Challenge* (Washington, D.C.: Island Press, 1992), 3.

7. Nathaniel P. Reed, "The Conservation Movement as a Political Force," in Snow, *Voices from the Environmental Movement*, 45.

8. Roush, "Conservation's Hour," 35.

9. Robert Gottlieb, *Forcing the Spring: The Transformation of the American Environmental Movement* (Washington, D.C.: Island Press, 1993), 105–108; Denis Hayes, interview by the author, Chattanooga, Tenn., October 16, 2012.

10. Snow, *Inside the Environmental Movement*, xvii; Nature Conservancy, "About Us," n.d., http://www.nature.org/about-us/index.htm (accessed April 3, 2015).

11. Mark Dowie, *Losing Ground: American Environmentalism at the Close of the Twentieth Century* (Cambridge, Mass.: MIT Press, 1995), index.

12. Dowie, "I Intend to Keep Shouting."

13. Quoted in Richard Severo, "David Brower, an Aggressive Champion of U.S. Environmentalism, Is Dead," *New York Times*, November 7, 2000.

Bibliography

Books

Abbey, Edward. *The Monkey Wrench Gang*. 1975. Reprint. New York: Perennial, 2000.

Adams, Ansel. *These We Inherit, the Parkland of America*. San Francisco: Sierra Club Books, 1962.

———. *This Is the American Earth*. San Francisco: Sierra Club Books, 1960.

Allin, Craig W. *The Politics of Wilderness Preservation*. Westport, Conn.: Greenwood Press, 1982.

Atkinson, Brooks, and W. Kent Olson. *New England's White Mountains: At Home in the Wild*. San Francisco: Friends of the Earth International, 1979.

Benson, Jackson J. *Wallace Stegner: His Life and Work*. New York: Viking, 1996.

Bohn, David. *Glacier Bay, the Land and the Silence*. San Francisco: Sierra Club Books, 1967.

Brooks, Karl B. *Public Power, Private Dams: The Hells Canyon High Dam Controversy*. Seattle: University of Washington Press, 2009.

Brower, David R. *For Earth's Sake: The Life and Times of David R. Brower*. Salt Lake City: Peregrine Smith Books, 1990.

———, ed. *Gentle Wilderness: The Sierra Nevada*. San Francisco: Sierra Club Books, 1964.

———, ed. *The Meaning of Wilderness to Science*. San Francisco: Sierra Club Books, 1960.

———, ed. *Not Man Apart*. San Francisco: Sierra Club Books, 1965.

———, ed. *The Sierra Club Wilderness Handbook*. San Francisco: Sierra Club Books, 1967.

———, ed. *Wilderness: America's Living Heritage*. San Francisco: Sierra Club Books, 1961.

———, ed. *Wildlands in Our Civilization*. San Francisco: Sierra Club Books, 1964.

——. *Work in Progress*. Salt Lake City: Peregrine Smith Books, 1991.

Brower, David, with Steve Chapple. *Let the Mountains Talk, Let the Rivers Run: A Call to Those Who Would Save the Earth*. San Francisco: HarperCollins West, 1995.

Brower, Kenneth. *Earth and the Great Weather: The Brooks Range*. San Francisco: Friends of the Earth and Seabury Press, 1971.

——, ed. *Guale, the Golden Coast of Georgia*. San Francisco: Friends of the Earth, 1974.

——. *The Wildness Within: Remembering David Brower*. Berkeley, Calif.: Heyday, 2012.

Brower, Kenneth, with photographs by William R. Curtsinger. *The Wake of the Whale*. New York: Dutton, 1979.

Brower, Kenneth, ed., and Eliot Porter. *Galápagos*. Vol. 1, *Discovery*. San Francisco: Sierra Club Books, 1968.

——. *Galápagos*. Vol. 2, *Prospect*. San Francisco: Sierra Club Books, 1968.

Brower, Kenneth, and Robert Wenkam. *Micronesia: Island Wilderness*. San Francisco: Friends of the Earth and Seabury Press, 1973.

Carson, Donald W., and James W. Johnson. *Mo: The Life and Times of Morris K. Udall*. Tucson: University of Arizona Press, 2001.

Cohen, Michael P. *The History of the Sierra Club, 1892–1970*. San Francisco: Sierra Club Books, 1988.

Cornell, Virginia. *Defender of the Dunes: The Kathleen Goddard Jones Story*. Carpinteria, Calif.: Manifest, 2001.

Cosco, Jon M. *Echo Park: Struggle for Preservation*. Boulder, Colo.: Johnson Books, 1995.

DeBell, Garrett, ed. *The Environmental Handbook*. New York: Ballantine and Friends of the Earth Books, 1970.

Desmann, Raymond F. *The Destruction of California*. New York: Macmillan, 1965.

Douglas, William O. *My Wilderness: The Pacific West*. New York: Doubleday, 1960.

——. *Of Men and Mountains*. New York: Harper, 1950.

Dowie, Mark. *Losing Ground: American Environmentalism at the Close of the Twentieth Century*. Cambridge, Mass.: MIT Press, 1995.

Duberman, Martin, Martha Vicinus, and George Chauncey Jr., eds. *Hidden from History: Reclaiming the Gay and Lesbian Past*. New York: New American Library, 1989.

Ehrlich, Paul. *The Population Bomb*. San Francisco: Sierra Club Books, 1969.

Flippen, J. Brooks. *Conservative Conservationist: Russell E. Train and the Emergence of American Environmentalism*. Baton Rouge: Louisiana State University Press, 2006.

Foreman, Dave. *Confessions of an Eco-Warrior*. New York: Harmony Books, 1991.

Gottlieb, Robert. *Forcing the Spring: The Transformation of the American Environmental Movement*. Washington, D.C.: Island Press, 1993.

Gussow, Alan. *A Sense of Place: The Artist and the American Land*. San Francisco: Friends of the Earth Books, 1972.

Halberstam, David. *The Fifties*. New York: Villard Books, 1993.

Harvey, Mark W. T. *A Symbol of Wilderness: Echo Park and the American Conservation Movement*. Albuquerque: University of New Mexico Press, 1994.

——. *Wilderness Forever: Howard Zahniser and the Path to the Wilderness Act*. Seattle: University of Washington Press, 2005.

Hirt, Paul W. *A Conspiracy of Optimism: Management of the National Forests Since World War Two*. Lincoln: University of Nebraska Press, 1994.

Hornbein, Thomas F., and Norman G. Dyhrenfurst. *Everest: The West Ridge*. San Francisco: Sierra Club Books, 1965.

Hyde, Philip, and François Leydet. *The Last Redwoods*. San Francisco: Sierra Club Books, 1964.

Jett, Stephen C., and Philip Hyde. *Navajo Wildlands: As Long as the Rivers Shall Flow*. San Francisco: Sierra Club Books, 1967.

Johnson, Rich. *The Central Arizona Project, 1918–1968*. Tucson: University of Arizona Press, 1977.

Johnston, Mirelle. *Central Park Country: A Tune Within Us*. San Francisco: Sierra Club Books, 1968.

Kauffman, Richard. *Headlands*. San Francisco: Friends of the Earth, 1976.

——. *The Primal Alliance: Earth and Ocean*. San Francisco: Seabury Press and Friends of the Earth Books, 1974.

Kilgore, Bruce, ed. *Wilderness in a Changing World*. San Francisco: Sierra Club Books, 1966.

Kline, Benjamin. *First Along the River: A Brief History of the U.S. Environmental Movement*. Lanham, Md.: Rowman & Littlefield, 2011.

Knight, Max. *Return to the Alps*. San Francisco: Friends of the Earth Books and Seabury Press, 1974.

Krutch, Joseph Wood, and Eliot Porter. *Baja California and the Geography of Hope*. San Francisco: Sierra Club Books, 1967.

Lane, L. W., Jr. *The Sun Never Sets: Reflections on a Western Life*. Stanford, Calif.: Stanford University Press, 2013.

Leydet, François. *Time and the River Flowing: Grand Canyon*. Edited by David Brower. San Francisco: Sierra Club Books, 1964.

——, ed. *Tomorrow's Wilderness*. San Francisco: Sierra Club Books, 1963.

Lien, Carsten. *Olympic Battleground: The Power Politics of Timber Preservation*. San Francisco: Sierra Club Books, 1991.

Lovins, Amory. *Eryri: The Mountains of Longing*. New York: McCall, 1972.

Manning, Harvey. *The Wild Cascades: Forgotten Parkland.* San Francisco: Sierra Club Books, 1965.

Martin, Russell. *A Story That Stands Like a Dam: Glen Canyon and the Struggle for the Soul of the West.* New York: Holt, 1989.

McCloskey, Maxine, ed. *Wilderness: The Edge of Knowledge.* San Francisco: Sierra Club Books, 1969.

McCloskey, Maxine, and James P. Gilligan, eds. *Wilderness and the Quality of Life.* San Francisco: Sierra Club Books, 1969.

McCloskey, Michael. *In the Thick of It: My Life in the Sierra Club.* Washington, D.C.: Island Press, 2005.

McPhee, John. *Encounters with the Archdruid.* New York: Farrar, Straus and Giroux, 1971.

Miles, John C. *Guardians of the Park: A History of the National Parks and Conservation Foundation.* Washington, D.C.: Taylor & Francis, 1995.

Miller, Char. *Gifford Pinchot and the Making of Modern Environmentalism.* Washington, D.C.: Island Press, 2001.

Miller, Char, and Hal Rothman, eds. *Out of the Woods: Essays in Environmental History.* Pittsburgh: University of Pittsburgh Press, 1997.

Mills, Stephanie. *Whatever Happened to Ecology?* San Francisco: Sierra Club Books, 1989.

Nash, Hugh. *Progress as If Survival Mattered.* San Francisco: Friends of the Earth Books, 1982.

Nash, Roderick, ed. *Grand Canyon of the Living Colorado.* San Francisco: Sierra Club and Ballantine Books, 1970.

——. *Wilderness and the American Mind.* New Haven, Conn.: Yale University Press, 1967.

Newhall, Nancy. *Ansel Adams.* Vol. 1, *The Eloquent Light.* San Francisco: Sierra Club Books, 1963.

Pearson, Byron E. *Still the Wild River Runs: Congress, the Sierra Club, and the Fight to Save Grand Canyon.* Tucson: University of Arizona Press, 2002.

Peattie, Roderick, ed. *The Cascades: Mountains of the Pacific Northwest.* New York: Vanguard, 1949.

Penalosa, Fernando. *Yosemite in the 1930s: A Remembrance.* Rancho Palos Verdes, Calif.: Quaking Aspen Book, 2002.

Porter, Eliot. *The Place No One Knew: Glen Canyon on the Colorado.* Edited by David Brower. San Francisco: Sierra Club Books, 1963.

——. *Summer Island: Penobscot Country.* San Francisco: Sierra Club Books, 1966.

Powell, James Lawrence. *Dead Pool, Lake Powell, Global Warming, and the Future of Water in the West.* Berkeley: University of California Press, 2008.

Reisner, Marc. *Cadillac Desert: The American West and Its Disappearing Water.* New York: Viking Penguin, 1986.

Robinson, Glen O. *The Forest Service: A Study in Public Land Management.* Baltimore: Johns Hopkins University Press, 1975.

Robinson, John. *San Gorgonio: A Wilderness Preserved.* San Bernardino, Calif.: San Gorgonio Volunteers Association, 1991.

Rome, Adam. *The Genius of Earth Day: How a 1970 Teach-in Unexpectedly Made the First Green Generation.* New York: Hill and Wang, 2013.

Russell, Renny. *Rock Me on the Water: A Life on the Loose.* Questa, N.M.: Animist Press, 2007.

Russell, Terry, and Renny Russell. *On the Loose.* San Francisco: Sierra Club Books, 1967.

Schneider, Paul. *The Adirondacks: A History of America's First Wilderness.* New York: Holt, 1997.

Schrepfer, Susan. *The Fight to Save the Redwoods: A History of Environmental Reform, 1917–1978.* Madison: University of Wisconsin Press, 1983.

Schwartz, William, ed. *Voices for the Wilderness.* San Francisco: Sierra Club Books, 1969.

Scott, Doug. *The Enduring Wilderness: Protecting Our Natural Heritage Through the Wilderness Act.* Golden, Colo.: Fulcrum, 2004.

Sellars, Richard West. *Preserving Nature in the National Parks: A History.* New Haven, Conn.: Yale University Press, 1997.

Shabecoff, Philip. *A Fierce Green Fire: The American Environmental Movement.* New York: Hill and Wang, 1993.

Snow, Donald. *Inside the Environmental Movement: Meeting the Leadership Challenge.* Washington, D.C.: Island Press, 1992.

——, ed. *Voices from the Environmental Movement: Perspectives for a New Era.* Washington, D.C.: Island Press, 1992.

Souder, William. *On a Farther Shore: The Life and Legacy of Rachel Carson, Author of "Silent Spring."* New York: Crown, 2012.

Spaulding, Jonathan. *Ansel Adams and the American Landscape: A Biography.* Berkeley: University of California Press, 1995.

Starr, Walter. *Guide to the John Muir Trail and the High Sierra Region.* San Francisco: Sierra Club Books, 1943.

Steen, Harold K. *The U.S. Forest Service: A History.* Seattle: University of Washington Press, 1976.

Stegner, Wallace. *Beyond the Hundredth Meridian: John Wesley Powell and the Second Opening of the West.* New York: Penguin, 1953.

——, ed. *This Is Dinosaur: Echo Park Country and Its Magic Rivers.* New York: Knopf, 1955.

Sturgeon, Stephen C. *The Politics of Western Water: The Congressional Career of Wayne Aspinall.* Tucson: University of Arizona Press, 2002.

Taylor, Joseph E., III. *Pilgrims of the Vertical: Yosemite Rock Climbers and Nature at Risk.* Cambridge, Mass.: Harvard University Press, 2010.

Thoreau, Henry David, with photographs by Eliot Porter. *In Wildness Is the Preservation of the World.* San Francisco: Sierra Club Books, 1962.

Turner, Tom. *David Brower: The Making of the Environmental Movement.* Berkeley: University of California Press, 2015.

———. *Sierra Club: 100 Years of Protecting Nature.* New York: Abrams, 1991.

Udall, Stewart L. *The Quiet Crisis and the Next Generation.* Layton, Utah: Gibbs Smith, 1988.

Wayburn, Edgar, and Allison Alsup. *Your Land and Mine: Evolution of a Conservationist.* San Francisco: Sierra Club Books, 2004.

Webb, Roy. *Riverman: The Story of Bus Hatch.* Rock Springs, Wyo.: Labyrinth Press, 1989.

Wenkam, Robert. *Kauai and the Park Country of Hawaii.* San Francisco: Sierra Club Books, 1967.

———. *Maui: The Last Hawaiian Place.* San Francisco: Friends of the Earth, 1971.

Wills, John. *Conservation Fallout: Nuclear Protest at Diablo Canyon.* Reno: University of Nevada Press, 2006.

Wirth, Conrad L. *Parks, Politics, and the People.* Norman: University of Oklahoma Press, 1980.

Wright, Cedric. *Words of the Earth.* San Francisco: Sierra Club Books, 1960.

Articles

"Actions of Board at December Meeting." *Sierra Club Bulletin,* January 1969, 5.

Adams, Ansel. "Tenaya Tragedy." *Sierra Club Bulletin,* November 1958, 1–13.

Adams, J. Donald. "Speaking of Books." *New York Times,* May 15, 1960.

"Adams Asks Ouster of Chief." *Monterey Peninsula Herald* (Monterey, Calif.), September 25, 1968.

"Agreement on Redwoods Park Plan." *San Francisco Chronicle,* October 20, 1967.

Anderton, Trish. "Soldiers Who Loved to Ski." *Appalachia,* Summer–Fall 2014, 64–70.

"Aspinall Gives Views on Wilderness Bill." *Grand Junction (Colo.) Daily Sentinel,* June 3, 1963.

"Aspinall Hits Sierra Club Lobby Tactics." *Grand Junction Daily Sentinel,* July 6, 1966.

Associated Press. "EPA Is Said to Plan a Curb on Dirty Air at Grand Canyon." *New York Times,* February 1, 1991.

"Atomic Plant Site Stirs Wide Coast Conservation Controversy." *New York Times,* April 4, 1967.

"Audubon Group Hears Brower." *Arizona Daily Star* (Tucson), November 10, 1964.

Avery, Ben. "Magazine's Motives Cause for Alarm." *Arizona Republic*, March 27, 1966.

"Background on Nipomo Dunes–Diablo Canyon Issue." *Sierra Club Bulletin*, February 1967, 12.

Balzar, David. "David Brower, Crusader for the Environment, Dies at 88." *Los Angeles Times*, November 7, 2000.

"Battle of the Wilderness." *Newsweek*, October 3, 1966, 108–108B.

Beier, Bob. "Aspinall Raps Opposition to Project." *Albuquerque Tribune*, November 18, 1966.

Bernstein, Adam. "David Brower Dies; Transformed Sierra Club into Powerful Force." *Washington Post*, November 7, 2000.

Bess, Donovan. "A New Attack by Sierra Club." *San Francisco Chronicle*, July 2, 1966.

Blakeslee, Sandra. "Drought Unearths a Buried Treasure." *New York Times*, November 2, 2004.

Blair, William M. "Feud Threatens Wilderness Bill." *New York Times*, March 1, 1963.

———. "Vast Water Plan Outlined by U.S." *New York Times*, January 22, 1963.

Bliven, Naomi. "Books." *New Yorker*, March 12, 1966, 173.

Bradley, Harold C. "Danger to Dinosaur." *Pacific Discovery*, January–February 1954, 5.

Bradley, Harold C., and David R. Brower. "Roads in the National Parks." *Sierra Club Bulletin*, June 1949, 31–54.

Bradley, Richard C. "Ruin for the Grand Canyon?" *Audubon*, January–February 1966, 34–41.

Brazil, Eric. "Environment Pioneer Brower Energizes S.F. State Students." *San Francisco Examiner*, April 28, 2000.

———. "Sierra Club's Squabble over Nuclear Arms." *San Francisco Examiner*, July 14, 1986.

Brooks, Paul. "Congressman Aspinall vs. the People of the United States." *Harper's*, March 1963, 60–63.

Brower, Anne. "Worm of Power Not Visible." *Palo Alto Times*, February 27, 1969.

Brower, David R. "Brower on the Yosemite Valley Plan." *San Francisco Chronicle*, November 20, 2002.

———. "David Ralph Simons 1936–1960." *Sierra Club Bulletin*, November 1960, 24–25.

———. "Editor's Mail Box, Diablo Power Site." *San Francisco Examiner*, July 9, 1968.

———. "Far from the Madding Mules: A Knapsacker's Retrospective." *Sierra Club Bulletin*, February 1935, 68–77.

———. Foreword to *Galápagos*, vol. 1, *Discovery*, edited by Kenneth Brower, with photographs by Eliot Porter, 24–26. London: Sierra Club Books, 1968.

———. Foreword to Eliot Porter, *The Place No One Knew: Glen Canyon on the Colorado*, edited by David Brower, 7–9. San Francisco: Sierra Club Books, 1963.

———. Foreword to *Wildlands in Our Civilization*, edited by David Brower, 11–19. San Francisco: Sierra Club Books, 1964.

———. "Glen Canyon: The Year of the Last Look." *Sierra Club Bulletin*, June 1962, 7.

———. "Grand Canyon Battle Ads." In *Grand Canyon of the Living Colorado*, edited by Roderick Nash, 130–131. San Francisco: Sierra Club and Ballantine Books, 1970.

———. "Headlands." *Not Man Apart*, September 1976, 7–10.

———. "How to Kill a Wilderness." *Sierra Club Bulletin*, August 1945, 2–5.

———. "It Couldn't Be Climbed." *Saturday Evening Post*, February 3, 1940, 24–25, 70–75.

———. "It's Healing Time on Earth." *Earth Island Journal*, Winter 1980, 51–52.

———. "Lake Powell and the Canyon That Was." *Sierra Club Bulletin*, April–May 1963, 2.

———. "Let the River Run Through It." *Sierra*, March–April 1997, 42–44.

———. "Mission 66 Is Proposed by Reviewer of Park Service's New Brochure on Wilderness." *National Parks Magazine*, January–March 1958, 1–4, 45–47.

———. "Pennington Glen Canyon Film Release." *Sierra Club Bulletin*, June 1965, 19.

———. "Please Keep Those Glen Canyon Tunnels Open Until Rainbow Bridge Protection Is Certain." *Sierra Club Bulletin*, March–April 1962, 2–3.

———. "Pursuit in the Alps." *Sierra Club Bulletin*, April 1946, 32–45.

———. "Rainbow Bridge and the Quicksands of Time." *Sierra Club Bulletin*, June 1961, 2.

———. "Rainbow Promise Breaking: New Threat at Echo Park." *Sierra Club Bulletin*, April–May 1960, 16.

———. "Scenic Resources for the Future." *Sierra Club Bulletin*, December 1956, 1–10.

———. "Should Atomic Energy Go Back to the Drawing Board?" *Not Man Apart*, August 1972, 2–8.

———. "Sierra High Trip." *National Geographic*, June 1954, 844–869.

———. "The Story Behind It: The Most Beautiful Book of Its Kind Ever Produced." *Sierra Club Bulletin*, October 1962, 6–7.

———. "Tioga Protest: What Happened Below Tenaya." *Sierra Club Bulletin*, October 1958, 1–7.

———. "A Tribute to Ansel Adams." *Sierra*, July–August 1984, 32.

———. "Wilderness and the Constant Advocate." *Living Wilderness*, Spring–Summer 1964, 42–47.

———. "Wilderness and the Constant Companion." *Sierra Club Bulletin*, September 1964, 3.

———. "The Wilderness Bill: Nobody Wants It but the People." *Sierra Club Bulletin*, March 1960, 2.

———. "Wilderness—Conflict and Conscience." In *Wildlands in Our Civilization*, edited by David Brower, 52–74. San Francisco: Sierra Club Books, 1964.

Brower, David, and Jim Harding, "Horsepower Sense." *Not Man Apart*, June 1974, 1–7.

Brower, David, and David Sive. "Two Davids, One Goliath." *Sierra Club Bulletin*, February 1966, 2.

Brower, Kenneth. "Climbing the Spiral Staircase." *California*, Spring 2013, 49.

"Brower Called a Dictator." *San Francisco Examiner*, February 5, 1969.

"Brower May Quit Post in Sierra Club." *San Francisco Examiner*, January 6, 1969.

Brown, Tom. "Alaskan Interview." *Not Man Apart*, January 1971, 12–14.

Caen, Herb. "In One Ear." *San Francisco Chronicle*, May 1, 1967.

Cain, Stanley A. "Address at the Diamond Jubilee Banquet." *Sierra Club Bulletin*, January 1968, 7–13.

"A Case for a Dam." *Sierra Club Bulletin*, February 1957, 3.

Chatfield, David. "That's FOE All Over." *Not Man Apart*, January–February 1979, 8–9.

Chrismer, Ellen. "Esteemed Leader Has Quit." *Modesto Bee*, May 20, 2000.

Cline, Alan. "Turmoil over Sierra Club Leadership." *San Francisco Examiner & Chronicle*, October 6, 1968.

"Controversy in the Grand Canyon: Beauty of Colorado River Dams?" *National Observer*, April 25, 1966.

Crowe, Harold E. "Announcement from the President." *Sierra Club Bulletin*, January 1953, 3–4.

"Dam Foes Twist Truth, Moss Says." *Phoenix Gazette*, June 22, 1966.

"Dam the Canyon?" *Newsweek*, May 30, 1966, 27.

"David Brower: Tireless Environmental Activist." *Mother Earth News*, May–June 1973, 3.

Davies, Lawrence E. "Fight Is on to Save Big Dinosaur Area." *New York Times*, December 21, 1953.

———. "Naturalists Get a Political Arm." *New York Times*, September 17, 1969.

———. "Udall Is Urged to 'Obey the Law.'" *New York Times*, January 12, 1963.

Davis, Mikhail. "Obituary: Anne Hus Brower (1913–2001)." *Berkeley Daily Planet*, November 28, 2001.

Deevey, Edward S. "Review." *Science*, December 9, 1960, 1759.

D'Emilio, John. "Capitalism and Gay Identity." In *The Material Queer: A LesBiGay Cultural Studies Reader*, edited by Donald Morton, 263–271. Boulder, Colo.: Westview Press, 1995.

Devall, Bill. "David Brower." *Environmental Review*, Autumn 1985, 243–244.

DeVoto, Bernard. "Let's Close the National Parks." *Harper's*, October 1953, 49–52.

——. "Shall We Let Them Ruin Our National Parks?" *Saturday Evening Post*, July 22, 1950, 42.

"Directors Launch Campaign for Expanded Grand Canyon Protection." *Sierra Club Bulletin*, June 1963, 6–7.

"Dominy Says River Project in Mess." *Farmington (N.M.) Times*, October 26, 1966.

Dowie, Mark. "I Intend to Keep Shouting." *Men's Journal*, February 2001, 80–81.

Duscha, Julius. "Wilderness Bill Not for Aspinall." *Washington Post*, April 12, 1963.

Dusheck, George. "Sierra's Chief Wins." *San Francisco Examiner*, May 7, 1967.

"The Earth's Defender." *San Jose Mercury News*, November 8, 2000.

"Echo Park Dam Not Needed." *New York Times*, June 16, 1955.

"An ECO from the Third World." *Not Man Apart*, May 1974, 7.

Emory, Jerry. "Condor Recovery: Hands-on or Hands-off?" *Not Man Apart*, November 1980, 15–17.

"Environmental Pioneer Hit by Heart Trouble." *San Francisco Examiner*, April 26, 1995.

Evanoff, Mark. "Boondoggle at Diablo: Saga of Greed, Deception, Ineptitude—and Opposition." *Not Man Apart*, September 1981, D1–D12.

Evans, Brock. "Showdown for the Wilderness Alps of Washington's North Cascades." *Sierra Club Bulletin*, April 1968, 7–16.

"Fairchild Award Given to Writer." *Rochester Times-Union*, November 10, 1960.

"Farewell to an Arch Druid." *Denver Post*, November, 8, 2000.

"FOE Book May Backfire on Maui." *Not Man Apart*, February 1979, 4.

"FOE's Branches." *Not Man Apart*, December 1984, 6.

"Free Speech Apparently Does Not Apply If Objecting to Action of Great Society." *Fort Lauderdale News*, June 28, 1966.

"Friend of the Earth." *St. Louis Post-Dispatch*, November 9, 2000.

Fritz, Emanuel. "Book Reviews." *Journal of Forestry*, September 1964, 641–643.

Gilliam, Harold. "The Sierra Club's First Century." *San Francisco Chronicle*, April 26, 1992.

——. "The Sierra Club: Time for a Reappraisal." *San Francisco Chronicle-Examiner*, April 27, 1969.

Gilroy, Harry. "Prize Is Awarded Crusading Book." *New York Times*, April 23, 1965.

Goodman, Jack. "How Wahweap Creek Became Wahweap Marina." *New York Times*, July 9, 1967.

———. "Taming the Colorado." *Saturday Evening Post*, September 15, 1962, 26–31.

Graham, Frank. "Dave Brower: Last of the Optimists?" *Audubon*, September 1982, 62–73.

"Grand Canyon." *Sierra Club Bulletin*, May 1966, 1–16.

"Grand Canyon 'Cash Registers.'" *Life*, May 7, 1965, 4.

Hager, Philip. "Executive Head of Sierra Club Resigns Post." *Los Angeles Times*, May 4, 1969.

"Handbook Edition." *Sierra Club Bulletin*, December 1967, 1–62.

Hanna, Bert. "Rivals Vow Congress Fight." *Denver Post*, April 3, 1966.

Harrison, Arthur E. "Mt. Rainier Tramway Plan Defeated." *Sierra Club Bulletin*, January 1955, 6.

Harvey, Mark W. T. "Defending the Park System: The Controversy over Rainbow Bridge." *New Mexico Historical Review*, January 1998, 45–67.

Heald, Weldon. "Cascade Holiday." In *The Cascades: Mountains of the Pacific Northwest*, edited by Roderick Peattie, 97–138. New York: Vanguard, 1949.

———. "San Gorgonio—Southern California's Rooftop." *Living Wilderness*, Summer 1963, 12–16.

Hendrick, Kimmish. "The Battle of Grand Canyon." *Christian Science Monitor*, May 19, 1966.

"Hickel Receives Approval from Senate." *Sierra Club Bulletin*, February 1969, 3.

Honan, William H. "Eliot Porter, Photographer, Is Dead at 88." *New York Times*, November 2, 1990.

Hope, Jack. "The King Besieged." *Natural History*, November 1968, 52–56, 72–82.

Humphries, Harrison. "Park Service Head Opposes Colorado River Power Plan." *Washington Post*, December 7, 1961.

"Inconsistent Club." *San Francisco Examiner*, July 3, 1968.

"IRS and the Grand Canyon." *New York Times*, June 17, 1966.

"IRS Threatens the Sierra Club." *New York Times*, June 12, 1966.

Kane, Joe. "The Friends of David Brower." *Outside*, December 1984, 39.

Kay, Jane. "Friends Recall Brower's Natural Gifts." *San Francisco Chronicle*, December 3, 2000.

Kearsh, Dean. "Debate at Rim on Grand Canyon." *Kansas City Star*, April 6, 1966.

Kenney, Harry C. "From the Bookshelf." *Christian Science Monitor*, November 29, 1962.

Kilgore, Bruce. "Rainbow Bridge: Final Act." *Sierra Club Bulletin*, June 1961, 8–9.

"Killer Mountain Goal of Alpinists." *New York Times*, June 9, 1935.

Linford, E. H. "Belated Crusade to Save the Redwoods." *Salt Lake Tribune*, February 15, 1964.

Logan, William. "Grand Canyon Battle in Depth." *Rocky Mountain News* (Denver), April 4, 1966.

Lombardi, Kate Stone. "Recalling the Glory Days of *Reader's Digest*." *New York Times*, October 1, 2010.

Martin, Glen. "Longtime Sierra Club Leader Quits." *San Francisco Chronicle*, May 9, 2000.

Matthews, Edwin. "Dreams Can Begin It." *Not Man Apart*, July 1979, 8–9, 14–15.

McCarthy, Justin. "About Half in U.S. Say Environmental Protection Falls Short." *Gallup Poll Briefing*, April 9, 2015, 1.

McCloskey, Michael. "Four Major New Conservation Laws: A Review and a Preview." *Sierra Club Bulletin*, November 1968, 4–10.

Meek, William W. "Barry, Swinging Late, Hits Hardest at Canyon Forum." *Arizona Republic*, April 1, 1966.

Meine, Curt, Michael Soule, and Reed F. Noss. "A Mission-Driven Discipline: The Growth of Conservation Biology." *Conservation Biology*, June 2006, 631–651.

Miller, Char, and V. Alaric Sample. "Gifford Pinochet and the Conservation Spirit." Introduction to Gifford Pinchot, *Gifford Pinchot: Breaking New Ground Commemorative Edition*, xi–xvii. Washington, D.C.: Island Press, 1998.

Miller, Scott. "Undamming Glen Canyon: Lunacy, Rationality, or Prophecy?" *Stanford Environmental Law Journal*, January 2000, 121–207.

Mitchell, John G. "On Environmental Publishing." *Not Man Apart*, November 1973, 3–11.

Montgomery, Paul L. "Tour of Hudson Led by Douglas." *New York Times*, March 7, 1966.

"Nash Raps Sierra Club Critic." *San Francisco Examiner*, March 21, 1969.

"The NCCC Proposes: A North Cascades National Park." *Sierra Club Bulletin*, October 1963, 7–13.

"1969 Annual Banquet, Claremont Hotel, Berkeley, May 3." *Sierra Club Bulletin*, May 1969, 4.

"No Barr Involved in Tax Wrangle." *Arizona Republic*, June 16, 1966.

North, Richard D. "Obituary: David Brower." *Independent* (London), November 9, 2000.

"North Cascades National Park." *Seattle Times*, January 30, 1967.

Oakes, John B. "Conservation: Fight for Parks." *New York Times*, October 2, 1960.

Onthank, Karl. "Not Yet a Lost Cause: Report on the Three Sisters." *Sierra Club Bulletin*, January 1958, 16–17.

O'Reilly, John. "Udall at the Bridge." *Sports Illustrated*, May 15, 1961, 26–27.

Ormes, Robert. "A Piece of Bent Iron." *Saturday Evening Post*, July 22, 1939, 13.

Pearlman, David. "Nation and State Take Inventory." *Sierra Club Bulletin*, January 1959, 3.

Pearson, Bruce. "We Have Almost Forgotten How to Hope: The Hualapai, the Navajo, and the Fight for the Central Arizona Project, 1944–1968." *Western Historical Quarterly*, Autumn 2000, 297–316.

Pearson, Drew. "Big Oil Man Visits the President Elect." *San Francisco Chronicle*, December 23, 1968.

"The Peripatetic Reader." *Atlantic Monthly*, May 1964, 134.

"Play Area Tend to Wilds Is Cited." *New York Times*, May 17, 1957.

Poore, Charles. "Books of the Times." *New York Times*, July 9, 1960.

Price, Ben. "Wilderness Bill Stirs Fuss." *Boston Herald*, December 21, 1958.

"Protective Measures for Rainbow Bridge." *Sierra Club Bulletin*, February 1957, 14.

Rauber, Paul. "With Friends Like These." *Mother Jones*, November 1986, 35–37, 47–49.

Reed, Christopher. "David Brower, Environmental Champion Whose Passion Founded Friends of the Earth." *Guardian* (London), November 7, 2000.

Reed, Nathaniel P. "The Conservation Movement as a Political Force." In *Voices from the Environmental Movement: Perspectives for a New Era*, edited by Donald Snow, 41–51. Washington, D.C.: Island Press, 1992.

Richardson, Elmo. "The Interior Secretary as Conservation Villain: The Notorious Case of Douglas 'Giveaway' McKay." *Pacific Historical Review*, August 1972, 336–339.

Robinson, Bestor. "The First Ascent of Shiprock." *Sierra Club Bulletin*, February 1940, 1–7.

Rogers, Paul. "Border Control Proposal Rejected." *San Jose Mercury News*, April 26, 1998.

——. "David Brower, Nature's Crusader Dead at 88: Mountaineer, Fiery Sierra Club Director Leaves a Legacy of Conservation from Coast to Coast." *San Jose Mercury News*, November 7, 2000.

Rourke, Mary. "August Fruge, 94; Publisher Transformed UC Press." *Los Angeles Times*, July 18, 2004.

Roush, G. Jon. "Conservation's Hour—Is Leadership Ready?" In *Voices from the Environmental Movement: Perspectives for a New Era*, edited by Donald Snow, 21–40. Washington, D.C.: Island Press, 1992.

Rugaber, Walter. "Hickel Sworn In at White House Ceremony and Hailed by Nixon." *New York Times*, January 25, 1969.

Saxon, Wolfgang. "Conrad L. Wirth, 93: Led National Parks Service." *New York Times*, July 28, 1993.

Saylor, John P. "Wilderness: The Outlook from Capitol Hill." In *Wilderness: America's Living Heritage*, edited by David Brower, 147–151. San Francisco: Sierra Club Books, 1961.

Schneider, Keith, and Cornelia Dean. "Stewart L. Udall, Conservationist in Kennedy and Johnson Cabinets, Dies at 90." *New York Times*, March 20, 2010.

Schrepfer, Susan. "Establishing Administrative Standing: The Sierra Club and the Forest Service, 1897–1958." *Pacific Historical Review*, February 1989, 55–81.

"Scorecard." *Sports Illustrated*, July 18, 1966.

Severo, Richard. "David Brower, an Aggressive Champion of U.S. Environmentalism, Is Dead at 88." *New York Times*, November 7, 2000.

Shabecoff, Philip. "Conservation Figure Ousted for Resisting Orders to Cut Staff." *New York Times*, July 7, 1984.

Singer, Ronald. "Reviews." *Natural History*, May 1963, 8.

"The Sierra Club Foundation." *Sierra Club Bulletin*, December 1968, 12.

"Sierra Club Gains in Fight on Taxes." *New York Times*, August 7, 1966.

"Sierra Club Keeps Brower." *Oakland Tribune*, May 7, 1966.

"Sierra Elections Should Not Signal Retreat, Discord." *Louisville Courier-Journal*, April 21, 1969.

"Silly Wilderness Preservation Proposal." *Salt Lake Tribune*, March 8, 1960.

Simons, David R. "These Are the Shining Mountains." *Sierra Club Bulletin*, October 1959, 1–13.

Sklarewitz, Norman. "Conservationists Turn from Guarding Nature to Internal Warfare." *Wall Street Journal*, February 13, 1969.

"The Smell of Retribution." *Wichita Eagle*, June 28, 1966.

Smith, Catherine. "Carter's Alaska Move: 100 Million Acres Saved." *Not Man Apart*, January–February 1979, 1.

Stegner, Wallace. "Backroads River." *Atlantic Monthly*, January 1948, 59–64.

——. Introduction to *Wildlands in Our Civilization*, edited by David Bower, 33–43. San Francisco: Sierra Club Books, 1964.

——. "Lake Powell." *Holiday*, May 1966, 64–68, 148–151.

——. "The Marks of Human Passage." In *This Is Dinosaur: Echo Park Country and Its Magic Rivers*, edited by Wallace Stegner, 3–17. New York: Knopf, 1955.

——. "Saga of a Letter: The Geography of Hope." *Living Wilderness*, December 1980, 42–47.

"Stop Glen Canyon Dam, Utahan Asks." *Salt Lake Tribune*, February 12, 1954.

Swain, Roger B. "The Archdruid Himself." *New York Times*, March 20, 1990.

Sylvester, Robert. "The Battle for the Grand Canyon." *Daily News* (New York), May 8, 1966.

Taugher, Mike. "David Brower, 1912–2000, Nature Loses a Best Pal." *Contra Costa Times*, November 7, 2000.

Taylor, Angela. "Fur Coats: Facing Extinction at Conservationists' Hand?" *New York Times*, December 30, 1969.

"32 Groups to Fight New Dam in West." *New York Times*, January 5, 1954.

"This is the American Earth." *Sierra Club Bulletin*, October 1959, 1.

Thornton, Gene. "Reappraising Eliot Porter." *New York Times*, December 16, 1979.

"Three Sisters Decision: Last Word or Just the Beginning?" *Sierra Club Bulletin*, February 1957, 10.

Thurber, Scott. "Brower Cut Off from Funds." *San Francisco Chronicle*, January 31, 1969.

——. "Ouster Plan That Failed." *San Francisco Chronicle*, May 8, 1967.

——. "Sierra Club Battle over Brower's Job." *San Francisco Chronicle*, May 6, 1967.

Turner, Tom, and David Chatfield. "Brower Still on Board." *Not Man Apart*, July–August 1984, 7.

Turner, Wallace. "Sierra Club Loses Exemption on Tax." *New York Times*, December 21, 1966.

United Press International. "Grand Debate: Flood Canyon?" *Deseret News* (Salt Lake City), April 1, 1966.

——. "Opponent of Dam Charges Suppression." *Dallas Chronicle*, May 13, 1966.

"United States: Endangered Species, the Environmental Movement." *Economist*, February 18, 2006, 50.

"Update: Dave Brower, the Environmentalist and the Bomb." *Mother Earth News*, September–October 1982, 94–96.

"U.S. Wilderness System Proposed." *Sierra Club Bulletin*, June 1956, 3.

Wayburn, Edgar. "A President's Message." *Sierra Club Bulletin*, April–May 1967, 4.

Wayburn, Peggy. "The Tragedy of Bull Creek." *Sierra Club Bulletin*, January 1960, 10–11.

"We Defend the Parks." *Sierra Club Bulletin*, January 1955, 3–5.

Weeks, Edward. "The Peripatetic Reviewer." *Atlantic Monthly*, May 1964, 134.

"What You Can Do for Dinosaur and the National Park System." *Sierra Club Bulletin*, January 1954, 3.

Wingo, Hal. "Close-up: California's David Brower, No. 1 Conservationist, Knight Errant to Nature's Rescue." *Life*, May 27, 1966, 37–42.

Woodbury, Angus. "Protecting Rainbow Bridge." *Science*, August 26, 1960, 519–528.

Woodward, Walt. "Sierra Club Will Survive War." *Seattle Times*, April 18, 1969.

Woolefenden, John. "Brower on Population Bomb." *Monterey Peninsula Herald* (Monterey, Calif.), October 15, 1968.

"World Conservation Stressed by Author." *Seattle Times*, March 8, 1969.

Wyant, Dan. "Battle Against Pumice Mining Bitterly Recalled." *Eugene Register-Guard*, October 4, 1971.

Ybarra, Michael J. "Exhibition: A Solitary, Singular Life." *Wall Street Journal*, September 10, 2009.

Oral Histories

Adams, Ansel. "A Conversation with Ansel Adams." An oral history conducted in 1972, 1974, and 1975 by Ruth Teiser and Catherine Harroun. Regional Oral History Office, Bancroft Library, University of California, Berkeley, 1978.

Berry, Phillip S. "Sierra Club Leader, 1960s–1980s: A Broadened Agenda, a Bold Approach." An oral history conducted in 1981 and 1984 by Ann Lage. Regional Oral History Office, Bancroft Library, University of California, Berkeley, 1988.

Brower, David R. "Environmental Activist, Publicist, and Prophet." An oral history conducted in 1974–1978 by Susan Schrepfer. Regional Oral History Office, Bancroft Library, University of California, Berkeley, 1979.

——. "Reflections on the Sierra Club, Friends of the Earth, and Earth Island Institute." An oral history conducted in 1999 by Ann Lage. Regional Oral History Office, Bancroft Library, University of California, Berkeley, 2012.

Clark, Lewis F. "Perdurable and Peripatetic Sierran: Club Officer and Outing Leader 1928–1984." An oral history conducted in 1975–1977 by Marshall Kuhn and included in "Sierra Club Reminisces III, 1910s–1970s." Regional Oral History Office, Bancroft Library, University of California, Berkeley, 1984.

Crowe, Harold E. "Sierra Club Physician, Baron, and President." An oral history conducted in 1973 by Richard Searle and included in "Sierra Club Reminisces II." Regional Oral History Office, Bancroft Library, University of California, Berkeley, 1975.

Dawson, Glen. "Pioneer Rock Climber and Ski Mountaineer." An oral history conducted in 1975 by Richard Searle and included in "Sierra Club Reminiscences II." Regional Oral History Office, Bancroft Library, University of California, Berkeley, 1975.

Dominy, Floyd. "Oral History Interviews." Conducted April 6, 1984, and April 8, 1986, by Brit Allan Story. U.S. National Archives and Records Administration, College Park, Md.

Dyer, Polly. "Preserving Washington's Parklands and Wilderness." An oral history conducted in 1983 by Susan Schrepfer and included in "Pacific Northwest Conservationists." Regional Oral History Office, Bancroft Library, University of California, Berkeley, 1986.

Evans, Brock. "Environmental Campaigner, from the Northwest Forests to the Halls of Congress." An oral history conducted in 1982 and 1984 by Ann Lage and included in "Building the Sierra Club's National Lobbying Program, 1967–1981." Regional Oral History Office, Bancroft Library, University of California, Berkeley, 1985.

Farquhar, Francis. "Sierra Club Mountaineer and Leader." An oral history conducted in 1974 by Ann Lage and Ray Lage and included in "Sierra Club Reminisces." Regional Oral History Office, Bancroft Library, University of California, Berkeley, 1974.

Fruge, August. "A Publisher's Career with the University of California Press, the Sierra Club, and the California Native Plant Society." An oral history conducted in 1997–1998 by Suzanne B. Riess. Regional Oral History Office, Bancroft Library, University of California, Berkeley, 2001.

Hildebrand, Joel. "Sierra Club Leader and Ski Mountaineer." An oral history conducted in 1974 by Ann Lage and included in "Sierra Club Reminiscences I, 1900–1960." Regional Oral History Office, Bancroft Library, University of California, Berkeley, 1974.

Horsehall, Ethel Rose Taylor. "On the Trail with the Sierra Club, 1920s–1960s." An oral history conducted in 1979 by George Baranoski and included in "Sierra Club Women III." Regional Oral History Office, Bancroft Library, University of California, Berkeley, 1982.

Jones, Kathleen Goddard. "Defender of California's Nipomo Dunes, Steadfast Sierra Club Volunteer." An oral history conducted in 1983 by Anne Van Tyne and included in "Sierra Club Nationwide II." Regional Oral History Office, Bancroft Library, University of California, Berkeley, 1984.

Leonard, Richard. "Mountaineer, Lawyer, Environmentalist, Volume I." An oral history conducted in 1976 by Susan R. Schrepfer. Regional Oral History Office, Bancroft Library, University of California, Berkeley, 1976.

——. "Mountaineer, Lawyer, Environmentalist, Volume II." An oral history conducted in 1976 by Susan R. Schrepfer. Regional Oral History Office, Bancroft Library, University of California, Berkeley, 1976.

McArdle, Richard E. "Dr. Richard E. McArdle: An Interview with the Former Chief, U.S. Forest Service, 1952–1962." An oral history conducted in 1973–1974 by Elwood R. Maunder. Forest History Society, Santa Cruz, Calif., 1975.

McCloskey, Michael. "Sierra Club Executive Director: The Evolving Club and the Environmental Movement, 1961–1981." An oral history conducted in 1981 by Susan Schrepfer. Regional Oral History Office, Bancroft Library, University of California, Berkeley, 1983.

McConnell, Grant. "Conservation and Politics in the North Cascades." An oral history conducted in 1983 by Rod Holmgren and included in "Sierra Club Nationwide I." Regional Oral History Office, Bancroft Library, University of California, Berkeley, 1983.

Moorman, James W. "Attorney for the Environment 1966–1981, Center for Law and Social Policy, Sierra Club Legal Defense Fund, Department of Justice, Division of Lands and Natural Resources." An oral history conducted in 1984 by Ted Hudson. Regional Oral History Office, Bancroft Library, University of California, Berkeley, 1994.

Ogilvy, Stewart. "Sierra Club Expansion and Evolution: The Atlantic Chapter, 1957–1969." An oral history conducted in 1978 by Jeri Nunn and included in "The Sierra Club Nationwide I." Regional Oral History Office, Bancroft Library, University of California, Berkeley, 1982.

Robinson, Bestor. "Thoughts on Conservation and the Sierra Club." An oral history conducted in 1974 by Ann Lage and Ray Lage and included in "Sierra Club Reminisces." Regional Oral History Office, Bancroft Library, University of California, Berkeley, 1974.

Searle, Richard. "Grassroots Sierra Club Leader." An oral history conducted in 1976 by Paul Clark and included in "Southern Sierrans." Regional Oral History Office, Bancroft Library, University of California, Berkeley, 1976.

Stegner, Wallace. "The Artist as Environmental Activist." An oral history conducted in 1982 by Ann Lage. Regional Oral History Office, Bancroft Library, University of California, Berkeley, 1983.

Torre, Gary. "Labor and Tax Attorney, 1949–1982; Sierra Club Foundation Trustee, 1968–1981, 1994–1998." An oral history conducted in 1998 by Carl Williams. Regional Oral History Office, Bancroft Library, University of California, Berkeley, 1999.

Udall, Stewart. "Stewart L. Udall Oral History Interview V." An oral history interview conducted on October 31, 1969, by Joe B. Frantz. Lyndon Baines Johnson Library, Austin, Tex.

Wayburn, Edgar. "Sierra Club Statesman, Leader of the Parks and Wilderness Movement: Gaining Protection for Alaska, the Redwoods, and Golden Gate Parklands." An oral history conducted in 1976–1981 by Ann Lage and Susan Schrepfer. Regional Oral History Office, Bancroft Library, University of California, Berkeley, 1985.

Wayburn, Peggy. "Author and Environmental Activist." An oral history conducted in 1990 by Ann Lage. Regional Oral History Office, Bancroft Library, University of California, Berkeley, 1992.

Archival Materials

Brower, David. *Grand Canyon Discussion at Silver Grotto and Marble Damsite, May 31 to June 13, 1977*. Audiocassettes, recorded by Ron Hayes. Grand Canyon National Park Museum Collection, GRCA 40237, Grand Canyon National Park, Ariz.

———. Interviewed by Tom Turner, March 30, 1983. Transcript, Carton 18, Folder 49. David Ross Brower Papers. BANC MSS 79/9c. Bancroft Library, University of California, Berkeley.

David Ross Brower Papers. BANC MSS 79/9c. Bancroft Library, University of California, Berkeley.

Friends of the Earth, U.S. Records, 1969–2009. American Heritage Center. University of Wyoming, Laramie.

Grand Canyon National Park Museum Collection. Grand Canyon National Park, Ariz.

Graves, C. Edward. "Opponent Addresses Memorandum." n.d. Superintendent's Office, Dinosaur National Monument.

Morris Udall Papers. University of Arizona Libraries, Special Collections. University of Arizona, Tucson.

Oral Histories. Regional Oral History Office. Bancroft Library, University of California, Berkeley.

Not Man Apart, 1970–1990. Alternative Press Collection. Thomas Dodd Center. University of Connecticut, Storrs.

Redwood National Park Establishment Papers Series. Redwood National Park, Orick, Calif.

Sierra Club Board of Directors Meeting Minutes, 1892–1995. BANC FILM 2945. Bancroft Library, University of California, Berkeley.

Sierra Club Bulletin, 1904–1967. Homer Babbidge Library. University of Connecticut, Storrs.

Sierra Club Bulletin, 1959–1963, 1964–1977. Sterling Memorial Library. Yale University, New Haven.

Sierra Club Members Papers. BANC MSS 71/295c. Bancroft Library, University of California, Berkeley.

Sierra Club Office of the Executive Director Records. BANC MSS 2002/230c. Bancroft Library, University of California, Berkeley.

Sierra Club Records. BANC MSS 71/103c. Bancroft Library, University of California, Berkeley.

Sierra Club Records, 1957–1980. NC1260. Special Collections, University of Nevada, Reno.

Sierra Club Records, 1969. William E. Colby Library. Sierra Club, San Francisco.

Stewart L. Udall Papers. AZ 372. University of Arizona Libraries, Special Collections, University of Arizona, Tucson.

Toney, Sharon. "Conservationists' Role in the Echo Park Dispute." Unpublished essay, n.d. Superintendent's Office. Dinosaur National Monument, Dinosaur, Colo.

Wayne Aspinall Papers. Penrose Special Collections and Archives. University of Denver.

Yosemite Park Museum and Library. Yosemite National Park, El Portal, Calif.

Government Documents

City of Berkeley Landmark Application. *Brower Houses and David Brower Redwood*, July 2008.

County of Alameda. Probate Court records of David Brower, 1971, and Anne Brower, 1972, Oakland, Calif.

Millar, Constance I. "Case Studies in Ecosystem Management: The Mammoth–June Ecosystem Management Project, Inyo National Forest." In *Sierra Nevada Ecosystem Project: Final Report to Congress*. Vol. 1, *Assessment Summaries and Management Strategies*, 146–150. Centers for Water and Wildland Resources, Report no. 37. Davis: University of California, 1996.

U.S. House of Representatives. Subcommittee on Irrigation and Reclamation. *Colorado River Basin Project: Hearing on H.R. 3300 and Similar Bills*. 90th Cong., 1st sess., March 13–17, 1967.

——. Subcommittee on Irrigation and Reclamation. *Colorado River Storage Project: Hearing on H.R. 4449, H.R. 4443, and H.R. 4463*. 83rd Cong., 2nd sess., January 18–23, 1954.

——. Subcommittee on Irrigation and Reclamation. *Lower Colorado River Basin Project: Hearing on H.R. 4671, 4672–4706, 9248*. 89th Cong., 1st sess., August 8–13, 1965, and September 1, 1965.

——. Subcommittee on Irrigation and Reclamation. *Lower Colorado River Basin Project: Hearing on H.R. 4671 and Similar Bills*. 89th Cong., 2nd sess., May 9–13 and 18, 1966.

——. Subcommittee on Public Lands. *National Wilderness Preservation Act: Hearing on S. 174, H.R. 293, H.R. 299, H.R. 496, H.R. 776, H.R. 1762, H.R. 1925, H.R. 2008, and H.R. 8237*. 87th Cong., 1st sess., November 6, 1961.

——. Subcommittee on Public Lands. *National Wilderness Preservation Act: Hearing on S. 174, H.R. 293, H.R. 299, H.R. 496, H.R. 776, H.R. 1762, H.R. 1925, H.R. 2008, and H.R. 8237*. 87th Cong., 2nd sess., May 9, 1962.

Interviews

Adams, Michael. Telephone interview by the author, October 4, 2012.

Alderson, George. Telephone interview by the author, November 19, 2012.

Berry, Phillip. Interview by the author, Lafayette, Calif., April 24, 2013.

Blackwelder, Brent. Telephone interview by the author, January 3, 2013.

Browder, Joseph. Telephone interview by the author, November 28, 2012.

Brower, Barbara. Interview by the author, Portland, Ore., May 24, 2013.

Brower, Joseph. Telephone interviews by the author, November 9 and 28, 2012, and August 23, 2013.

Brower, Joseph and Gayle Brower. Interview by the author, Santa Rosa, Calif., March 26, 2013.

Dobson, Ed. Telephone interview by the author, June 14, 2010.

Dullen, Deanna. Telephone interview by the author and e-mail correspondence, August 30, 2012.

Evans, Brock. Telephone interview by the author, September 11, 2012.

Fahn, Larry. Interview by the author, San Francisco, March 29, 2013.

Gabel, Dan. Telephone interview by the author, January 10, 2013.

Gediman, Scott. Interview by the author, Yosemite National Park, Calif., August 2, 2012.

Hamilton, Bruce. Interview by the author, San Francisco, August 7, 2012.

Hayes, Denis. Interview by the author, Chattanooga, Tenn., October 16, 2012.

Hoops, Herm. Interview by the author, Vernal, UT , June 19, 2011.

Ingram, Jeffrey. Interview by the author, Tucson, Ariz., June 11, 2013; telephone interviews by the author, November 12, 2012, and July 8, 2014.

Leahy, Steve. Telephone interview by the author, December 18, 2012.

Litton, Martin. Interview by the author, May 26, 2013, Portola Valley, Calif.; telephone interview by the author, October 8, 2012.

McCloskey, Michael. Interview by the author, Portland, Ore., January 22, 2013, and May 24, 2013; telephone interview by the author, July 10, 2014.

Millar, Constance I. Interview by the author, Inyo National Forest, Calif., July 30, 2012.

Mills, Stephanie. Telephone interview by the author, August 5, 2014.

Moore, Avis Ogilvy. Telephone interview by the author, February 19, 2013.

Norton, Boyd. Telephone interview by the author, January 16, 2013

Phillips, David. Interview by the author, Berkeley, Calif., January 18, 2013.

Pope, Carl. Telephone interview by the author, July 8, 2013.

Reis, Greg. E-mail correspondence with the author, September 12, 2012.

Risser, Mary. Interview by the author, Dinosaur National Monument, Colo., June 21, 2010.

Soucie, Gary. Interview by the author, Williamstown, Mass., October 26, 2012; telephone interview by the author, August 26, 2013.

Steiger, Wayne. Telephone interview by the author, March 5, 2013.

Stolow, David. E-mail correspondence with the author, March 1, 2013.

Tamhane, Mark. E-mail correspondence with the author January 5, 2013.

Vinyard, Lucille. Interview with the author, Trinidad, Calif., March 27, 2013.

Werbach, Adam. Telephone interview by the author, August 20, 2013.

Woolfenden, Wallace B. E-mail correspondence with the author, July 20–August 29, 2012.

Unpublished Documents

Devall, William B. "The Governing of a Voluntary Organization: Oligarchy and Democracy in the Sierra Club." Ph.D. diss., University of Oregon, 1970.

Mills, Stephanie. "An Archdruid—or Some Damn Thing." Manuscript, n.d. Copy in author's files.

Electronic Documents

Arnow, Ed. "David Brower Farewell Speech at Sierra Club." KPIX, San Francisco, May 3, 1969. http://diva.sfsu.edu/collections/sfbatv/bundles/189379.

Asahi Glass Foundation. "1998 Blue Planet Prize." Press release, n.d. http://www.af-info.or.jp/en/blueplanet/list.html.

Bagley, Katherine. "Field Guide to the Modern U.S. Environmental Movement." *Inside Climate Change*, April 7, 2014. http://insideclimatenews.org.

Brower, David. Interview by Ken Verdoia. KUED, University of Utah, aired October 1999. http://www.kued.org/productions/glencanyon/interviews/brower/html.

Coyle, Daniel. "The High Cost of Being David Brower." *Outside*, December 1995. http://outsideonline.com/outdoor-adventure/The-High-Cost-of-Being-David Brower.html?page=all.

"Creation and Growth of the National Wilderness Preservation System." Wilderness.net, n.d. http://www.wilderness.net/NWPS/AtoZ.

Dominy, Floyd. Interviewed by Ken Verdoia. KUED, University of Utah, aired October 1999. http://www.kued.org/productions/glencanyon/interview/Dominy/html.

Fedarko, Kevin. "Ain't It Just Grand." *Outside*, May 31, 2005. http://www.outsideonline.com/templates/Outside_Print_Template?content=123280513.

Friends of the Earth International at 40. Amsterdam: Friends of the Earth International, 2011. http://www.foei.org/wp-content/uploads/2011/06/10-foei-at-40-lr.pdf.

Gallup. "Climate Change, Environment" (1989–2015). n.d. http://gallup.com /pll/1615/environment.aspx.

Hayden, Laura. "Q&A: Kenneth Brower Remembers His Father." *Sierra*, June 21, 2012. http://sierraclub.typepad.com/greenlife/2012/06/kenneth-brower-remembers -father-an-interview-with-david-browers-son.html.

Ingram, Jeff. "Ingram Events Journal, 1966–68, Part 1, Added to 6/3/11." http:// gcfutures.blogspot.com/2011/06/ingram-events-journal-1966-8-part-1.html.

Kupfer, David. "Final Interview: David R. Brower." August 5, 2000. http://www .wildnesswithin.com/kupfer/.html.

McKeever, Brian and Sarah L. Pettijohn. "The Nonprofit Sector in Brief 2014." Urban Institute, October 2014. http://www.urban.org/UploadedPDF/413277- Nonprofit-Sector-in-Brief-2014.pdf.

National Center for Charities Statistics. "Public Charities: NTEE = C (Environ- ment)." 2013. http://nccs.urban.org/database/overview.cfm.

The Nature Conservancy. "About Us." n.d. http://www.nature.org/about-us /index.htm.

Rogers, Jedediah. "Glen Canyon Unit." U.S. Bureau of Reclamation, 2006. http:// www.usbr.gov/projects/ImageServer?imgName=Doc_1232657383034.pdf.

Sierra Club. "LeConte Memorial Lodge." n.d. http://www.sierraclub.org.education /leconte/.

Sleight, Ken. Interviewed by Ken Verdoia. KUED, University of Utah, aired October 1999. http://www.kued.org/productions/glencanyon/interviews/sleight /html.

"Three Sisters Wilderness." Wilderness.net, n.d. http://www.wilderness.net /NWPS/wildView?WID=602.

Turner, Tom. "From Heresy to Conventional Wisdom at Blinding Speed: A History of Earth Island Institute's 25 Years." *Earth Island Journal*, Spring 2008. http://www.earthisland.org/journal/index.php/eij/article/from_heresy_to _conventional_wisdom_st_blinding_speed.

U.S. National Park Service. "Olmsted Point Plowing." June 2, 2005. http://www .nps.gov/yose/photosmultimedia/olmstedplowing.htm.

Wayburn, Edgar and Allison Alsup. "Dr. Edgar Wayburn, MD: 1906–2010." Sierra Club, n.d. http://www.sierraclub.org/history/wayburn/.

Videos

DeGraaf, John, prod., dir., and ed. *For Earth's Sake: The Life and Times of David Brower*. KCTS, 1989.

Duane, Kelly, dir. *Monumental: David Brower's Fight for Wild America*. DVD. KEET-TV, First Run Features, 2004.

Bibliography

Reynolds, Claire, exec. prod., and Sam Greene, prod. and dir. *Redwood National Park: Preserving Ancient Forests*. DVD. KEET-TV, 2009.
Sierra Club. *Two Yosemites*. Sierra Club Films, 1955.
——. *Wilderness Alps of Stehekin*. Sierra Club Films, 1957. http://content.sierraclub.org/brower/video.
——. *Wilderness River Trail*. Dawson Productions, 1953.

Acknowledgments

David Brower was a saver. He kept documents, memos, letters, and other memorabilia during his entire lifetime. Much of it was donated to the Bancroft Library at the University of California in Berkeley. The Sierra Club and many of its leaders also supplied similar material to the library. I spent several months in Berkeley mining those resources. I thank the entire staff for their assistance, in particular Susan Snyder, now retired. Although the Bancroft records were the primary resource for this book, they were often missing one thing: any comment by or introspection from Brower about events as they were unfolding. That gap is unfortunate because I was striving to understand such a complex individual as David Brower.

I also thank the many individuals, libraries, archive centers, and government agencies that helped make this book possible. They include Sterling Memorial Library, Yale University; Ellen Bryne with the William Colby Memorial Library at the Sierra Club in San Francisco; Linda Eade of the Yosemite Park Museum and Library; Mary Risser at Dinosaur National Monument; Deana Dulen at Devils Postpile National Monument; Berkeley Architectural Heritage Association; James O'Barr at Redwoods National Park; the Special Collections Office at the University of Nevada, Reno; the Grand Canyon National Park Museum and Library; the Library Manuscript Collection at the University of Arizona; the Penrose Special Collection at the University of Denver; the American Heritage Center at the University of Wyoming; the Lyndon Baines Johnson Presidential

Library at the University of Texas at Austin; the archives of the U.S. Bureau of Reclamation; and the Forest History Society. I owe special thanks to my home library and the Thomas Dodd Research Center at the University of Connecticut. Both were invaluable, as were Lana Babji in the Interlibrary Loan Office and Steve Batt. Finally, I thank the many people who took the time to allow me to interview them, sometimes more than once, to help me better understand Brower.

The University of Connecticut and its Journalism Department were generous in providing me both the time and the financial resources to research and write this book, and I especially want to thank Maureen Croteau, the longtime department director. Both Rebecca Ortinez and Paul Lyzun provided assistance within the department. In addition, a number of colleagues in the department and at the university were supportive in various ways, including Margaret Breen, Kent Holtsinger, Gail MacDonald, Micki McElya, Helen Rowzadowski, Steve Smith, Mike Stanton, Bruce Stave, and Wayne Worcester. Outside the university, help was provided by Janet Cusack, Paul and Sheila Ekberg, Milton and Mary Fuji, Herm Hoops, Brian Jones, Tony Lioce, Connie Millar, Greg Reis, and Wally Woolfenden.

I thank Patrick Fitzgerald, Bridget Flannery-McCoy, Kathryn Schell, and others at Columbia University Press as well as Jonathan Cobb, Stephanie Mills, and Mark T. W. Harvey.

A longtime colleague and eminent journalist, Tim Kenny, deserves special commendation for his assistance with the manuscript.

Finally and most importantly, my thanks go to my colleague and partner, Dianne Sprague, who assisted with the research, read every word of the book several times, listened patiently, and tirelessly offered advice through the long process of creating a book.

Index